WIRELESS PERSONAL COMMUNICATIONS: *Trends and Challenges*

THE KLUWER INTERNATIONAL SERIES IN ENGINEERING AND COMPUTER SCIENCE

COMMUNICATIONS AND INFORMATION THEORY

Consulting Editor

Robert Gallager

Other books in the series:

ELLIPTIC CURVE PUBLIC KEY CYRPTOSYSTEMS, Alfred Menezes
ISBN: 0-7923-9368-6
SATELLITE COMMUNICATIONS: Mobile and Fixed Services, Michael Miller, Branka Vucetic and Les Berry
ISBN: 0-7923-9333-3
WIRELESS COMMUNICATIONS: Future Directions, Jack M. Holtzman and David J. Goodman
ISBN: 0-7923-9316-3
DISCRETE-TIME MODELS FOR COMMUNICATION SYSTEMS INCLUDING ATM, Herwig Bruneel and Byung G. Kim
ISBN: 0-7923-9292-2
APPLICATIONS OF FINITE FIELDS, Alfred J. Menezes, Ian F. Blake, XuHong Gao, Ronald C. Mullin, Scott A. Vanstone, Tomik Yaghoobian
ISBN: 0-7923-9282-5
WIRELESS PERSONAL COMMUNICATIONS, Martin J. Feuerstein, Theodore S. Rappaport
ISBN: 0-7923-9280-9
SEQUENCE DETECTION FOR HIGH-DENSITY STORAGE CHANNEL, Jaekyun Moon, L. Richard Carley
ISBN: 0-7923-9264-7
DIGITAL SATELLITE COMMUNICATIONS SYSTEMS AND TECHNOLOGIES: Military and Civil Applications, A. Nejat Ince
ISBN: 0-7923-9254-X
IMAGE AND TEXT COMPRESSION, James A. Storer
ISBN: 0-7923-9243-4
VECTOR QUANTIZATION AND SIGNAL COMPRESSION, Allen Gersho, Robert M. Gray
ISBN: 0-7923-9181-0
THIRD GENERATION WIRELESS INFORMATION NETWORKS, Sanjiv Nanda, David J. Goodman
ISBN: 0-7923-9128-3
SOURCE AND CHANNEL CODING: An Algorithmic Approach, John B. Anderson, Seshadri Mohan
ISBN: 0-7923-9210-8
ADVANCES IN SPEECH CODING, Bishnu Atal, Vladimir Cuperman, Allen Gersho
ISBN: 0-7923-9091-1
SWITCHING AND TRAFFIC THEORY FOR INTEGRATED BROADBAND NETWORKS, Joseph Y. Hui
ISBN: 0-7923-9061-X
ADAPTIVE DATA COMPRESSION, Ross N. Williams
ISBN: 0-7923-9085
SOURCE CODING THEORY, Robert M. Gray
ISBN: 0-7923-9048-2
AN INTRODUCTION TO ERROR CORRECTING CODES WITH APPLICATIONS, Scott A. Vanstone, Paul C. van Oorschot
ISBN: 0-7923-9017-2
FINITE FIELDS FOR COMPUTER SCIENTISTS AND ENGINEERS, Robert J. McEliece
ISBN: 0-89838-191-6
AN INTRODUCTION TO CRYPTOLOGY, Henk C. A. van Tilborg
ISBN: 0-89838-271-8

WIRELESS PERSONAL COMMUNICATIONS:
Trends and Challenges

edited by

Theodore S. Rappaport
Brian D. Woerner
Jeffrey H. Reed
Virginia Polytechnic Institute

KLUWER ACADEMIC PUBLISHERS
Boston / Dordrecht / London

Distributors for North America:
Kluwer Academic Publishers
101 Philip Drive
Assinippi Park
Norwell, Massachusetts 02061 USA

Distributors for all other countries:
Kluwer Academic Publishers Group
Distribution Centre
Post Office Box 322
3300 AH Dordrecht, THE NETHERLANDS

Library of Congress Cataloging-in-Publication Data

A C.I.P. Catalogue record for this book is available
from the Library of Congress.

Printed on acid-free paper.

Printed in the United States of America

TABLE OF CONTENTS

Preface

"Well informed people know it is impossible to transmit the voice over wires, and that were it possible to do so, the thing would be of no practical value."

from an editorial in the Boston Post - 1865

Fortunately for the telecommunications industry, the unknown author of the above statement turned out to be very mistaken indeed. Even as he spoke, Alexander Graham Bell was achieving the impossible, with a host of competing inventors close behind. The communications revolution which ensued has changed the way in which we live and work, and the way in which we view the world around us. Wired telephone lines now encircle the globe, allowing instantaneous transmission of voice and data. Events from Times Square to Red Square are now as accessible as events on the local courthouse lawn.

The advent of wireless communications has extended Bell's revolution to another domain. Personal communications promises voice, data and images which are accessible everywhere. Although predictions are dangerous, a look back over the last decade reveals spectacular growth. In the United States alone, there are now over 50 million cordless phones in use throughout the country - at least one cordless phone for every 3 households - and nearly 20 million pocket pagers. U.S. Cellular telephone service, launched commercially in 1984, has experienced 30-40% annual growth rates despite a sluggish economy. The service has proved so popular that the Federal Communications Commissions has announced plans this year to free new spectrum for expanded Personal Communications Services (PCS) and tentatively awarded two licenses based on pioneer's preferences. Events in Europe, Canada and Japan are unfolding at an equally hectic pace. With the hindsight of a century of changes, we can only be certain of more change on the horizon. Considerable fortunes are being staked on the continued growth of personal communication in all its forms.

The Mobile and Portable Radio Research Group (MPRG) was founded at Virginia Tech in 1990 as a resource to the personal communications industry. The MPRG has grown with the industry, developing a broad program of teaching and research in personal communications. Beginning with the 1993 academic year, the MPRG will be a part of the newly established Center for Wireless Telecommunications on the Virginia Tech campus, supported by Virginia's Center for Innovative Technology.

Since the 1991, the MPRG has sponsored an annual Symposium on Wireless Personal Communications each spring on the Blacksburg campus of Virginia Tech. At the 1993 Symposium, nearly 200 wireless professionals and students from around the world gathered here for two days of lively discussions on the future of wireless communications. This book consists of twenty one papers presented at the 3rd Virginia Tech Symposium, as well as one additional paper. The papers cover a diverse range of current research activity in wireless communications.

The research presented in this book reflects the rapid growth and maturity of the wireless industry. Although overcoming the propagation characteristics of the wireless channel will remain the unique problem confronted by wireless system designers, a plethora of innovative techniques for equalization and interference rejection have emerged. New applications of wireless systems have also been proposed and implemented, ranging from high data rate paging to video transmission. Research in Code Division Multiple Access (CDMA) has begun to move from performance evaluation to system design. Research also continues on a broad range of simulation, coding and multiple access techniques. And as the wireless community has expanded, there has been the need to draw upon expertise in seemingly disjoint disciplines such as semiconductor design and fabrication, and human factors. In this text, the large number of papers with multiple authors and from multiple institutions is indicative of the collaborative nature of wireless research, necessitated by the increasing complexity and breadth of wireless systems.

The book is organized into five sections, each focusing on a distinct set of issues in wireless communications. Section I covers radio propagation and recent approaches to compensate for propagation characteristics. Section II focuses on a range of applications for personal communications systems. Recently there has been a heavy research emphasis in the field of Spread Spectrum, and new Code Division Multiple Access (CDMA) techniques are covered in Section III. Section IV presents a range of novel ideas in simulation, error correction and multiple access. Finally, Section V presents a range of perspectives on personal communications systems.

The wireless channel remains an important constraint on system performance. Mountains present one of the greatest challenges for wide-area radio system designers, as they cause large multipath delays. R. L. Kirlin, W. Du, Y. Cui, G. Robertson, Y. Zhang, and P. F. Driessen at the University of Victoria present a study of channel impulse responses in mountainous environments. Increasing attention has focused on efforts to compensate for the effects of the channel. Adaptively steerable antenna arrays which focus the radio beam are one such technique. Stephan Schell of Penn State along with William Gardner and Peter Murphy of the University of California at Davis describe a blind antenna steering method which requires no training sequences. Joe Liberti and Ted Rappaport quantitatively demonstrate capacity improvements in CDMA systems using adaptive antennas, and present new analytical results which tie RF propagation and antenna pattern into the bit error rate of CDMA systems. Equalization is another technique for overcoming the multipath characteristics. Michel Fattouche of the University of Calgary and Hatim Zaghloul of Wi-LAN Inc. have developed equalization algorithms for systems employing frequency division multiplexing.

Interference rejection for wideband systems plays a role which is analogous to equalization for narrowband systems. There has been great interest in interference rejection techniques as a means of overcoming the well known near/far problem in CDMA systems. I. Howitt, V. Vemuri, T. C. Hsia of UC-Davis and Jeff Reed of Virginia Tech present an excellent tutorial on the use of neural networks for interference rejection. John Doherty of Iowa State examines interference rejection in terms of vector spaces. This perspective has proved useful in the development of modulation techniques. Finally, Brian Agee of Radix Technologies explores the use of both spatial and spectral diversity techniques for interference rejection.

Section II includes papers on emerging PCS applications. A common theme throughout the PCS industry is the demand for increasing data rates. As evidenced by the papers in this section, this demand is present in both low and high cost applications. Rade Petrovic of the University of Mississippi, Walt Roehr of Telecommunication Networks Consulting, and Dennis Cameron of MTEL Technologies have prepared a pair of papers on high data rate paging systems. The first of these papers examines the crucial design issues for increasing the throughput in a simulcast paging environment. The second of these papers describes the use of permutation modulation, an ingenious generalization of conventional FSK signalling, suitable for use in a simulcast environment. At the other end of the data rate spectrum, a Stanford University research group consisting of Teresa Meng, Ely Tsern, Andy Hung, Sheila Hemami and Benjamin Gordon have pushed back the frontiers of video compression to the point where video over wireless is now possible. The low power video compression technique described remains robust to channel errors.

Intense interest surrounds the use of CDMA techniques, and these techniques are investigated by the research described in Section III. Over the past year, CDMA research has moved beyond interminable comparisons with TDMA, to focus on the design of CDMA systems. At UCLA, the research group of Jonathan Min, Ahmadreza Rofougaran, Victor Lin, Michael Jensen, Henry Samueli, Asad Abidi, Gregory Pottie, and Yahya Rahmat-Samii is designing a low power CDMA transceiver which employs frequency-hopping. The other CDMA papers focus primarily on direct-sequence spread-spectrum. Gary Lomp and Donald Schilling of Interdigital Communications Corporation along with Larry Milstein of UC-San Diego present an overview of the Broadband CDMA techniques which have been proposed for use in PCS systems. Scott Miller of the University of Florida describes one method for applying trellis coding techniques to CDMA systems. Trellis coding techniques have successfully improved data rates over wireline modems and hold the potential for similar improvements in wireless systems. The final two papers in the CDMA section deal with improved synchronization in CDMA systems. Alternative techniques are presented by Peter Schelbert, W. J. Burmeister and M. A. Belkerdid of University of Central Florida and by Daeho Kim and Hoyoung Kim of the Electronics and Telecommunications Research Institute (ETRI) in South Korea.

Section IV contains continuing wireless research along many fronts. The first two papers describe advances in simulation techniques for performance evaluation of wireless systems. Wael Al-Qaq, Micheal Devetsikiotis and Keith Townsend of North Carolina State University discuss the application of importance sampling techniques to wireless communications. By focusing on key events, these techniques hold the potential to reduce simulation times by orders of magnitude. Meanwhile, S. Srinivas and Sam Shanmugan of the University of Kansas attempt to model the bursty errors of wireless channels with Markov models. In the next paper, Young-Ok Park of ETRI considers the use of variable rate convolutional codes to increase the efficiency of wireless systems. J. L. Sobrino and

J. M. Brazio of CAPS institute in Portugal compare reservation and demand assignment strategies for multiplexing in packet voice systems.

Wireless communications are touching an ever wider segment of the economy. Fittingly, the final section of this book offers a variety of fresh perspectives. George Hagn and E. Lyon of SRI International begin this section by looking back to the first wireless PCS system. Interestingly enough, the experiments which they describe took place in the Piedmont region of Virginia, only a two hour drive from the Virginia Tech campus. In the next paper, Mike Schwartz of National Semiconductor examines wireless communications from the perspective of the semiconductor industry, foreseeing a steadily increasing volume of business. James Proffitt of PacTel describes the problems and business opportunities which will be created by networking wireless systems with the existing phone network. Finally, Larry Dworkin and Louis Taylor of The MITRE Corporation look ahead and ask the question "Whither Personal Communications?" Speaking from the perspective of military consumers of PCS, the authors articulate the need for standardization in wireless systems.

In a field which is changing as rapidly as wireless, no collection of papers can convey a comprehensive survey of the state of the art. Perhaps the best that can be attained is a snapshot which samples the best work in a broad range of areas. The editors hope that this book represents such a snapshot.

We would like to thank the many people whose work has made this book possible. The quality of research presented in this book is due to the diligent efforts of the individual authors. The MPRG Industrial Affiliates companies continue to provide generous financial support for our Symposium on Wireless Personal Communications, upon which this book is based. Jenny Frank has worked tirelessly to coordinate the 1993 symposium and assemble the papers which appear in this book. The symposium would not have been possible without the dedicated efforts of Prab Koushik and Annie Wade. Lastly, the efforts of a great many MPRG students have contributed mightily to the Symposium and to this book. We hope the community finds this journal useful for continued work in wireless, and invite the readers to attend our future symposia at the Virginia Tech campus.

1

Measured 900 MHz Complex Impulse Responses in Mountainous Terrain: Relationship to Topographical Map Data

R.L. Kirlin, W. Du, Y. Cui, G. Robertson, Y. Zhang, P.F. Driessen
Department of Electrical and Computer Engineering
University of Victoria
Victoria, B.C. Canada V8W 1P6
e-mail: peter@sirius.uvic.ca

Abstract

900 MHz complex impulse response data in mountainous terrain is measured at closely spaced locations, and is processed as data from a synthetic aperture array. Experimental data from linear and crossed arrays with 50 or 100 elements is considered. The direction of arrival for each delayed component is identified, and contour plots of the receiver power at bearings and distances are produced. These coutour plots closely match the topography of the region, and clearly indicate that the strongest mountain reflections come from the steepest mountain slopes.

These results are used to establish a relationship between the mountain reflection coefficients and the topography, thus making it possible to invert the problem and estimate the impulse response (multipath delay profile) in mountainous terrain directly from topographical map data. Such estimates can help to select cell site locations and antenna configurations to minimize the delay spread.

1 Introduction

Our objective is to determine a method that allows topographic data bases to be converted to coverage maps usable for digital cellular radio systems. For such digital systems, two parameters (the multipath delay profile as well as the signal strength) will affect the received signal quality (bit error rate) and thus a useful coverage map will need to specify both parameters versus location. If the terrain is described as a collection of scatterers with known scattering coefficients (e.g. triangular reflectors from a topographic data base), and all paths from transmitter via each scatterer to receiver are defined, then in principle, both parameters may be determined, and may also be characterized statistically [1][2]. However, it is necessary (and may be difficult) to accurately define the scattering coefficients (or cross-section) at the frequency of interest (e.g. 910 MHz) for each triangular region using only coutour, groundcover and mineral composition data.

Thus in this work, we seek to determine these scattering coefficients using measured complex impulse response data, and to locate the significant reflectors which contribute to the multipath delay profile. Measured impulse response data was collected at two sites in Vancouver, Canada (Figure 1) which include steep mountain slopes to the north, causing significant power in long delay multipaths at 900 MHz [3]. The data was recorded at complex baseband, using a 10 Mb/sec sliding correlator, thereby yielding complex impulse responses for the multipath channel with time resolution of 50 nsec.

A synthetic array was implemented using a precision positioner. This allows inversion of the data to correlate major reflections with topographic features. This is accomplished with a high-resolution spectrum or Direction-of- Arrival (DOA) estimator and knowledge of the delay time for each arrival, thereby creating the intersection of a DOA radial and an elliptical delay-time distance locus with transmitter and receiver as foci.

Results show clearly that significant reflectors can be located this way, and therefore their properties can be used in classification algorithms to classify topographic features for trial cell-sites at other locations. Although not tested, suggestions on how this might be accomplished are included.

The steps involved in achieving these results are as follows:

1. Direction-of-arrival estimation of measured echo signals.
2. Location of principal reflectors.
3. Correlation of reflectors with topography.
4. Channel identification and modeling.
5. Echo prediction based on the stochastic model of the channel.

Array processing began with tests of two schemes, conventional beamforming and minimum variance distortionless response (MVDR). MVDR (also referred to as Linear Constrained Minimum Variance (LCMV)) was subsequently selected as the preferred method due to its proper dimensioning to estimate the signal power from a given direction. The qualitative correlation of response data (specifically reflected power and reflector location) with topographic features results are very good. Overlays of reflected power vs x-y location correspond highly with steep topography gradients.

2 Array Processing Fundamentals

2.1 Problem Formulation

The signal received by a receiver in a mountainous area can be considered as a superposition of direct path signal and reflected radio waves. Let $s(t)$ be the base-band signal transmitted through a channel with impulse response $h(t)$, so that the received signal

$$r(t) = \int_{-\infty}^{\infty} s(t-\tau)h(\tau)d\tau \tag{1}$$

is determined by

$$h(t) = \sum_{k=0}^{D} \rho_k e^{j\phi_k} \delta(t - \tau_k) \tag{2}$$

where $\phi_k = 2\pi f_0 \tau_k$, f_0 is the rf carrier frequency, and $\tau_k = (r_{TS_k} + r_{S_k R})/c$ is the absolute delay of the path from T via the k-th scatterer S_k to R. A typical measured $|h(t)|^2$ is shown in Figure 2. For general terrain characterized by the normalized radar cross-section σ^0,

$$|\rho_k|^2 = \frac{\lambda^2}{(4\pi)^3} \int_A \frac{\sigma^0 dA}{r_{TS_k}^2 r_{S_k R}^2} \tag{3}$$

where the integral is evaluated as a sum by defining the elements of area dA for each scatterer S_k using a topographical data base, and an estimate of σ_k^0 for each scatterer. The sum is performed over all scatterers S_k for which $(r_{TS_k} + r_{S_k R})$ is constant within the distance resolution [3]. Assuming that $r(t)$ contains reflected signals from D scatterers, $r(t)$ can be modeled as the sum of D components

$$s_k(t) = \rho_k e^{j\phi_k} s(t) \qquad k = 1, 2, \ldots, D, \tag{4}$$

where ρ_k is the reflection coefficient and ϕ_k is the phase shift; both of them can be treated as independent random variables. Assume that data measured at M adjacent locations are used to form a synthetic aperture or an M sensor array. The array receives $D + 1$ plane waves from D scatterers and one direct path. We further assume that the baseband signals, $s_k(t)$ have approximate constant amplitude during the time it takes for the wave to travel across the array. This assumption (narrow-bandness) enables modeling the time delays between receivers as phase shifts. Let $x_i(t)$ denote the complex representation of the signal at sensor i and $x(t)$ denote the array output vector with the ith element $x_i(t)$. Then the array output vector can be written by

$$x(t) = \sum_{k=0}^{D} a(\theta_k) s_k(t) + n(t), \tag{5}$$

where θ_k denotes the incident direction of the kth signal with $k = 0$ corresponding to the direct path, $a(\theta_k)$ denotes array manifold vectors

$$a(\theta_k) = [1, e^{j2\pi\Delta sin(\theta) f_0/c}, \ldots, e^{j2\pi\Delta(M-1) sin(\theta) f_0/c}]^T \tag{6}$$

with Δ is the distance between adjacent sensing elements, the direction θ is measured relative to the normal of the array (broadside), and $n(t)$ represents the additive measurement noise. A more compact expression of (5) is given by

$$x(t) = A(\theta) s(t) + n(t), \tag{7}$$

where

$$\theta = [\theta_0, \theta_1, \theta_2, \ldots, \theta_D]^T \tag{8}$$

$$A(\theta) = [a(\theta_0), a(\theta_1), a(\theta_2), \ldots, a(\theta_D)] \tag{9}$$

$$s(t) = [s_0(t), s_1(t), s_2(t), \ldots, s_D(t)]^T \tag{10}$$

Assuming that the noise is independent of the signal waveforms, the covariance matrix of the array output vectors is

$$R = E[x(t)x^H(t)] = A(\theta)SA(\theta) + N, \tag{11}$$

where S and N are the covariance matrices of the signals and noise, respectively. If only a finite observation is available, the sample covariance matrix can be used in place of the true covariance matrix.

2.2 DOA Estimation

More recently, the eigenvectors associated with the array covariance matrix have been employed to obtain high resolution DOA estimates. When using these methods, one must first form an estimate of the array covariance matrix from the sampled array outputs and then a generalized eigen-analysis of the matrix pair $(R,\ N)$ is made, that is

$$Rv_i = \lambda_i N v_i \qquad \text{for} \quad 1 \leq i \leq M, \tag{12}$$

where v_i and λ_i are the eigenvector and eigenvalue of the matrix pair $(R,\ N)$.

Subspace-based methods are those which explicitly or implicitly partition the vector space into signal and noise subspaces. The number of signals is determined by considering the distribution of the generalized eigenvalues. No matter which algorithm is used, the fact that the individual source steering vectors are orthogonal to each of the noise level eigenvectors is utilized to obtain the DOA estimates of the echo signals. For example, the MUSIC spectrum is given by

$$P(\theta) = \frac{1}{\sum_{i=D+2}^{M} ||a^H(\theta)v_i||^2}, \tag{13}$$

the peaks of which represent the MUSIC estimates of DOA's. The peaks correspond to trial θ that give $a(\theta)$ orthogonal to the M-D noise-space eigenvectors.

2.3 Spatial Smoothing

For subspace-based high resolution algorithms, accurate partition of the signal and noise subspace is crucial. If such a partition is not correctly done, these algorithms will totally fail to function. When there are some coherent signals present, or there is at least one signal which is a scaled version of another signal, the dimension of signal subspace will be reduced according to the number of coherent signal present. Thus subspace partition will inevitably be erroneous. In order to overcome this problem, spatial smoothing was proposed as a preprocessing scheme for subspace-based algorithms for a uniform linear array. Later, another spatial smoothing algorithm was proposed for arrays of any geometrical structures using interpolated arrays. However the spatial smoothing method using an interpolated array will cause some transformation errors , and special care has to be taken to make sure that these errors are much smaller than the errors introduced by observation noise. Since some echoes from mountainous scatterers may show strong coherence, spatial smoothing should be implemented as a preprocessing procedure. Further array processing information appears in [4]-[8].

2.4 Waveform Estimation

In addition to the DOA information, we are also interested in range, scatterer's signature and channel impulse response due to each scatterer. If the signal waveform from each individual scatterer or scattering region can be extracted, the information required should be easily obtained. The traditional means of doing this is beamforming [9], where a beam is formed artificially by electronic steering. The output of a beamformer is the linear combination of array output

$$y(t) = w^H x(t). \tag{14}$$

Beamformer algorithms differ only in the method of choosing the weighting vector w. With the DOA estimation available, the linearly constrained minimum variance beamformer (LCMV) is suitable for waveform estimation. To design a LCMV beamformer to extract the ith signal, a constraint matrix C is formed such that the array response to the remaining D signals are nulls while allowing the ith signal to pass without attenuation or

$$C^H w = f, \tag{15}$$

where

$$C = [a(\theta_0), a(\theta_1), \ldots, a(\theta_{i-1}), a(\theta_i), a(\theta_{i+1}), \ldots, a(\theta_D)] \tag{16}$$

$$f = [0, 0, \ldots, 0, 1, 0, \ldots, 0]^T. \tag{17}$$

The determination of the optimal weighting vector becomes the constrained minimization problem of

$$w = \text{argmin}_w w^H R w \qquad \text{subject to} \qquad C^H w = f, \tag{18}$$

and it is found to be

$$w = R^{-1} C [C^H R C]^{-1} f. \tag{19}$$

An alternative approach for waveform estimation is to employ least squares (LS) methods and waveforms corresponding to all signals can be obtained simultaneously. We have successfully applied the LS methods to interference cancellation in seismic signal processing. Specifically, there are two LS solutions; stochastic and deterministic LS solutions. The stochastic LS estimate of $s(t)$ is given by

$$s_{sls}(t) = SA^H(\theta)R^{-1}x(t), \tag{20}$$

where S is signal covariance matrix, and it can be estimated by

$$S = A^{\#}(\theta)(R - N)A^{\#*}(\theta), \tag{21}$$

with $A^{\#}(\theta)$ denoting the pseudo-inverse of $A(\theta)$.

The deterministic LS solution is

$$s_{dls}(t) = A^{\#}x(t), \tag{22}$$

2.5 Preliminary Tests

The MUSIC method outlined in Section 2.2 was tested on the array data. Since the recorded data is complex demodulated to base band, the phase information retained in the demodulated signal contains the carrier phase information; i.e., the phase change between each trace is actually the relative carrier phase shift at 910 MHz. This makes the application of simple narrow- band MUSIC possible.

It can be shown that the covariance of the demodulated PN sequence at the ith and jth sensors (traces) is:

$$E[s(t - \tau_i)s(t - \tau_j)\exp(j2\pi f_0(\tau_i - \tau_j))] \tag{23}$$

where τ_i and τ_j are the delay at the ith and jth traces, and f_0 is carrier frequency. This verifies that the direction vector should incorporate the carrier frequency (910 MHz). That is, the phase difference between two sensors should be $2\pi f_0 d_{ij}$ with f_0 being the carrier frequency and d_{ij} the delay between ith and jth traces. The complex number in the i, jth element of cov $[x]$ will have phase angle (for a single wavefront) equal to $2\pi f d_{ij}$. Knowing the carrier frequency, we can deduce the delay and arrival angle from the sample.

Using MUSIC, we form one covariance matrix for each time window position (short around each trial path delay), and do not change that window location for different trial angles of arrival. That is, we may determine multiple DOA's near each delay time with only one window, utilizing a number of time slices $\underline{x}_j$ to get cov $[x]$.

3 Reflector Location Estimation

A direction finding algorithm like MVDR or MUSIC is used for DOA. The delay time between transmitter and receiver can be used to determine an ellipse with foci at the transmitter and receiver locations. Together the DOA and time delay ellipse determine the reflector location. Time delay of course translates to distance. A change of coordinates eventually relates the reflector position to that of the transmitter.

3.1 Reflector Coordinates

The next step is to convert the angle vs distance results into distance vs distance plots, in order to coordinate the actual points of reflection with a topographical map. Assume that the angle α between the transmitter and the line of the receiver array and the distance D between the transmitter and the receiver can be measured. With the total distance T traveled by the signal and the angle of arrive ϕ_1 at the receiver known, the following equations can be derived from the cosine law (derivation for a West to East receiver array)

$$b^2 = r^2 + D^2 - 2rD\cos(\phi_2) \tag{24}$$

where

D = distance between the transmitter and the receiver

r = the radial distance from the receiver to the reflective point

ϕ_2 = the angle between the transmitter and the plane of the receiver array plus the angle of arrival (from 0 to 180 instead of -90 to 90): $\phi_2 = \alpha + \phi_1$

T = total distance travelled by the signal: $T = b + r$.

With $b^2 = T^2 - 2rT + r^2$, the value of r becomes,

$$r = \frac{T^2 - D^2}{2T - 2D\cos(\phi_2)} \tag{25}$$

The location of the receiver with respect to the transmitter, in x-y coordinates (representing West to East and North to South coordinates), can be expressed as

$$R_x = -D\cos(\alpha) \qquad R_y = +D\sin(\alpha) \tag{26}$$

Finally the reflector location with respect to the transmitter site can be written as

$$x = r\cos(\phi_1) + R_x \qquad y = r\sin(\phi_1) + R_y \tag{27}$$

There are slight differences in the derivation of the x-y coordinates for different orientations of the receiver and transmitter locations.

3.2 Linear Array Results Summary

The results of BF processing for the Capilano/Park Royal data have been plotted to the same scale as a topographic map of the region. These plots (Figures 3, 4) are then made into overlays to observe the regions of reflectance from the topographical map, Figure 1. Interesting responses follow significant topo contour gradients even back into a canyon due north. NS array results are similar, but show the expected front-back ambiguity.

When one positions the overlays on top of the provided topographical map, making sure that the North-South directions are aligned properly such that the point (0, 0) on the overlay lies on the center of the "cross" on the map, one can see the major sources of the reflection. One of the more impressive regions to notice is the area near Grouse Mountain. Near the bottom of Grouse Mountain, beside the lake, there are very steep and jagged cliff formations, most likely causing the concentration of the power in that region. Looking at the "W to E" overlay, one can observe how well the contours of the surrounding topography are closely mapped to the source reflectance contours.

4 Summary

The results reported above are very encouraging. The method and results of processing of cross-array data (two orthogonal linear arrays), phase drift correction, center element location are omitted for lack of space.

Howver, much work remains to be done. It is the long term goal of this research to produce an algorithm which predicts channel response in a coverage area. Most literature on the subject deals with deterministic approaches using direct paths and paths of similar delay from known or hypothesized objects, surfaces or structures, for example [1]. These approaches involve extensive calculations, incorporating numerical solutions to equation like those in Section 2.

We are suggesting that all potential reflecting incremental surfaces may be preclassified by its features, thereby excluding all but a small proportion from the final estimation of the channel response. A suggested methodology for quickly determining regions of reflecting surfaces which are likely to cause significant reflections is omitted for lack of space. How large a training set should be found to instruct the classifier and what parameters should be used, aside from the obvious ones, needs further study. We also expect that a number of refinements must be added to the method we have outlined. Nevertherless we are confident that further research merging the array inversion methods with classification can lead to quick solutions to cell site design and coverage prediction. Lastly, we point out that we have not yet associated complex impulse responses with the reflectors we have located. However this may straightforwardly be done with procedures of Section 2.4.

References

[1.] Lebherg, M., W. Wiesbeck and W. Krank, "A versatile wave propagation model for the VHF/UHF range considering three-dimensional terrain", IEEE Trans. Antennas and Propagation, vol. 40, pp. 1121-1131, 1992.

[2.] Braun, Walter R. and Ulrich Dersch, "A physical mobile radio channel model", IEEE T-Vehic. Tech., 40, No. 2, pp. 472-482, May 1991.

[3.] Driessen, P.F., "Multipath delay characteristics in mountainous terrain at 900 MHz", *Proc. IEEE Vehic. Tech. Conf.*, Denver, May 1992, pp. 520-523.

[4.] Drosopoulos, A. and S. Haykin, "Experimental characterization of diffuse multipath at 10.2 GHz using method of multiple windows", *IEE Electronics Letters*, Vol. 27, No. 10, pp. 798-799, May 1991.

[5.] Marple, S.L., Jr., *Digital spectral analysis with applications*, Prentice-Hall, INC., 1987.

[6.] Evans, J.E., J.R. Johnson and D.F. Sun, "Application of advanced signal processing techniques to angle of arrival estimation in ACT navigation and surveillance systems", M.I.T. Lincoln Lab., Lexington, MA, Tech. Rep. 582, June 1982.

[7.] Friedlander, "Direction finding using an interpolated array", IEEE Proceedings, ICASSP, 1990.

[8.] Kirlin, R.L. and W. Du, "Design of transformation matrices for array processing", Proc. IEEE Int. Conf. Acoustics, Speech & Signal Processing, ICASSP-91, pp. 1389-1392, May 1991.

[9.] Neidell, N.S. and M.T. Taner, "Semblance and other coherency measures for multi-channel data", vol. 26, No. 3, pp. 482-497, 1971.

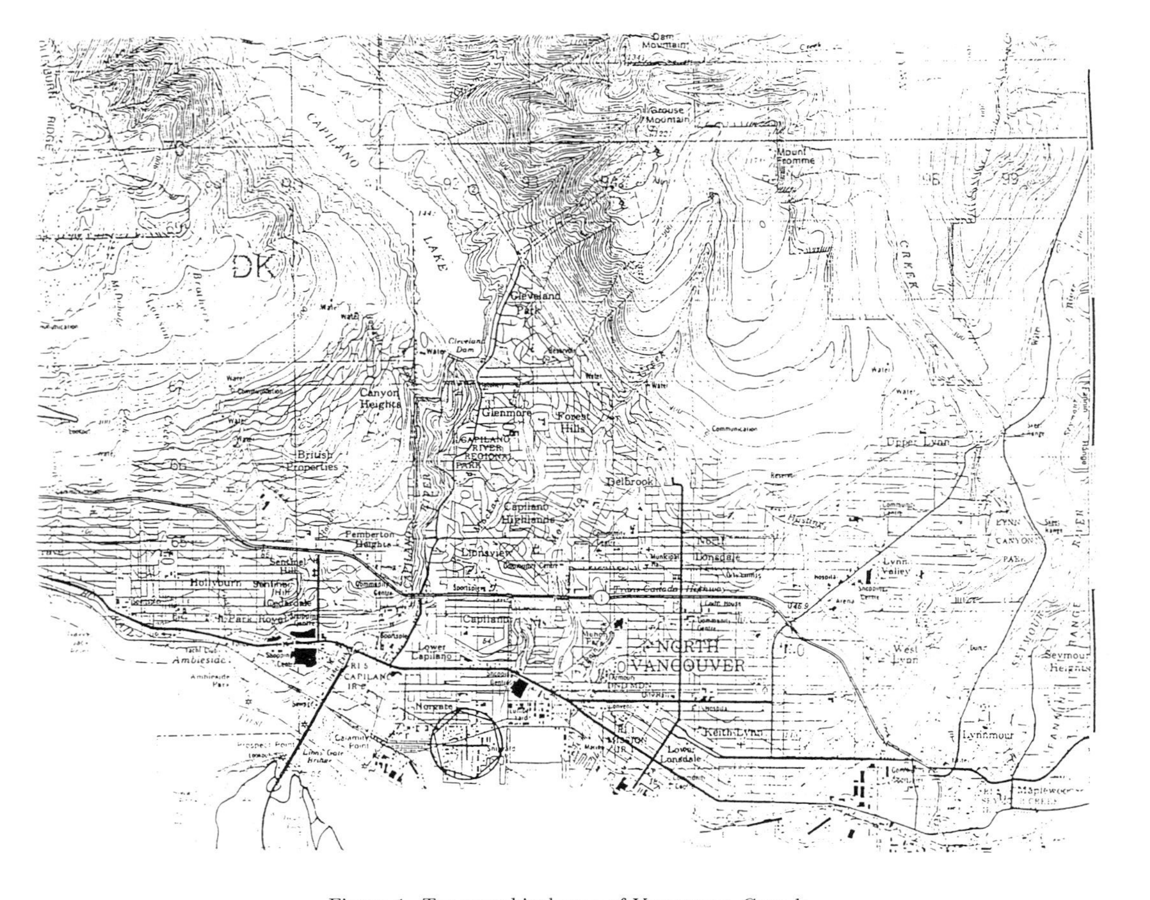

Figure 1. Topographical map of Vancouver, Canada.

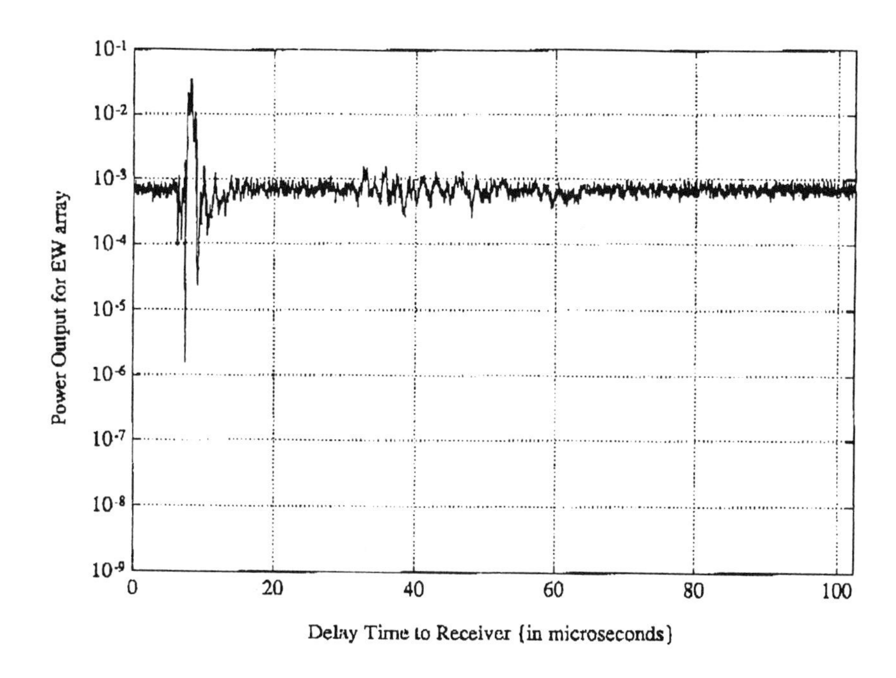

Figure 2. Typical data arrival, baseband complex demodulation power vs delay.

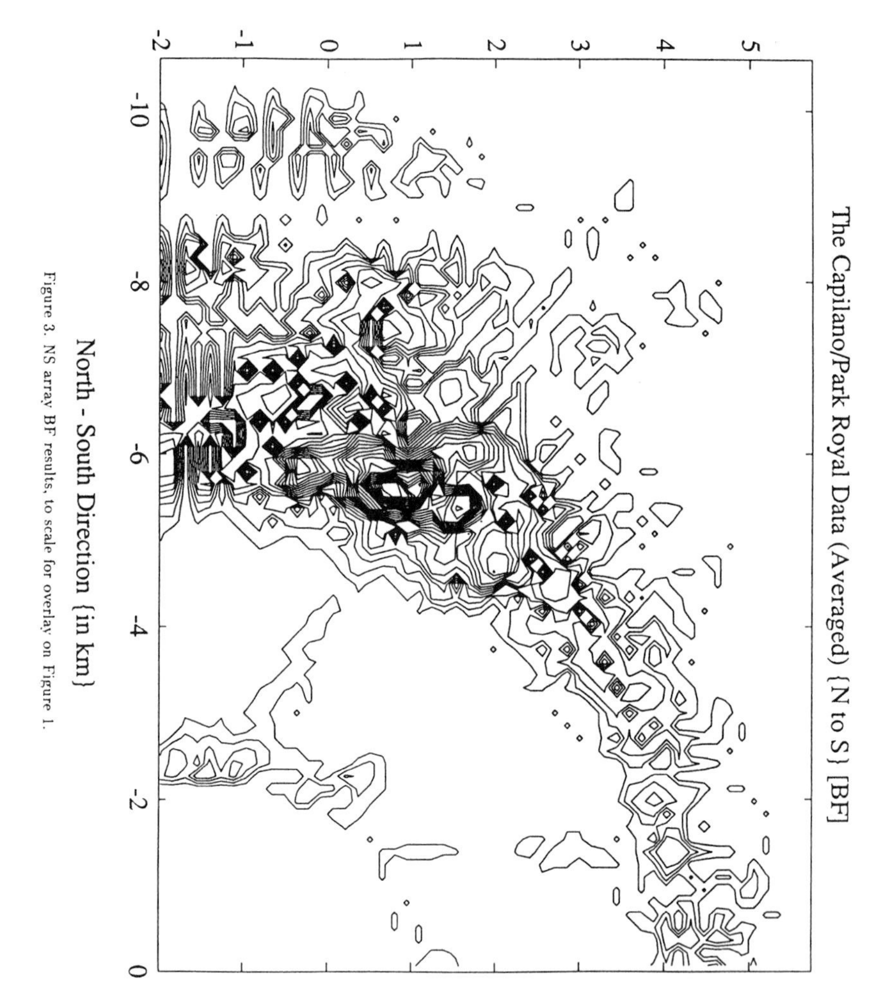

Figure 3. NS array BF results, to scale for overlay on Figure 1.

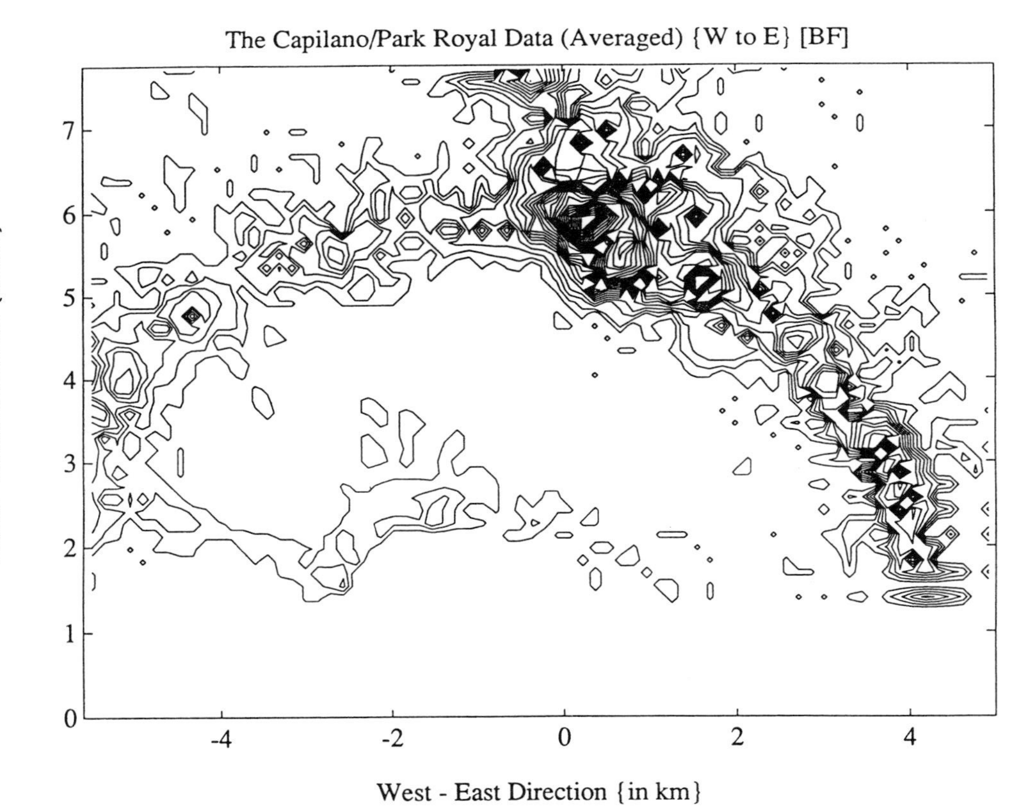

Figure 4. EW array BF results, to scale for overlay on Figure 1.

2

Blind Adaptive Antenna Arrays for Increased Capacity in Cellular Communications*

Stephan V. Schell
Dept. of Electrical Eng.
The Pennsylvania State University
University Park, PA 16802, USA

William A. Gardner
Dept. of Elec. & Comp. Eng.
University of California
Davis, CA 95616, USA

Email: schell@gauss.ece.psu.edu

Abstract

In this paper we describe a new cellular communication scheme, based on blind adaptive spatial filtering, for increasing spectral efficiency relative to existing and proposed systems. Depending on propagation conditions and the number of antennas in the array, between 2 and 64 times as many users can be accommodated as in existing and proposed systems. In this scheme, an antenna array at the base station is very rapidly adapted to separate spectrally overlapping (but individually bandwidth-efficient) signals received from multiple users. The adapted array is also used to transmit signals to the users. Transmission and reception are time-division multiplexed with each other, and are synchronized among all users; thus, antenna arrays are not required on the mobile units for co-channel interference suppression. Conventional adaptive array schemes require training signals and thus are limited in the allowable number of antenna elements by the tradeoff between training-signal length and message capacity that must arise in the rapidly varying propagation environment typical in land mobile cellular radio. In the new scheme, the Spectral Coherence Restoral (SCORE) algorithm, which adapts the antenna array to separate the signals of different users on the basis of their different cyclostationarity properties, does not require any training signals or calibration data. It is shown that the bandwidth efficiency of the scheme can be substantially higher than that of other schemes, primarily because a large number of antennas can be used in the array at the base station, and thus a large number of users can occupy the same frequency band. The performance of the new method is evaluated in computer simulations as a function of the number of antenna elements, the number of active spectrally overlapping users, and the spatio-temporal properties of the multipath.

---

*This work was supported in part by the Office of Naval Research under contract N00014-92-J-1218.

1 Introduction

Demand for mobile communication continues to increase as it becomes easier to use, is more widely available, and offers a greater variety of services. The need for new mobile communication systems having increased spectral efficiency relative to current systems is compounded by increasing demand for radio spectrum allocations from other communication services. Various multiple access schemes have been proposed for increasing spectral efficiency, including time division multiple access (TDMA) and frequency division multiple access (FDMA), although code division multiple access (CDMA) might offer the greatest potential increase in capacity (e.g., see [1],[2],[3]) in addition to inherently mitigating the effects of multipath. Sectorization, in which fixed multi-beam or multi-sector antennas are used, can increase capacity at the expense of having to manage the handoff between sectors within the same cell. However, the multiplicity of spatial channels, that arises because each mobile user occupies a unique spatial location, is not fully exploited by any of the aforementioned schemes.

All schemes use a large-scale form of space division multiple access (SDMA) by dividing a large geographic area into cells. The mobile users within each cell are served by a base station, shown in Figure 1 as being at the center of each cell.

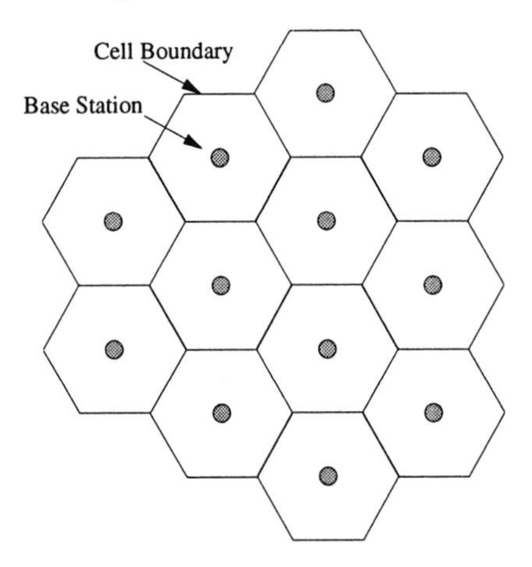

Figure 1: Typical cell structure in a cellular radio system.

Some schemes which use SDMA within each cell by spatially filtering to separate spectrally overlapping signals from different users have been proposed with potential increases in spectral efficiency over conventional analog FM-FDMA schemes of a factor of 30 (cf. [4, 5]). These schemes adapt the antenna array either by estimating the directions of arrival of the spectrally overlapping signals and then using these estimates to compute appropriate weights for the spatial filter [6], or by minimizing the time-averaged squared error between a known training signal and the output of the spatial filter [4, 7, 8, 9]. In either case, the properly adapted array at the base station can spatially separate the spectrally overlapping users as shown in Figure 2, where multipath is neglected for clarity.

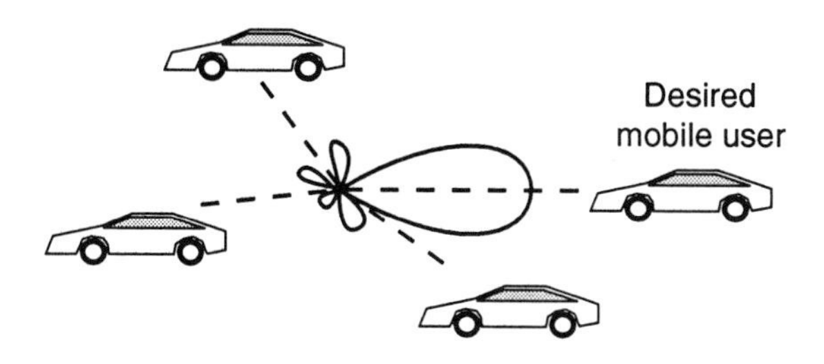

Figure 2: Antenna pattern of array at base station extracting one of several spectrally overlapping signals from mobile users.

The schemes based on direction estimation have numerous disadvantages, including computationally intensive algorithms, poor performance in the presence of multipath signals arriving from different directions, the need to measure, store, and update array calibration data, and considerable sensitivity to errors in the array calibration data. The schemes that require a training signal have different disadvantages, including the need to use capacity to periodically transmit the training signal, the need to synchronize the received and locally generated copies of the training signal, and the need to adaptively increase or decrease the duration of the training signal to accommodate varying levels of interference.

2 The SCORE Algorithm

An alternative means of adapting the array, which requires neither a training signal nor any type of array calibration data, is summarized here. As explained in [10, 11, 12], the Spectral Coherence Restoral (SCORE) algorithm adaptively combines the output signals of an antenna array so as to maximize the spectral coherence (or degree of cyclostationarity) exhibited by the signal at the output of the combiner. This scheme is appropriate for many man-made communication signals, which exhibit cyclostationarity due to the periodic gating, keying, sampling, and mixing operations used in modulators [13, 14, 15]. Of particular relevance to the scheme proposed in this paper is the fact that the L most dominant eigenvectors defined by the following equation,

$$\boldsymbol{R}_{\boldsymbol{x}\boldsymbol{x}^*}^{2f} \boldsymbol{R}_{\boldsymbol{x}\boldsymbol{x}}^{-*} \boldsymbol{R}_{\boldsymbol{x}\boldsymbol{x}^*}^{2f\,H} \boldsymbol{w}_m = \lambda_m \boldsymbol{R}_{\boldsymbol{x}\boldsymbol{x}} \boldsymbol{w}_m, \quad \text{for } m = 1, \ldots, M$$

each extracts some linear combination of the L multipath signals of the user having carrier frequency f and rejects interference (including the signals from other users), where M is the number of antennas in the array, $\boldsymbol{x}(n)$ is the sampled complex envelope of the vector of signals at the outputs of the antennas, the correlation matrices are defined by

$$\boldsymbol{R}_{\boldsymbol{x}\boldsymbol{x}^*}^{2f} = \left\langle \boldsymbol{x}(n)\boldsymbol{x}^T(n)e^{-j2\pi(2f)n} \right\rangle \quad \text{and} \quad \boldsymbol{R}_{\boldsymbol{x}\boldsymbol{x}} = \left\langle \boldsymbol{x}(n)\boldsymbol{x}^H(n) \right\rangle,$$

$\langle\rangle$ denotes the time-average over N samples of data, and superscripts $*$, T, H, and $-*$ denote conjugation, transposition, conjugate transposition, and matrix inversion and conjugation, respectively. Thus, the mth output of SCORE is $\boldsymbol{w}_m^H \boldsymbol{x}(n)$, and can be shown to be a linear combination of the multipath signals from the user having carrier frequency f, plus residual contributions from other users and noise.

This version of SCORE can be interpreted as a multidimensional generalization of a standard carrier-recovery device for scalar BPSK signals, which squares the signal to regenerate a sine wave at the doubled carrier frequency. SCORE adapts the most dominant weight vector $\boldsymbol{w}_1$ so that the ratio of power in the regenerated sine wave at the doubled carrier to the power in the output ignal is maximized. This action necessarily minimizes the contributions of other signals that do not have the chosen carrier frequency.

3 Overview of STFDMA Scheme

The space-time-frequency division multiple access (STFDMA) scheme first proposed in [5] and described here has some things in common with that proposed in [8], but uses SCORE to eliminate the need for a reference (training) signal to adapt the array. An adaptive antenna array at the base station separates the temporally and spectrally overlapping received signals of different users in the cell and transmits directively to each user, exploiting multipath when present. Unlike schemes that rely solely on frequency, time, or code division multiplexing and thus use only one spatial channel, the proposed scheme exploits space as well as partial time and frequency division multiplexing and thus uses multiple spatial, temporal, and spectral channels. If the signals from a group of users are spatially separable at the base station, then they can be assigned to spectrally overlapping bands. Also, if the signals from a group of users are spatially *in*separable at the base station, then they must be assigned to spectrally disjoint bands.

Also, signals coming from the individual users can be assigned to time intervals that are interleaved with those assigned to signals coming from the base station. Under the assumption that users are sufficiently well distributed throughout the cell, all available spatial and spectral channels can be used effectively. Since the number of multiple spatial channels that can be separated from each other by the antenna array is approximately equal to the number of antenna elements in the array (which can be quite large), overall capacity can be much greater than schemes using a single spatial channel. Also, unlike adaptive array schemes that require direction estimation processors or known training signals, the proposed scheme uses the SCORE algorithm, and thus does not require array calibration data or computationally intensive multidimensional searches nor does it waste channel capacity by transmitting a training signal.

4 Details of STFDMA Scheme

As shown in Figure 3, the scheme uses TDM of reception (Rx, or uplink) and transmission (Tx, or downlink) frames for a given user so that a spatial filter adapted during reception can be used for transmission. Thus, spatial directivity for Rx is preserved for Tx. Up to 10 users share the same carrier frequency, and are separable by the TDM scheme shown in Figure 3. The data rate of each user's vocoded speech is 8 kb/s, which must then be doubled due to TDM of Rx and Tx, and multiplied further by a factor of 10 due to TDM of 10 users, to yield an instantaneous data rate of 160 kb/s. The signals are BPSK using 100% excess-bandwidth Nyquist-shaped pulses, for a signal bandwidth of 320 kHz. This data rate is high enough that a sufficient number of independent time samples can be collected for adaptation of the spatial filter over a period of roughly 500 μs, which is short enough that the propagation environment is approximately stationary, given a typical fast fading rate of 100 fades/s for land mobile radio.

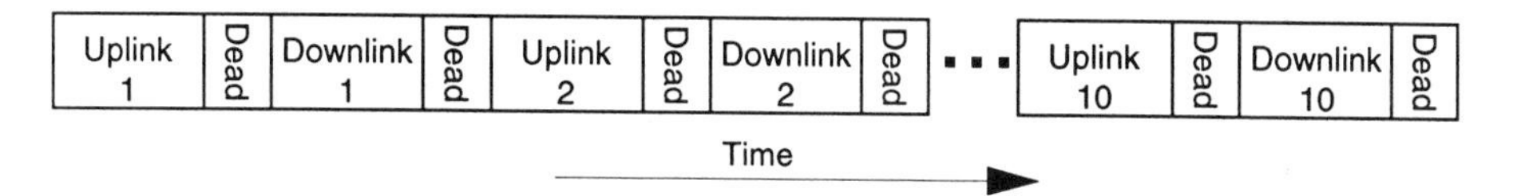

Figure 3: One complete TDM cycle for the uplink (reception) and downlink (transmission) phases at the base station for 10 mobile users sharing the same carrier frequency.

As shown in Figure 4, carrier assignments for different groups of 10 TDM'd users can be such that their signals are almost completely spectrally overlapping, provided that the users assigned to the first TDM time slot in different groups are spatially separable, and similarly for the users assigned to the second through the tenth TDM time slots. Since only one user having a particular carrier frequency is active at any given time, SCORE is used to adapt a spatial filter to separate that user from all other users. The number of spatially separable users is limited by the number of antennas in the array, which in turn is limited by convergence-time considerations.

In general, for a total system-bandwidth B_t, single-user channel-bandwidth B_c, frequency-reuse factor r, and minimum carrier separation f_{sep}, the maximum number L of users that can

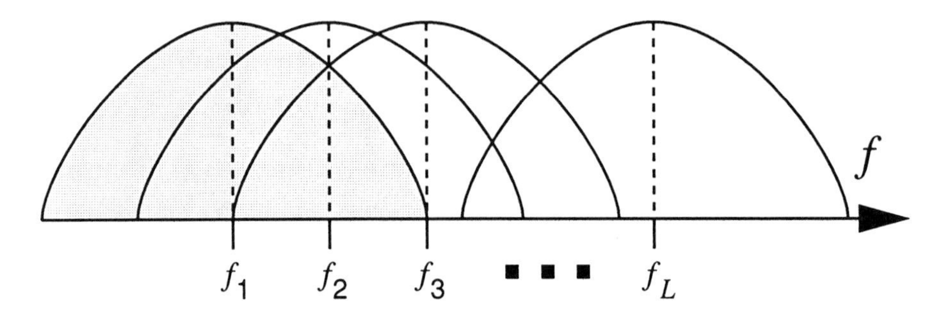

Figure 4: Frequency allocation in the proposed scheme. The spectrum of user number one is shaded to enhance clarity.

be accommodated in one cell by the frequency allocation scheme in Figure 4 is $L = 10(B_t/r - B_c)/f_{sep} + 10$. The number M of antenna elements required to separate these signals is bounded from below by the number K of users whose signals are spectrally overlapping with any given user's signal, where $M > K = 2(B_c/f_{sep} - 1)$. Using $B_c = 320$ kHz and $f_{sep} = 10$ kHz, at least 63 antenna elements (which can be omnidirectional) are needed to separate the signals of all users, assuming that the energy from each user arrives at the base station from a single direction, and assuming that the users are uniformly distributed throughout the cell. In practice, more antenna elements might be required to achieve adequate performance at full capacity in the presence of spatially separable multipath, although fewer antenna elements can suffice if a lower capacity is opted for and channels are appropriately allocated. Also, adaptive equalization should follow the spatial filtering to mitigate the time-smearing effects of multipath.

5 Capacity Example

For the purpose of comparing the potential increase in capacity due to the proposed SCORE-STFDMA scheme relative to the analog FM-FDMA, TDMA, and CDMA schemes as considered in [1], consider a total system-bandwidth of $B_t = 12.5$ MHz. In the following comparison, the number of channels needed by each user in the FM-FDMA, TDMA, and CDMA schemes is two (one for transmission and one for reception), and one channel is needed by each user in the proposed scheme because transmission and reception are multiplexed in time. With FM-FDMA, using a channel bandwidth of 30 kHz, 2 channels per user, and a frequency reuse factor of 7 yields 30 users per cell. With TDMA, using a channel bandwidth of 30 kHz with three time slots for TDMA, 2 channels per user, and a frequency reuse factor of 4 yields 150 users per cell. With CDMA, using 2 channels per user, a frequency reuse factor of 1, sectorization of 3, and voice activity factor of 3/8 yields 1200 channels per cell [1] or 600 users per cell. With the proposed SCORE-STFDMA scheme, using a channel bandwidth of 320 kHz, 1 channel per user, a frequency reuse factor of 3, and a carrier separation of 10 kHz yields up to 3850 users per cell. Decreasing the carrier separation to 5 kHz allows up to 7700 users per cell at the expense of doubling the number of antennas. If future evaluations of SCORE-STFDMA show that the spatial directivity at the base station allows the frequency reuse factor to drop from three to one, then the capacity of SCORE-STFDMA would triple.

These results are summarized in Table 1.

Scheme	FM-FDMA	TDMA	CDMA	SCORE-STFDMA $f_{sep} = 10$ kHz
Users/cell	30	150	600	3850
Rel. Eff.	1	5	20	128

Table 1: Summary of capacity and relative efficiency.

6 Performance Evaluation

In this section, the performance of the proposed method is investigated via computer simulations. The bit error rate (BER) of the mobile users' messages received at the base station is estimated for different numbers of active users, antennas, and propagation conditions. Since the same propagation paths and frequency bands are used in both reception and transmission under the assumptions that the channel is linear and time-invariant over a TDM slot, it is reasonable to assume that this BER is also a good estimate of the BER of the base station's messages received by the mobile users.

6.1 Modeling the Environment

Unlike the temporal (delay) and angular distribution of received signals at the mobile unit, which has been studied extensively [16]-[21] (and references therein), few results on the distribution at the base station are available. Obviously, the standard technique of measuring the delay-Doppler characteristics of a coded pulse is inapplicable to the base station, which is immobile. The alternative of using a modern array-based high-resolution direction-finding system based on MUSIC [22], ESPRIT [23], or Weighted Subspace Fitting [24] can be costly.

Consequently, the true properties of the spatio-temporal channel at the base station must be inferred from the existing (though not directly applicable) studies [16]-[21] and some common sense. Three types of propagation conditions are considered: 1) a pessimistic choice, in which significant reflectors exist around the base station, such that the signal from each mobile user arrives from up to 10 angles randomly dispersed throughout the range from 0 to 360 degrees, regardless of the true line of bearing between the base station and the mobile users; 2) a more optimistic choice, in which the signal from each mobile user arrives from 20 angles randomly dispersed in a 10-degree sector centered on the true line of bearing, where the mobile users are uniformly spaced throughout the range 0 to 360 degrees; and 3) a third choice in which the signal from each mobile user arrives from only three angles (three-ray model) randomly dispersed in a 5-degree sector centered on the true line of bearing. Environment 1 could correspond to a dense urban environment whereas Environment 2 could correspond to a less cluttered suburban environment and Environment 3 could correspond to an even less cluttered environment. In all three cases, Doppler shifts of up to 90 Hz (for vehicle speeds up to 70 MPH or 110 KPH) are present, the carrier phase for each signal path is a uniformly distributed random variable on the interval $[0, 2\pi)$ (radians), and the antenna array is circular with half-wavelength spacing between adjacent elements.

6.2 Environment 1

In the simulated urban environment, the BER is evaluated after spatial filtering is performed with a 32-element array adapted by SCORE, for in-band SNRs ranging from -12 dB to 33 dB. In-band SNR is defined here as the ratio of power of a line-of-sight signal to the power of the noise in that signal's spectral support. For each value of in-band SNR, 1000 independent trials are performed. Ten spectrally overlapping users are present, and the multipath delays range from 0 to 15 μs, where the strength of a path is an exponentially decaying function of the delay. As shown in Figure 5, the BER decreases as the SNR rises to 3 dB but then remains at approximately 0.015 even as the SNR increases to 33 dB.

6.3 Environment 2

In this simulated suburban environment, the BER is evaluated for a 16-element array as a function of the number of users, which ranges from 1 to slightly less than the number of array elements. The delay on each path is a uniformly distributed random variable in the range 0 to 6 μs (one baud period), and the strength of each reflected path is a randomly scaled version of the strength of the direct path, where the scale factor is uniformly distributed on the interval $[0, 1)$. As shown in Figure 6, the BER remains low (10^{-3} or less) when the number of users is less than half the number of array elements but rises sharply for more users. From one to three percent of the independent trials for each choice of the number of users and the maximum delay yielded spurious results that were excluded from the overall average. Even for choices that yielded higher average BER, many of the trials resulted in no bit-errors at all.

6.4 Environment 3

In this simulated sparse suburban environment, the BER is evaluated for a 64-element array as a function of the number of users, which ranges from 1 to slightly less than the number of array elements. The delays and strengths of each path are the same as in En-

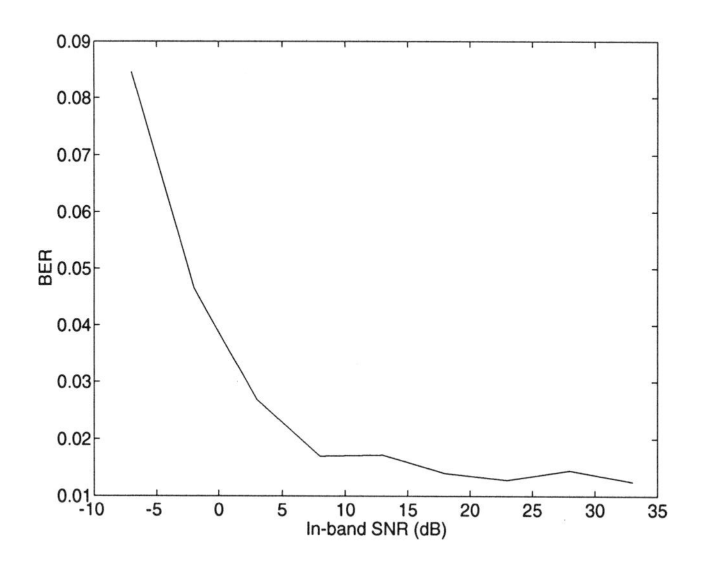

Figure 5: BER in simulated urban environment for 32-element array.

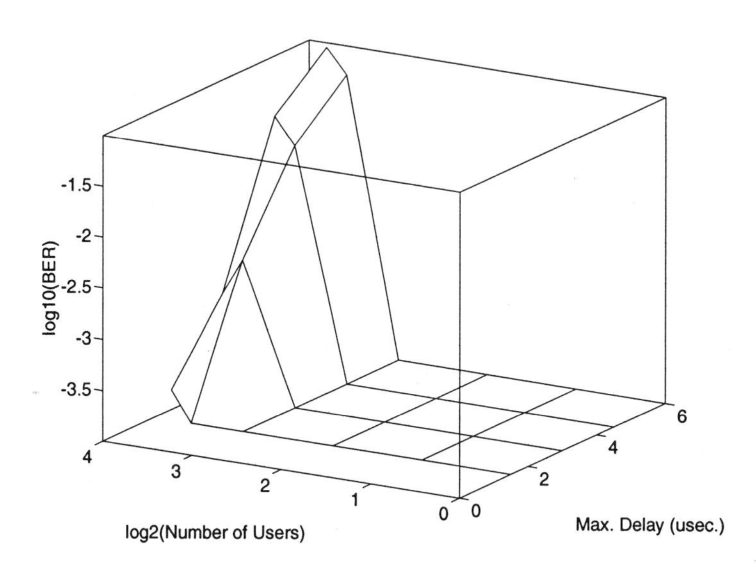

Figure 6: BER in simulated suburban environment for 16-element array.

vironment 2. As shown in Figure 7, the BER remains low (10^{-3} or less) when 32 or fewer users are present but rises sharply for more users. As before, from one to three percent of the independent trials yielded spurious results that were excluded from the overall average.

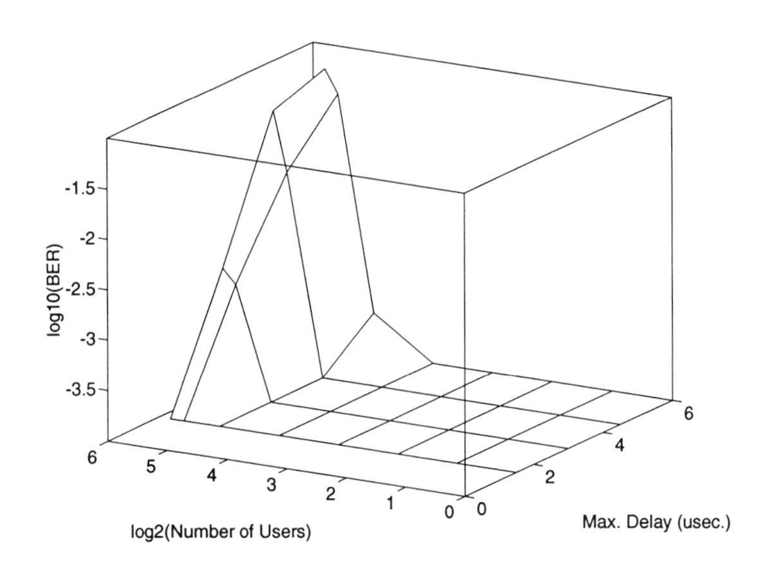

Figure 7: BER in simulated sparse suburban environment for 64-element array.

7 Conclusion

As indicated in the example of potential capacity and results of computer simulations in this paper, the scheme proposed in this paper could be a high-capacity alternative to existing schemes based on TDMA or CDMA in the mobile cellular environment. The average BER obtained in the computer simulations ranged from 10^{-2} in a dense urban environment down to below 10^{-4} in a suburban environment, without the use of adaptive temporal equalization or coding. Based on these promising results, a more extensive performance evaluation and an experimental characterization of the spatio-temporal channel at the base station are warranted. Also, it is unclear from the present results whether the signal is degraded more by its own multipath or by residual interference from other users. In the former case, adaptive temporal equalization and/or coding would be especially beneficial, and the the Constant Modulus Algorithm [25, 26] could also be used to adaptively linearly combine the multiple SCORE output signals. If residual interference from other users is significant, then more antennas may be needed in the array.

8 Acknowledgement

The authors gratefully acknowledge the assistance of Peter A. Murphy with the computer simulations.

References

[1] W. C. Lee, "Overview of cellular CDMA," *IEEE Trans. Vehic. Tech.*, vol. VT-40, no. 2, pp. 291–302, May 1991.

[2] K. S. Gilhousen, I. M. Jacobs, R. Padovani, A. J. Viterbi, L. A. Weaver, Jr., and C. E. Wheatley III, "On the capacity of cellular CDMA system," *IEEE Trans. Vehic. Tech.*, vol. VT-40, no. 2, pp. 303–312, May 1991.

[3] R. L. Pickholtz, L. B. Milstein, and D. L. Schilling, "Spread spectrum for mobile communications," *IEEE Trans. Vehic. Tech.*, vol. VT-40, no. 2, pp. 313–322, May 1991.

[4] S. C. Swales, M. A. Beach, D. J. Edwards, and J. P. McGeehan, "The performance enhancement of multibeam adaptive base-station antennas for cellular land mobile radio systems," *IEEE Trans. Vehic. Tech.*, vol. VT-39, no. 1, pp. 56–67, Feb. 1990.

[5] W. A. Gardner, S. V. Schell, and P. A. Murphy, "Multiplication of cellular radio capacity by blind adaptive spatial filtering," in *Proc. of IEEE Int'l. Conf. on Selected Topics in Wireless Comm.*, (Vancouver, B.C., Canada), pp. 102–106, June 1992.

[6] S. Anderson, M. Millnert, M. Viberg, and B. Wahlberg, "An adaptive array for mobile communication systems," *IEEE Trans. Vehic. Tech.*, vol. VT-40, no. 1, pp. 230–236, Feb. 1991.

[7] J. H. Winters, "Optimum combining in digital mobile radio with cochannel interference," *IEEE Trans. Vehic. Tech.*, vol. VT-33, no. 3, pp. 144–155, Aug. 1984.

[8] P. S. Henry and B. S. Glance, "A new approach to high-capacity digital mobile radio," *Bell Syst. Tech. J.*, vol. 60, no. 8, pp. 1891–1904, Oct. 1981.

[9] Y.-S. Yeh and D. O. Reudnik, "Efficient spectrum utilization for mobile radio systems using space diversity," *IEEE Trans. Comm.*, vol. COM-30, no. 3, pp. 447–455, Mar. 1982.

[10] B. G. Agee, S. V. Schell, and W. A. Gardner, "Spectral self–coherence restoral: A new approach to blind adaptive signal extraction," *Proc. IEEE*, vol. 78, no. 4, pp. 753–767, Apr. 1990.

[11] S. V. Schell and W. A. Gardner, "Maximum likelihood and common factor analysis-based blind adaptive spatial filtering for cyclostationary signals," in *Proc. IEEE Int. Conf. Acoust., Speech, Signal Processing*, (Minneapolis, Minnesota), pp. IV:292–295, Apr. 1993.

[12] S. V. Schell, "An overview of sensor array processing for cyclostationary signals," in *Cyclostationarity in Communications and Signal Processing* (W. A. Gardner, ed.), IEEE Press, 1993.

[13] W. A. Gardner, "Spectral correlation of modulated signals: Part I — analog modulation," *IEEE Trans. Comm.*, vol. COM–35, no. 6, pp. 584–594, June 1987.

[14] W. A. Gardner, W. A. Brown, III, and C.-K. Chen, "Spectral correlation of modulated signals: Part II — digital modulation," *IEEE Trans. Comm.*, vol. COM–35, no. 6, pp. 595–601, June 1987.

[15] W. A. Gardner, "Exploitation of spectral redundancy in cyclostationary signals," *IEEE Signal Processing Mag.*, vol. 8, no. 2, pp. 14–37, Apr. 1991.

[16] J. Shapira, "Channel characteristics for land cellular radio, and their systems implications," *IEEE Ant. and Prop. Mag.*, vol. 34, no. 4, pp. 7–16, Aug. 1992.

[17] D. C. Cox, "Delay doppler characteristics of multipath propagation at 910 MHz in a suburban mobile radio environment," *IEEE Trans. Antennas Propagat.*, vol. AP-20, no. 5, pp. 625–635, Sept. 1972.

[18] D. C. Cox, "A measured delay-doppler scattering function for multipath propagation at 910 mhz in an urban mobile radio environment," *Proc. IEEE*, pp. 479–480, Apr. 1973.

[19] A. S. Bajwa and J. D. Parsons, "Small-area characterisation of UHF urban and suburban mobile radio propagation," *Proc. IEE, Pt. F*, vol. 129, no. 2, pp. 102–109, Apr. 1982.

[20] W. R. Braun, "A physical mobile radio channel," *IEEE Trans. Vehic. Tech.*, vol. VT-40, no. 2, pp. 472–482, May 1991.

[21] H. J. Thomas, T. Ohgane, and M. Mizuno, "A novel dual antenna measurement of the angular distribution of received waves in the mobile radio environment as a function of position and delay time," in *Proc. of IEEE 42nd Vehic. Tech. Conf.*, (Denver, CO), pp. 546–549, May 1992.

[22] R. O. Schmidt, "Multiple emitter location and signal parameter estimation," *IEEE Trans. Antennas Propagat.*, vol. AP-34, no. 3, pp. 276–280, Mar. 1986.

[23] R. Roy and T. Kailath, "ESPRIT–estimation of signal parameters via rotational invariance techniques," *IEEE Trans. Acoust., Speech, Signal Processing*, vol. ASSP-37, no. 7, pp. 984–995, July 1989.

[24] M. Viberg, B. Ottersten, and T. Kailath, "Detection and estimation in sensor arrays using weighted subspace fitting," *IEEE Trans. Signal Processing*, vol. SP–39, no. 11, pp. 2436–2449, Nov. 1991.

[25] J. Treichler and M. Larimore, "New processing techniques based on the constant modulus adaptive algorithm," *IEEE Trans. Acoust., Speech, Signal Processing*, vol. ASSP-32, no. 2, pp. 420–431, Apr. 1985.

[26] R. Gooch and J. Lundell, "The CM array: An adaptive beamformer for constant modulus signals," in *Proc. IEEE Int. Conf. Acoust., Speech, Signal Processing*, (Tokyo), pp. 2523–2526, 1986.

3

Reverse Channel Performance Improvements in CDMA Cellular Communication Systems Employing Adaptive Antennas

Joseph C. Liberti and Theodore S. Rappaport

Mobile and Portable Radio Research Group
Bradley Department of Electrical Engineering
Virginia Tech, Blacksburg, Virginia 24061-0111

E-mail: liberti@mprg1.mprg.ee.vt.edu

Abstract

In this paper, we examine the performance enhancements that can be achieved by employing adaptive antennas in Code Division Multiple Access (CDMA) cellular radio systems. The goal is to determine what improvements are possible using narrowbeam antenna techniques, assuming that adaptive algorithms and the associated hardware to implement these systems can be realized. Simulations and analytical results are presented which demonstrate that adaptive antennas at the base station can dramatically improve the reverse channel performance of multi-cell radio systems, and new analytical techniques for characterizing mobile radio systems which employ frequency reuse are described. We also discuss the effects of using adaptive antennas at the portable unit.

1. Introduction

Current day mobile radio systems are becoming congested due to growing competition for spectrum. Many different approaches have been proposed to maximize data throughput while minimizing spectrum requirements for future wireless personal communications services [Coo78, Sal90].

The reverse link presents the most difficulty in CDMA cellular systems for several reasons. First of all, the base station has complete control over the relative power of all of the transmitted signals on the forward link, however, because of different radio propagation paths between each user and the base station, the transmitted power from each portable unit must be dynamically controlled to prevent any single user from driving the interference level too high for all other users [Rap92]. Second, transmit power is limited by battery consumption at the portable unit, therefore there are limits on the degree to which power may be controlled. Finally to maximize performance, all users on the forward link may be delay synchronized much more easily than users on the reverse link [Pur77].

Adaptive antennas at the base station and at the portable unit may mitigate some of these problems. In the limiting case of infinitesimal beamwidth and infinitely fast tracking ability, adaptive antennas can provide a unique channel that is free from interference for most users. All users within the system would be able to communicate at the same time using the same frequency channel, in effect providing Space Division Multiple Access (SDMA) [Gar92]. In addition, a perfect adaptive antenna system would be able to track individual multipath components and combine them in an optimal manner to collect all of the available signal energy [Swa90].

The perfect adaptive antenna system described above is not feasible since it requires infinitely large antennas (or alternatively, infinitely high frequencies). This raises the question of what gains might be achieved using reasonably sized arrays with moderate directivities.

For interference limited asynchronous reverse channel CDMA over an AWGN channel, operating with perfect power control with no interference from adjacent cells and with omnidirectional antennas used at the base station, the bit error rate (BER), P_b,is approximated by [Pur77]

$$P_b \approx Q\left(\sqrt{\frac{3N}{K-1}}\right) \qquad \text{Eq. 1.1}$$

where K is the number of users in a cell and N is the spreading factor. $Q(Y)$ is the standard Q-function. Eq. 1.1 assumes that the signature sequences are random and that K is sufficiently large to allow the Gaussian approximation, described in [Pur77], to be applied.

In order to develop simple bit error rate expressions for simultaneous asynchronous interference limited CDMA users when directional antennas are used we assume that the bit error rate expression of Eq. 1.1 can be expressed as

$$P_b \approx Q\left(\sqrt{3N \times CIR}\right) \qquad \text{Eq. 1.2}$$

where CIR is the ratio of the power of the desired signal to the total interference. This approximation is known to be inaccurate when the received power levels from different users are widely different and when the number of users is small [Mor89], however, it provides a reasonable approximation when K is large.

Let us assume that the base station is able to form a directional beam with power pattern $G(\varphi)$, whose two-dimensional directivity is defined as:

$$D = \frac{2\pi}{\int_0^{2\pi} G(\varphi)\, d\varphi} \qquad \text{Eq. 1.3}$$

We assume that the beam is steered such that, in the direction of the desired user, $G(\varphi_0) = 1$.

If we assume that K users are uniformly distributed throughout a cell with radius R, the expected value of interference received at the base station on the reverse link is given by

$$I = (K-1)\int_0^R \int_0^{2\pi} \frac{rP_r}{\pi R^2} G(\varphi)\, dr d\varphi = \frac{(K-1)P_r}{D} \qquad \text{Eq. 1.4}$$

Using the fact that the desired signal power, weighted by the antenna pattern, is simply P_r, and, substituting Eq. 1.4 into Eq. 1.2 we obtain,

$$P_b = Q\left(\sqrt{\frac{3DN}{K-1}}\right) \qquad \text{Eq. 1.5}$$

Eq. 1.5 is useful in showing that the probability of error for a CDMA system is related to the beam pattern of a receiver.

2. Reverse Channel Performance with Adaptive Antennas at the Base Station

The use of adaptive antennas at the base station receiver is a logical first step in improving capacity for several reasons. First of all, space and power constraints are not nearly as critical at the base station as they are at the portable unit. Second, the physical size of the array does not pose difficulty at the base station [Rap92].

Equation 1.5 is only valid when a single cell is considered. To consider the effects of adaptive antennas when CDMA users are simultaneously active in several adjacent cells, we must first define the geometry of the cell region. For simplicity, we consider the geometry proposed in [Rap92] with a single layer of surrounding cells, as illustrated in Figure 2.1.

Let $d_{i,j}$ represent the distance from the ith user to base j. Let $d_{i,0}$ represent the distance from the ith user to base station 0, the center base station.

Assume that the received power, at base j from the transmitter of user i, $P_{r;j,i}$, is given by a simple distance dependent path loss relationship

$$P_{r;j,i} = P_{T;i}\left(\frac{\lambda}{4\pi d_{ref}}\right)^2 \left(\frac{d_{ref}}{d_{i,j}}\right)^n \qquad \text{Eq. 2.1}$$

where n is the path loss exponent typically ranging between 2 and 4, and d_{ref} is a close-in reference distance [Rap92].

If we assume perfect power control such that the power received by each base from a user within that cell is P_r, then the interference power, $P_{r;0,i}$, received at the central base station from a mobile unit i in adjacent cell j is given by:

$$P_{r;0,i} = P_r\left(\frac{d_{i,j}}{d_{i,0}}\right)^n \qquad \text{Eq. 2.2}$$

which may be expressed as

$$P_{r;0,i} = P_r\left(1 + \left(\frac{2R}{d_{i,0}}\right)^2 - \frac{4R}{d_{i,o}}\cos\varphi_{i,0}\right)^{n/2} \qquad \text{Eq. 2.3}$$

where $\varphi_{i,0}$ is the angle of the mobile unit relative to a line drawn between the central cell base station and base j.

Let χ represent the expected value of the interference power, received at the central base station, from a single user in one of the adjacent cells when omnidirectional base station antennas are used. We can express the expected value of central cell interference power from a single adjacent cell user as

$$\chi = \beta P_r \qquad \text{Eq. 2.4}$$

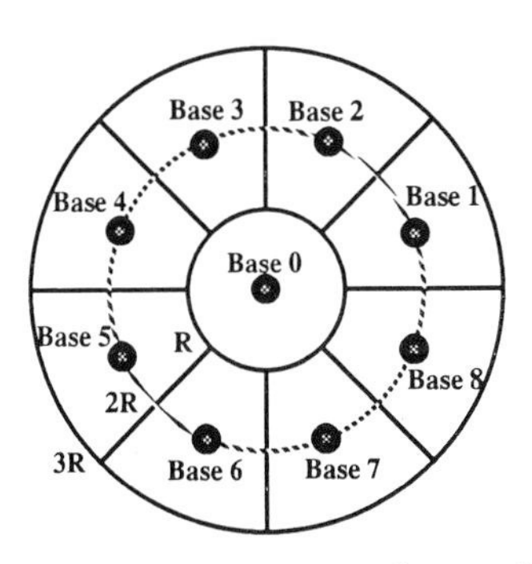

Figure 2.1 The nine cell wedge geometry proposed in [Rap92].

Table 2.1. Values of β as a function of the path loss exponent, n as determined by Eq. 2.5.

n	β
2	0.14962
3	0.08238
4	0.05513

where β, the average ratio of interference received from a single adjacent cell user to the interference received from a single in-cell user, is given by:

$$\beta = \int_{R}^{3R} \int_{-\pi/8}^{\pi/8} f_u(r, \varphi) \left(\left(1 + \left(\frac{2R}{r} \right)^2 - \frac{4R}{r} \cos\varphi \right) \right)^{n/2} dr d\varphi$$

Eq. 2.5

$f_u(r, \varphi)$ is the probability density function describing the geographic distribution of users in a single adjacent cell. Table 2.1 lists values of β for several values of n, assuming that users are uniformly distributed throughout the nine cell region.

β is related to the reuse factor, f, which is defined in [Rap92], for a single layer of eight adjacent cells, as

$$f = \frac{N_0}{N_0 + 8N_{a1}}$$ Eq. 2.6

where N_0 is the interference from users in the central cell, received at the central base station, on the reverse link. N_{a1} is the total interference seen by the desired central cell user from all users in a single adjacent cell.

When perfect power control is applied, then equation 2.6 may be expressed as:

$$f = \frac{(K-1)P_r}{(K-1)P_r + 8K\beta P_r} \approx \frac{1}{1+8\beta}$$ Eq. 2.7

where we have assumed that there are K users in each of the nine cells. For n=4, from Eq. 2.7, we obtain $f = 0.693$, implying that 31% of the interference power received at the central base station is due to users in adjacent cells. Note that, when omnidirectional antennas are used at both the base station and the portable unit, the value of the reuse factor, f, is determined by the cell geometry, power control scheme, and the path loss exponent.

When omnidirectional antennas are used at both the base station and the portable unit, the total interference seen on the reverse link by the central base station is the sum of the interference from users within the central cell, $(K-1)P_r$, and interference from users in adjacent cells, $8K\beta P_r$,

$$I = (K-1)P_r + 8K\beta P_r$$ Eq. 2.8

Let us assume that for the mth user in the central cell, an antenna beam from the base station, with pattern $G(\varphi)$, may be formed by the base station with maximum gain in the direction of user m. From Eq. 1.4, the average interference power contributed by a single user in the central cell is given by

$$E[P_{r0,i} | (0<r<R)] = \frac{P_r}{D}$$ Eq. 2.9

where D is the directivity of the beam with pattern $G(\varphi)$. When $G(\varphi)$ is piecewise constant over $(2p-1)(\pi/8) < \varphi - \varphi_d < (2p+1)(\pi/8)$, for $p = 0\ldots7$, for any angle φ_d between $-\pi/8$ and $\pi/8$, the average interference power at the central base station receiver (after the array weighting), due to a single user in an adjacent cell, can be shown to be:

$$E[P_{r0,i} | (R<r<3R)] = \frac{P_r\beta}{D}$$ Eq. 2.10

Using Eq. 2.10 with Eq. 2.9, the total interference power received at the center base station receiver is given by

$$I = \frac{(K-1)P_r + 8KP_r\beta}{D} \quad \text{Eq. 2.11}$$

Substituting Eq. 2.11 into the Eq. 1.6, using the fact that the desired signal power at the array port is P_r, we obtain an average bit error probability for the CDMA system employing a piece-wise constant directional beam is approximated by:

$$P_b \approx Q\left(\sqrt{\frac{3ND}{K(1+8\beta)}}\right) \quad \text{Eq. 2.12}$$

Eq. 2.12 relates the probability of error to the number of users per cell, the directivity of the base station antenna, and the propagation path loss exponent through the value of β. It is assumed that perfect power control is applied as described earlier in this section.

3. Simulation of Adaptive Antennas at the Base Station for Reverse Channel

To explore the utility of Eq. 2.12 and to verify its accuracy, we considered five base station antenna patterns which are illustrated in Figure 3.1. These antenna patterns are assumed to be directed such that maximum gain is in the direction of the desired mobile user. The first base station antenna pattern, shown in Figure 3.1(a) is an omnidirectional pattern similar to that used in traditional cellular systems.

The second pattern, illustrated in Figure 3.1(b), used 120° sectorization at the base station. In our model, the base station used three sectors, one covering the region from 30° to 150°, the second covering the region from 150° to 270°, and the third covering the region from −90° to 30°.

The third simulated base station pattern, shown in Figure 3.1(c), was a "flat-topped" pattern. The main beam was 30 degrees wide with uniform gain in the main lobe. A uniform side lobe

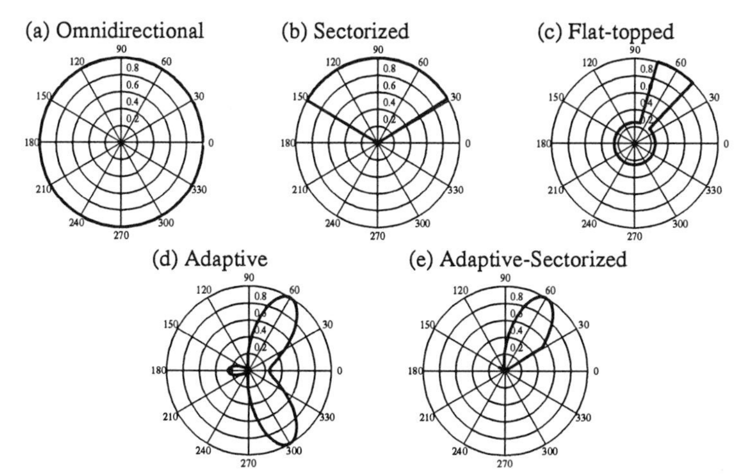

Figure 3.1 The five base station antenna patterns used in this study. Shown here are (a) the omnidirectional pattern, (b) the 120° sectorized pattern, (c) the flat-topped pattern, (d) the

level, which was 6 dB below the main beam gain, was assumed.

The fourth pattern, which used a simple three element linear array, is illustrated in Figure 3.1(d). This is the beam pattern formed by a binomial phased array with elements spaced a half wavelength apart. While the three dimensional gain of a binomial phased array is constant at 4.3 dB regardless of scan angle, the two-dimensional directivity, defined by Eq. 1.3, varies between 2.6 and 6.0 dB, depending on scan angle.

The pattern for the fifth simulated base station configuration, a sectorized adaptive antenna, is shown in Figure 3.1(e). Beginning with the sectorized system whose pattern is illustrated in Figure 3.1(b), we added a three element linear phased array to each sector. This pattern has a significantly higher gain than the beam pattern shown in Figure 3.1 (d).

To evaluate the link performance when these patterns are used at base stations, a simulation was designed using the cell geometry illustrated in Figure 2.1. Users were randomly placed throughout the region with an average of K users per cell.

Each user was assigned to one of the nine cells based on geographical location. Path loss was assumed to follow the model described in Eq. 2.1, and perfect power control was applied as described in Section 2.

We define $P_{i,j,k}$ as the component of the received power at the base station receiver (weighted by the array pattern) at the kth base station from the ith user associated with cell j. The

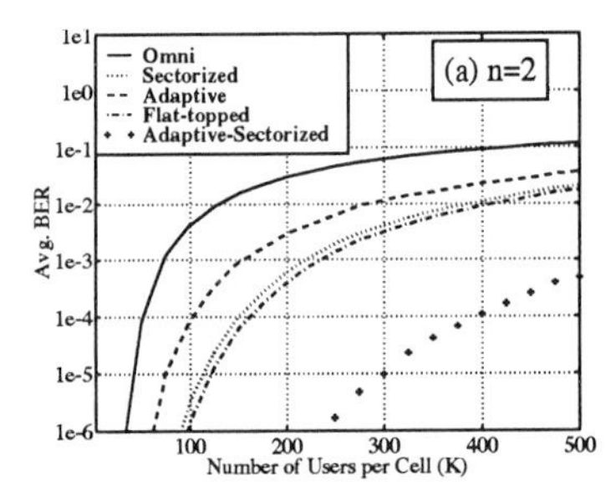

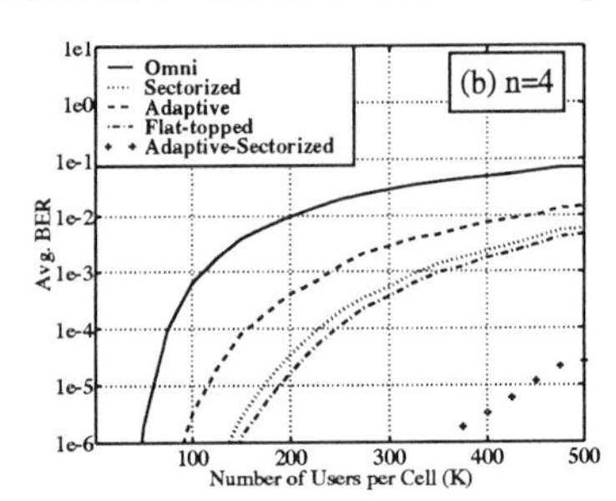

Figure 3.2 BER using adaptive antennas at the base station for (a) n=2, and (b) n=4. These results were developed through simulation by averaging the BER of every user in the center cell.

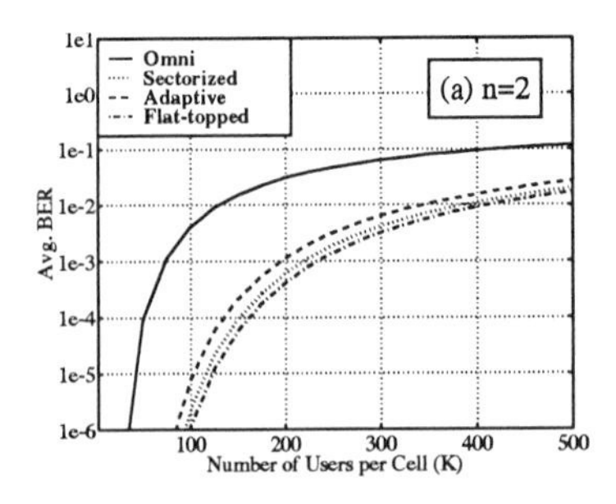

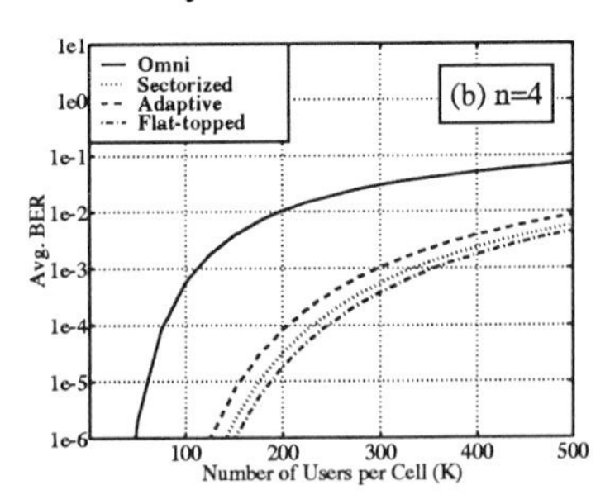

Figure 3.3 Plots of analytical results using equation 2.12 with two-dimensional directivities of 1.0, 2.67, 3.0, and 3.2 for the omni, adaptive, sectorized and flat-topped patterns, respectively, for (a) n=2, and (b) n=4.

CIR for the ith user in the central cell was calculated from

$$CIR_i = \frac{P_{i,0,0}}{\sum_{\substack{n=0 \\ n \neq i}}^{K-1} P_{n,0,0} + \sum_{m=1}^{8} \sum_{n=0}^{K-1} P_{n,m,0}} \qquad \text{Eq. 3.1}$$

The bit error rate for the ith user in cell 0 on the reverse link was determined by first calculating the CIR for the ith user from Eq. 3.1 then using that value in Eq. 1.2. A spreading factor of $N = 511$ was assumed.

This calculation was carried out for every user in the central cell and the resulting bit error rates were averaged to obtain an average bit error rate for the cell. Figure 3.2 shows bit error rates resulting from the simulation.

Figure 3.3 shows results calculated using the analytical result of Eq. 2.12 for four of the antenna patterns shown in Figure 3.1. By comparing Figures 3.2 and 3.3, it can be seen that for omnidirectional antennas, 120° sectorization, and the flat-topped pattern, the calculated bit error rates from Eq. 2.12 match the simulation results exactly. For the case of the adaptive binomial phased array, the analytical results for P_b are somewhat different from the simulation results. Unlike the omnidirectional, sectorized, and flat-topped patterns, the adaptive binomial phased array did not exhibit constant two-dimensional gain as a function of scan angle. Nevertheless, these figures demonstrate the accuracy of Eq. 2.12 when compared with simulations.

4. Simulation of Adaptive Antennas at the Portable Unit and at the Base Station

In this section, we examine how the reverse channel is affected by using adaptive antennas at a portable transmitter. A flat-topped beam was used to model an adaptive antenna at the portable transmitter. It is assumed that through retrodirectivity, the portable unit is able to adapt a beam pattern based on the received signal and transmit using that same pattern. Since space is extremely limited on the portable unit, the gain achievable by the portable unit antenna will be considerably less than that at the base station. For this study, it was assumed that the portable unit could achieve a beamwidth of 60 degrees with a uniform side lobe level that was 6 dB down from the main beam. This corresponds to an antenna with a gain of 4.3 dB. The pattern is similar to that shown in Figure 3.1(c) except that the beamwidth is wider in this case.

It was assumed that each portable unit was capable of perfectly aligning the boresight of its

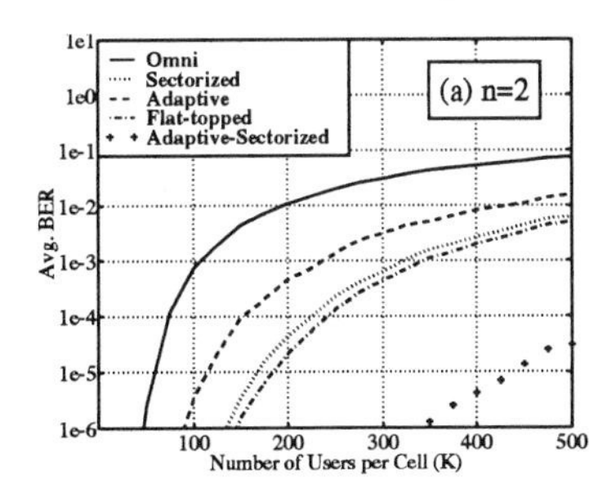

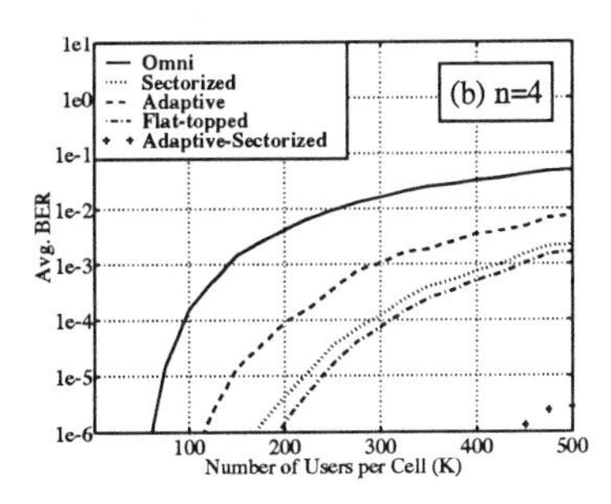

Figure 4.1 BER for five different base station configurations using adaptive antennas at the portable unit for (a) n=2 and (b) n=4. These results were obtained by simulation.

adaptive antenna with the base station associated with that portable unit. In this manner, portable units could radiate maximum energy to the desired base station.

Portable units with adaptive antennas were simulated for each of the five base station patterns described in Section 3. As in Section 3, average values of P_b were found by averaging the bit error rates of each user in the central cell, subjected to interference from the central cell and all immediately adjacent cells. The resulting bit error rates for these systems are shown in Figure 4.1. Note that, comparing Figure 3.2 and Figure 4.1, the bit error rates for the reverse channel are improved when directional antennas are used at the portable unit. For omnidirectional base stations, the average BER is only decreased by a small amount (20% or less) for K>200 when steerable directional antennas are used at the portable unit. However, for highly directional base station antenna patterns such as the adaptive-sectorized pattern, the average BER was decreased by an order of magnitude for K>300.

The relatively small improvements obtained by using adaptive antennas at the portable unit can be explained by the fact that when omnidirectional antennas are used at the mobile unit, no more than 1-0.455, or 0.545, of the total interference power is due to users in adjacent cells (see Table 4.1). When using adaptive antennas at the mobile unit, all users in the central cell will appear no different to the central base station than if they had used omnidirectional antennas. Thus, adaptive antennas at the portable unit will only reduce out-of-cell interference levels. Therefore, the maximum improvement in CIR, on the reverse link, that can be achieved by using adaptive antennas rather than omnidirectional antennas at the portable unit is only 3.5 dB.

Table 4.1 shows several values of the reuse factor, f, defined in Eq. 2.6 as the ratio of in-cell interference to total interference, for several base station patterns when omnidirectional antennas are used at the portable unit. Similarly, Table 4.2 shows values of f when steerable, directional antennas, with directivities of 4.3 dB, are used at the portable units.

Comparing Tables 4.1 and 4.2, it can be concluded that the use of adaptive antennas at the base station does nothing to improve the reuse factor, f, however the use of adaptive antennas at the portable unit does allow f to be improved. When omnidirectional antennas are used at the portable unit, f is entirely determined by the cell geometry, the power control scheme, and path loss

Table 4.1. Ratio of in-cell interference to total interference, f, as a function of path loss exponent when omnidirectional antennas are used at the portable units.

Base station pattern	Path Loss Exponent		
	n=2	n=3	n=4
Omni	0.454	0.601	0.693
Sectorized	0.453	0.601	0.692
Adaptive	0.452	0.600	0.692
Flat-topped	0.453	0.601	0.693
Adaptive-sectorized	0.453	0.601	0.692
$\frac{1}{1+8\beta}$ (Eq. 2.7)	0.455	0.603	0.694

Table 4.2. Ratio of in-cell interference to total interference, f, as a function of path loss exponent when antennas with directivities of 4.3 dB are used at the portable units.

Base station pattern	Path Loss Exponent		
	n=2	n=3	n=4
Omni	0.675	0.816	0.883
Sectorized	0.675	0.815	0.882
Adaptive	0.675	0.815	0.882
Flat-topped	0.675	0.815	0.883
Adaptive-sectorized	0.675	0.815	0.882

exponent, n, which is a function of propagation and not easily controlled by system designers. Using adaptive antennas at the portable unit, it is possible to tailor f to a desired value. Ideally, driving f to unity would allow system design to much less sensitive to the inter-cell propagation environment, when perfect power control is assumed.

This is important in CDMA cellular systems, since it indicates that use of adaptive antennas at the portable unit could help to allow for more frequent reuse of signature sequences.

5. Conclusions

It was shown in this study that adaptive antennas at the base station, with relatively modest beamwidth requirements, and no interference nulling capability, can provide large improvements in BER, as compared to omnidirectional systems.

Simple analytical expressions which relate the average BER of a CDMA user to the antenna directivity and propagation environment were derived and used to determine bit error rate improvements offered by a number of base station antenna patterns. In terms of capacity, the results of Section 3 indicate that using adaptive antennas at the base station can allow the number of users to increase by 200% to 400%, while maintaining an average BER of 10^{-3} on the reverse link.

The bit error rate on the reverse channel is further improved by adding adaptive antennas at the portable unit. Using a 4.3 dB gain antenna at the portable, the bit error rate for the directional base station antenna patterns was less than half of the bit error achieved without directional antennas at the portable unit.

In Section 4, it was demonstrated that while only modest improvements in bit error rate were achieved by using adaptive antennas at the portable unit, adaptive antennas at the portable unit allow the reuse factor to be altered. In this way, the impact of adjacent cell interference may be controlled.

In short, adaptive antennas at the base station can have a major effect on bit error rate performance, but cannot impact the reuse factor, f. Conversely, it has been shown in this paper that adaptive antennas at the portable unit can provide no more than a 3.5 dB improvement in reverse channel CIR, however, they allow the reuse factor, f, to be altered.

Neither interference nulling or the multipath channel were considered in this study, however, both of these factors will be significant in developing algorithms for successfully adapting array elements. Furthermore, work is currently underway to establish more accurate expressions for bit error rate in systems employing adaptive antennas.

6. References

[Coo78]G.R. Cooper and R.W. Nettleton, "A spread-spectrum technique for high-capacity mobile communications," *IEEE Trans. on Vehicular Technology,* vol. VT-27, no. 4, Nov. 1978.

[Sal90]A. Salmasi, "An overview of advanced wireless telecommunication systems employing code division multiple access," *Conf. on Mobile, Portable & Personal Communications,* Kings College, England, Sept., 1990.

[Gar92]W.A. Gardner, S.V. Schell, and P.A. Murphy, "Multiplication of cellular radio capacity by blind adaptive spatial filtering," *IEEE Int. Conf. on Sel. Topics in Wireless Commun. Mobile,* Vancouver, B.C., Can. Jun 1992.

[Swa90]S.C. Swales, M.A. Beach, D.J. Edwards, and J.P. McGeehan, "The performance enhancement of multibeam adaptive base-station antennas for cellular land mobile radio systems," *IEEE Trans. on Vehicular Technology,* vol. 39, no. 1, Feb. 1990.

[Pur77]M.B. Pursley, "Performance evaluation for phase-coded spread spectrum multiple-access communications with random signature sequences," *IEEE Trans. Commun.,* vol. COM-25, Aug. 1977.

[Rap92]T.S. Rappaport and L.B. Milstein, "Effects of radio propagation path loss on DS-CDMA cellular frequency reuse efficiency for the reverse channel," *IEEE Trans. on Veh. Tech.,* vol. 41, no. 3, Aug. 1992.

[Mor89]R.K. Morrow, and J.S. Lehnert, "Bit-to-Bit Error Dependence in Slotted DS/SSMA Packet Systems with Random Signature Sequences," *IEEE Trans. Comm.*, Vol 37, No. 10., Oct, 1989.

7. Acknowledgment

This work was supported by the Bradley Fellowship in Electrical Engineering and the MPRG Industrial Affiliates Program at Virginia Tech.

4

A New Equalizer for Wideband OFDM over a Frequency-Selective Fading Channel

Michel Fattouche [1] and *Hatim Zaghloul* [2]
[1] Department of Electrical & Computer Engineering
The University of Calgary, Calgary, Alberta,
T2N 1N4
[2] VP Research & Development, Wi-LAN Inc.
Calgary, Alberta

Abstract: *The paper proposes two algorithms for equalizing Wideband Orthogonal Frequency Division Multiplexing (W-OFDM) transmitted over a frequency-selective fading channel. The equalization techniques use a Hilbert transform as well as a parametric two-ray model to estimate the channel's group delay from the channel's amplitude response. Based on such estimates, the equalization techniques substantially reduce the irreducible Bit Error Rate (BER) caused by the channel's group delay. This is demonstrated on an indoor radio channel.*

1. Introduction:

The Eureka project [1,2] which is the emerging standard for the European Digital Audio Broadcast (DAB) project calls for a frequency domain differential phase modulation, namely: Differential Quadrature Phase Shift Keying (DQPSK) using Orthogonal Frequency Division Multiplexing (OFDM) [3-5], DQPSK-OFDM for short. One can show that the effect of the group delay of the frequency-selective fading channel on DQPSK-OFDM is to create an irreducible BER. Traditionally, research aiming at removing such an irreducible BER was confined either to using Forward Error Correction (FEC) coding [1,2,5] or to sending a pilot tone to monitor the fading channel [4]. FEC coding is complex and costly, while sending a pilot tone represents a waste of power and is effective only over narrowband signals. For this reason, we aim in this paper at reducing the irreducible BER by estimating the channel's group delay using the channel's amplitude response.

In [6], we show that it is possible to use the Hilbert transform to estimate the absolute value of the group delay of a frequency-selective fading channel from the amplitude response of the channel. The Hilbert transform is based on Voelcker's phase-envelope relationship [7]. Research aiming at estimating the phase differential using the amplitude of the received signal is not new. We have successfully estimated in [8] the random FM associated with a flat fading channel using the envelope of a narrowband DMPSK signal transmitted over the channel. Also, Poletti and Vaughan have estimated in [9] the phase differential of a fading channel from the amplitude of a narrowband signal transmitted over the channel.

In section 2, we use a frequency-domain version of Voelcker's phase-envelope relationship to estimate the absolute value of the group delay of a selective-fading channel from the amplitude response of the channel. In section 3, we replace the Hilbert-based phase estimate by a parametric estimate which is simpler to use and which offers a smaller Relative Mean Square Error (RMSE). In section 4, we use the resulting estimate of the absolute value of the group delay of the fading channel to reduce the irreducible BER for Wideband-OFDM (W-OFDM) transmitted over a frequency-selective fading channel. Section 5 concludes the paper with a summary of results. A wideband signal is defined throughout this paper as a signal with a bandwidth wide enough to contain at least one entire frequency domain null.

2. Application of Voelcker's Relationship to Selective Fading

When a system is Minimum Phase (MP) [13], we have a frequency domain version of Voelcker's relationship

$$\frac{d\Psi(f)}{df} = -H[\frac{d}{df}\ln|G(f)|] \quad (1)$$

between the amplitude response $|G(f)|$ and the phase response $\Psi(f)$ of the complex frequency response $G(f)$ of the system where $H[\cdot]$ denotes Hilbert transform and f denotes frequency. Such a relationship is used in [1] to estimate $\frac{d\Psi(f)}{df}$ for an indoor radio channel (with a carrier frequency centered at 1700MHz) [30] from the amplitude response $|G(f)|$.

In Fig-1a, the amplitude response $|G(f)|$ is shown for $-100\text{MHz} < f < 100\text{MHz}$. Two curves are displayed in Fig-1b. The first curve displays the true phase differential $\{\delta\Psi_k\}_{k=1}^{N}$ of $G(f)$ while the second curve displays its estimate $\delta\hat{\Psi}_k = -H[\ln|G_k| - \ln|G_{k-1}|]$ obtained using (1), where $\delta\Psi_k = \Psi_k - \Psi_{k-1}$, $\Psi_k = \Psi(kB_o)$, $G_k = G(kB_o)$, B_o is the sampling interval and $N = 511$ is the number of observed samples.

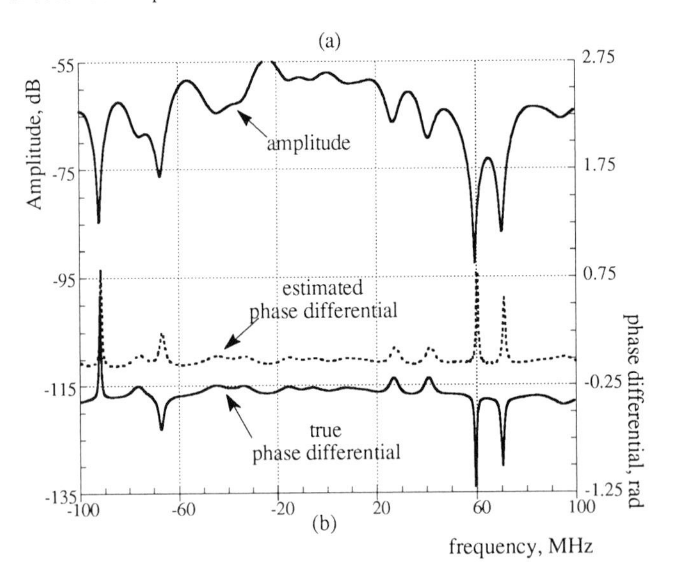

Fig-1 (a) The amplitude |G(f)| of G(f); (b) The true phase differential of G(f) compared with its estimated phase differential.

A measure of the goodness of fit between a time series: $\{x_k\}_{k=1}^{N}$ and its estimated series $\{\hat{x}_k\}_{k=1}^{N}$ is the Relative Mean Square Error (RMSE) which is defined as

$$\text{RMSE} = \frac{\sum_{k=1}^{N}[x_k - \hat{x}_k]^2}{\sum_{k=1}^{N}[x_k]^2}$$

where N is the length of each series. Note that the RMSE is not necessarily less than one. In Fig-1b, the RMSE between $|\delta\Psi_k|$ and its unbiased estimate $|\delta\hat{\Psi}_k + \mu|$ is 0.16, where $\mu = \frac{1}{N}\sum_{k=1}^{N}\delta\Psi_k$.

The effect of the sampling rate on estimate (1) is shown in Fig-2 which displays three RMSE curves corresponding to the three sampling rates: 256, 512 and 1024 samples per 200MHz. From Fig-2 one can see that the RMSE value is decreased by increasing the sampling rate. This is due to the fact that even though all of the above rates are larger than the Nyquist rate which corresponds to twice the time spread of the impulse response $g(t)$, the function $\frac{d\Psi(f)}{df}$ possesses a much larger time spread. For example, during a null of 20dB, $\frac{d\Psi(f)}{df}$ has ten times the time spread of $g(t)$ [14].

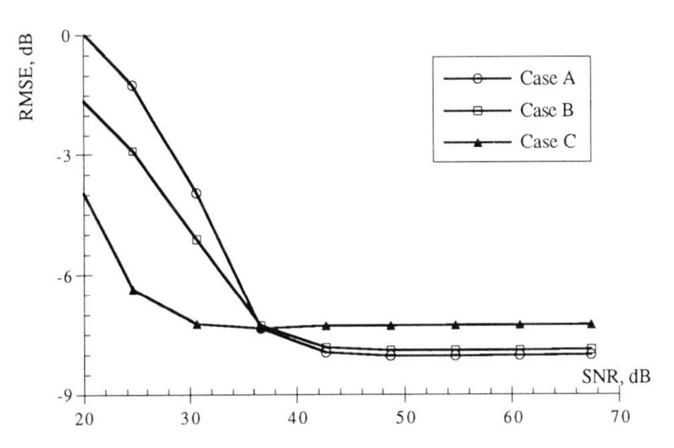

Fig-2 The RMSE between the true phase differential and its unbiased estimate using (1) for a number N per 200MHz: (A) N=1024, (B) N=512 and (C) N=256.

3. Theoretical Justification

In order to justify using (1) to estimate $\frac{d\Psi(f)}{df}$, we first sample the complex impulse response $g(t)$ corresponding to the complex transfer function $G(f)$, then we z-transform the result. The sampling is achieved by observing $G(f)$ over a finite band $[-B, B]$ then, by forcing $G(f)$ to be periodic with a period $2B$ one can find its Fourier coefficients $\{c_i\}$ spaced with a period

$1/2B$ time spacing. The discrete impulse response of the channel $g(i/2B)$ is displayed in Fig-3 which shows that $g(t)$ is nonzero only over a finite time interval $[T_1, T_2]$, i.e.

$$g(i/2B) = \begin{cases} c_i & \text{for } i = p, \ldots, p+n \\ 0 & \text{otherwise} \end{cases} \tag{2}$$

where $p = 2BT_1$ and $n = 2B(T_2 - T_1)$. Taking the z-transform of $g(i/2B)$, we have

$$G(z) = \sum_{i=p}^{p+n} c_i z^{-i} \tag{3}$$

By factorizing $G(z)$ into n zeros we obtain

$$G(z) = c_p z^{-p} \prod_{i=1}^{n} (1 - \alpha_i z^{-1}) \tag{4}$$

where α_i is the ith root of $G(z)$ in the z-plane.

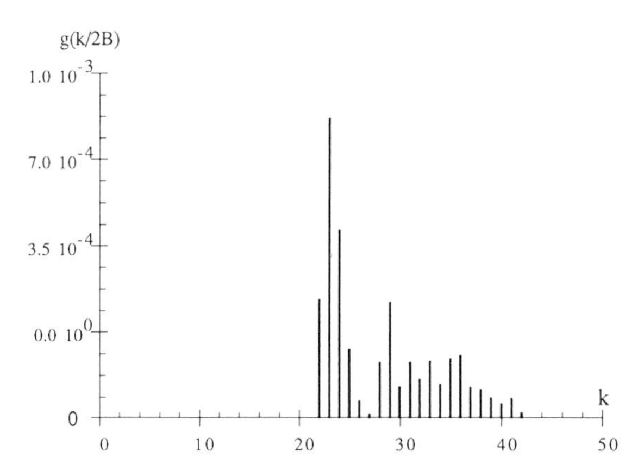

Fig-3 The amplitude of the discrete impulse response g(t).

Since the calculation of $G(z)$ on the unit circle reduces to $G(f)$, then the amplitude response $|G(f)|$ and the phase response $\Psi(f)$ are

$$|G(f)| = c_p \prod_{i=1}^{n} |G_i(f)| \tag{5a}$$

and

$$\Psi(f) = \sphericalangle c_p - \frac{2\pi p f}{2B} + \sum_{i=1}^{n} \sphericalangle G_i(f) \tag{5b}$$

where $G_i(f) = (1 - \alpha_i e^{-j2\pi f/2B})$ and $\sphericalangle c_p$ is the phase of c_p. In order to justify using (1) to estimate $\frac{d\Psi(f)}{df}$, we have to compare $H[\frac{d}{df}\ln|G(f)|]$ with $\frac{d\Psi(f)}{df}$. From (5a&b), we have

$$H[\frac{d}{df}\ln|G(f)|] = \sum_{i=1}^{n} H[\frac{d}{df}\ln|G_i(f)|] \tag{6a}$$

and

$$\frac{d\Psi(f)}{df} = +\sum_{i=1}^{n} \frac{d}{df} \sphericalangle G_i(f) - \frac{2\pi p}{2B} \tag{6b}$$

When $\alpha_i < 1$, $G_i(f) = (1 - \alpha_i e^{-j2\pi f/2B})$ is a MP function and

$$\frac{d}{df}\triangleleft G_i(f) = -H[\frac{d}{df}\ln|G_i(f)|] \qquad (7a)$$

When $|\alpha_i|>1$, $G_i(f)=(1-\alpha_i e^{-j2\pi f/2B})$ is a Maximum Phase (MaxP) function and can be written as $G_i(f)=-\alpha_i e^{-j2\pi f/2B}(1-1/\alpha_i e^{j2\pi f/2B})$. Hence,

$$\frac{d}{df}\triangleleft G_i(f) = +H[\frac{d}{df}\ln|G_i(f)|] - \frac{2\pi}{2B} \qquad (7b)$$

From the above, $G_i(f)$ can be either MP or MaxP. By ordering $\{G_i(f)\}_{i=1}^{n}$ such that $\{G_1(f),\ldots,G_{n_1}(f)\}$ are MaxP and $\{G_{n_1+1}(f),\ldots,G_n(f)\}$ are MP (where n_1 is the number of MaxP terms and $n-n_1$ is the number of MP terms in (4)) $\frac{d\Psi(f)}{df}$ in (6b) reduces to

$$\frac{d\Psi(f)}{df} = \sum_{i=1}^{n_1} H[\frac{d}{df}\ln|G_i(f)|] - \sum_{i=n_1+1}^{n} H[\frac{d}{df}\ln|G_i(f)|] - \frac{2\pi(n_1+p)}{2B} \qquad (8)$$

Comparing $\frac{d\Psi(f)}{df}$ in (8) with $H[\frac{d}{df}\ln|G(f)|]$ in (6b), one can see that they are equal except for two discrepancies. The first discrepancy is the extra term "$\frac{-2\pi(n_1+p)}{2B}$" in (8). The second discrepancy is the positive sign in "$+\sum_{i=1}^{n_1} H[\frac{d}{df}\ln|G_i(f)|]$" in (8) when $G_i(f)$ is MaxP. Both discrepancies are removed in the following section, but first an example.

Example: In order to illustrate the above theoretical justification of using the frequency domain version of Voelcker's phase-envelope relationship to estimate $\frac{d\Psi(f)}{df}$ from the amplitude response $|G(f)|$ of $G(f)$, we first z-transform $g(t)$ shown in Fig-3 and then display its zeros in the z-plane in Fig-4 for n=20, p=22 and B=100MHz.

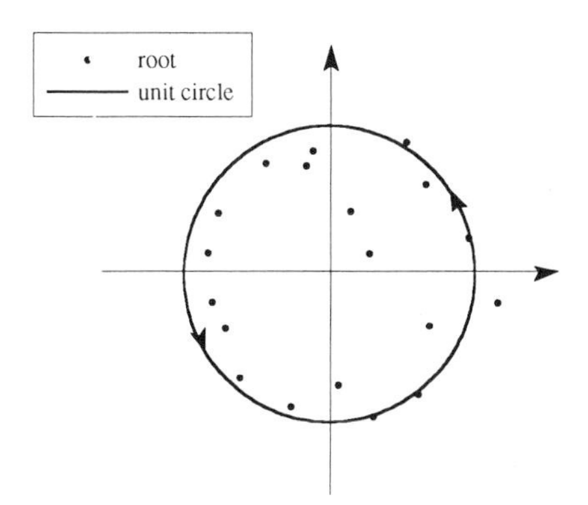

Fig-4 The roots of G(z) shown relative to the unit circle.

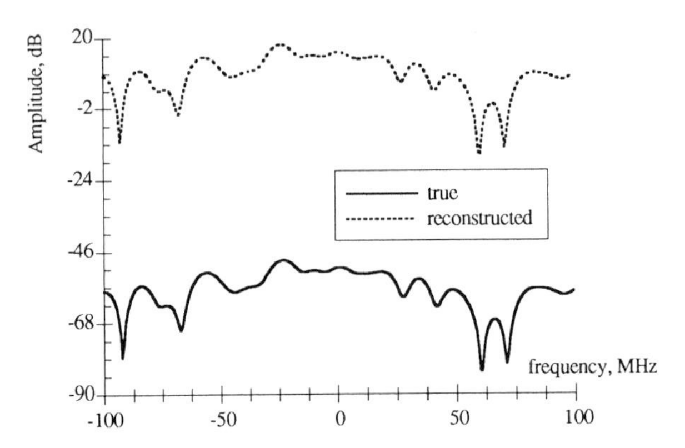

Fig-5 Comparison between the true amplitude of G(z) and the one reconstructed from the zeros of G(z) in the z-plane.

In Fig-5 we show the amplitude response $|G(f)|$ in conjunction with the product $\prod_{i=1}^{n}|G_i(f)|$ in (5a). In Fig-6 we show the phase differential $\Psi(f)-\Psi(f-B_0)$ in conjunction with the sum $\sum_{i=1}^{n}(\sphericalangle G_i(f)-\sphericalangle G_i(f-B_0))$. In Fig-7 we show the phase differential $\sum_{i=1}^{n_1}(\sphericalangle G_i(f)-\sphericalangle G_i(f-B_0))$ due to the zeros of $G(z)$ outside the unit circle, in conjunction with the phase differential $\sum_{i=n_1+1}^{n}(\sphericalangle G_i(f)-\sphericalangle G_i(f-B_0))$ due to the zeros of $G(z)$ inside the unit circle. From Fig-7 one can conclude that the first phase differential is due to a system which is MaxP while the second phase differential is due to a system which is MP. In Fig-7, $n_1=5$ and $n-n_1=15$.

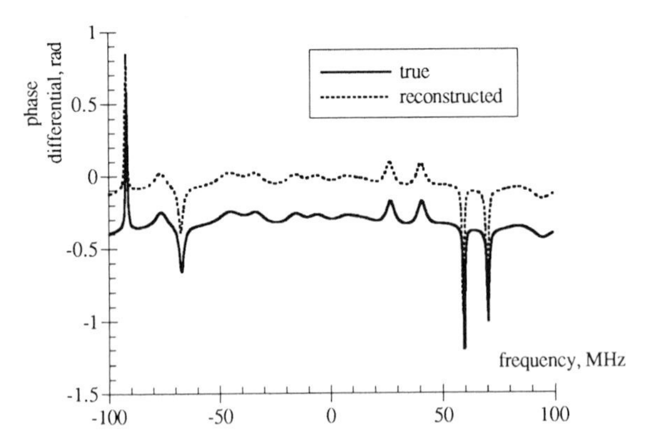

Fig-6 Comparison between the true phase differential of G(z) and the one reconstructed from the zeros of G(z) in the z-plane.

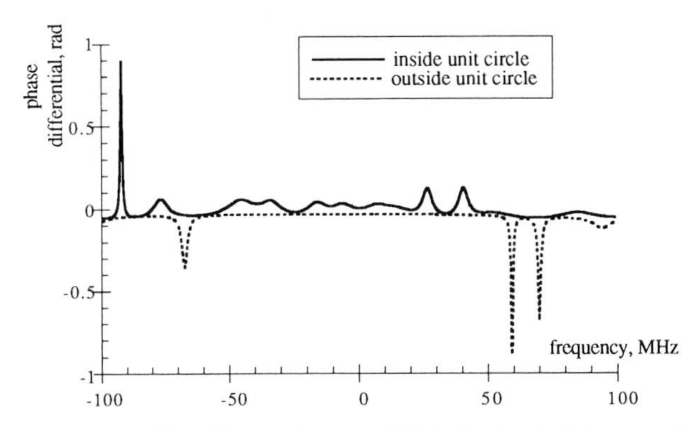

Fig-7 The phase differential due to the roots of G(z) inside the unit circle compared with the phase differential due to the roots of G(z) outside the unit circle.

In Fig-8 we show the reconstructed amplitude of $|G(f)|$ (the same as in Fig-1a) shifted by -45dB. Also Fig-8 shows the function: $20\log_{10}|1.-|\alpha_i||$ versus the phase $\sphericalangle\alpha_i$ of α_i as one travels from 0 to 2π around the unit circle, in an anticlockwise rotation. Since the function $20\log_{10}|1.-|\alpha_i||$ indicates the closeness of the zeros of $G(z)$ to the unit circle, one can see from Fig-8 that deep nulls in the amplitude response $|G(f)|$ of $G(f)$ are caused by distinct zeros and that the closer the zero to the unit circle the deeper the null. This implies that in the neighborhood of a deep null $G(f)$ can be well modeled by a single zero in the z plane. The neighborhood of a deep null is defined here as the frequency band $[f_1, f_2]$ over which $G(f)$ is at least 3dB below its running mean. Mathematically, the single zero model implies that

$$G(f) \cong A_k e^{-j\Omega' f}(1-\alpha_k e^{-j2\pi f/2B}) \qquad f_1 < f < f_2 \tag{9}$$

where Ω' is a constant, A_k is slowly varying for $f_1 < f < f_2$ and α_k is the zero in the z-plane closest to the arc between $2\pi f_1/2B$ and $2\pi f_2/2B$. The single zero model is a localized version of Rummler's Two-Ray model [11] of the multipath channel.

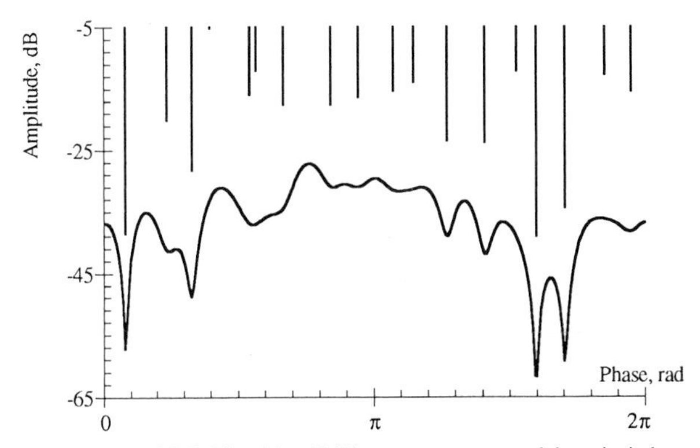

Fig-8 The envelope of G(f) (biased by -45dB) as one moves around the unit circle compared with the distance in dB between each root of G(z) and the unit circle.

4. Parametric Modeling

Based on (9), equation (1) can be approximated as

$$H[\frac{d}{df}\ln|G(f)|] \cong 1/f_0 \frac{1}{1+(f'/f_0)^2} \tag{10}$$

where $f' = f - f_{\min}$, $f_{\min}$ is the frequency in $[f_1, f_2]$ when $|G(f)|$ reaches its minimum, $f_0 = \frac{|1-|\alpha_k||}{\sqrt{|\alpha_k|\Omega}}$ and $\Omega = \frac{2\pi}{2B}$. Fig-9 displays two curves. The first curve (Curve A) corresponds to the phase differential $\{\delta\Psi_k\}_{k=1}^{N}$ shown in Fig-1b, while the second curve (Curve B) corresponds to the phase differential $\{|\delta\Psi_k'|\}_{k=1}^{N}$ obtained through equation (10) with f_o found by summing the frequency corresponding to the maximum value of $\ln|G_k| - \ln|G_{k-1}|$ over the same interval, then halving the sum. In Fig-9, the RMSE between $\{|\delta\Psi_k|\}_{k=1}^{N}$ and $\{|\delta\hat{\Psi}_k + \mu|\}_{k=1}^{N}$ using (10) is 0.07 which is smaller than the RMSE between $\{|\delta\Psi_k|\}_{k=1}^{N}$ and $\{|\delta\hat{\Psi}_k + \mu|\}_{k=1}^{N}$ using (1) which was 0.16.

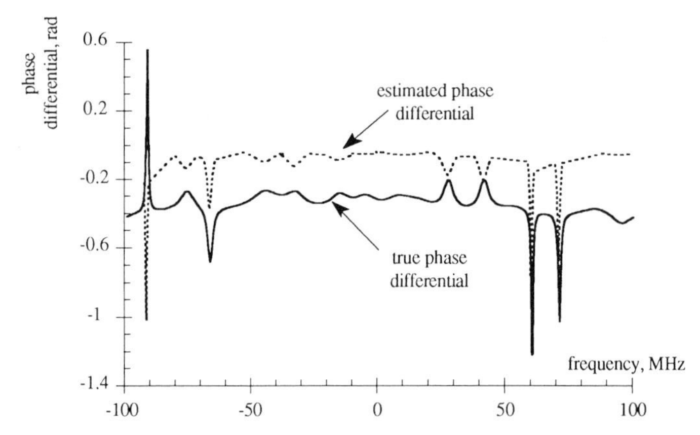

Fig-9 The true phase differential of G(f) compared with the estimated phase differential using (10).

In order to examine the effects of Additive White Gaussian Noise (AWGN) on the estimates (1) and (10), we generate a noisy system $G'(f)$ from $G(f)$ as: $G'(f) = G(f) + N_0(f)$ where $N_0(f)$ is a zero mean complex AWGN. The corresponding amplitude and phase of $G'(f)$ are denoted $|G'(f)|$ and $\Psi'(f)$ respectively. Fig-10 displays three RMSE curves versus the SNR. The first curve (Curve A) displays the RMSE between $\{|\delta\Psi_k|\}_{k=1}^{N}$ and its noisy version $\{|\delta\Psi_k'|\}_{k=1}^{N}$. The second curve (Curve B) displays the RMSE between $\{|\delta\Psi_k|\}_{k=1}^{N}$ and its unbiased estimate $\{|\delta\hat{\Psi}_k + \mu'|\}_{k=1}^{N}$ using (1), i.e. $\delta\hat{\Psi}_k = -H[\ln|G_k'| - \ln|G_{k-1}'|]$ where $\mu' = \frac{1}{N}\sum_{k=1}^{N}\delta\Psi_k'$. The third curve (Curve C) displays the RMSE between $\{|\delta\Psi_k|\}_{k=1}^{N}$ and its unbiased estimate $\{|\delta\hat{\Psi}_k + \mu'|\}_{k=1}^{N}$ using (10), i.e. $\delta\hat{\Psi}_k = 1/f_0 \frac{1}{1+((kB_0 - f_{\min})/f_0)^2}$ during the nulls that are at least 3dB below the running mean of $G'(f)$. Two remarks can be made based on Fig-10. First, one can observe from

Fig-10 that the effects of the AWGN on the estimates of $\frac{d\Psi(f)}{df}$ are negligible as long as the SNR is larger than 35dB. Also, one can see that for an SNR value larger than 35dB estimate (10) has a smaller RMSE value than estimate (1).

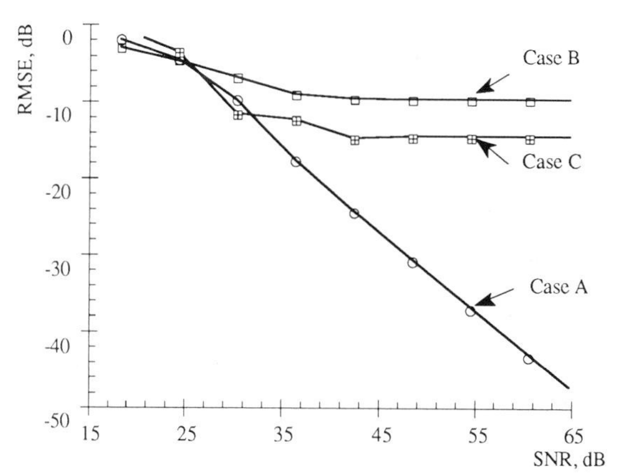

Fig-10 The RMSE between the true phase differential and (A) its noisy version, (B) its unbiased estimate from (1) and (C) its unbiased estimate from (10).

5. Reducing the effects of the Group Delay due to Selective Fading

The complex envelope $Z_0(f)$ of a W-OFDM signal transmitted over a selective fading channel over the frequency band $[-B,B]$ can be written as the complex envelope of the W-OFDM signal $U(f)$, multiplied by the complex frequency response $G(f)$ of the selective-fading channel, contaminated with AWGN. Mathematically, we have

$$Z_0(f) = U(f)G(f) + N_0(f) \tag{11}$$

where

$$U(f) = \sqrt{E_b / N_0}\exp[j\sum_{i=1}^{N}(2\pi / M)\alpha_i P(f - iB_0)] \tag{12}$$

$\alpha_i = \beta_i + \alpha_{i-1}$, β_i takes one of the M values in the alphabet $\{0,1,\cdots,M-1\}$, $P(f)$ is a shaping pulse, E_b is the average transmitted bit energy and $N_0(f)$ is the zero mean white Gaussian noise. β_i is the information digit between adjacent frequencies separated by B_0 where $NB_0 = 2B$.

The proposed algorithm consists of five steps. The first step consists of obtaining an estimate $|\hat{G}(f)|$of the amplitude $|G(f)|$. This is accomplished by first obtaining the amplitude $|Z_0(f)|$ of $Z_0(f)$, then timelimiting it to the interval $[(p-n/2)/2B,(p+3n/2)/2B]$. The reason for the time limitation is to reduce the power of the thermal noise which is directly related to the signal time spread. As an example, let us assume that a DMPSK-OFDM signal with $B_0 = 2B/N = 200\text{MHZ}/512$ is transmitted over an indoor radio selective-fading channel with a time spread $n/2B = 100\text{ns}$. In this case, the noise reduction due to the time limitation of $|Z_0(f)|$ is $10\log_{10}(512/200\text{MHz}/200\text{ns}) \cong 11\text{dB}$. In other words, when the SNR of the received signal is

24dB, the SNR of the received amplitude response $|G(f)|$ is 24+11=35dB which corresponds to a noise level with little or no effect on either (1) or (10), as shown in Fig-10.

The second step consists of obtaining an estimate for $|\frac{d\Psi(f)}{df}|$ using either (1) or (10), i.e. obtaining either

$$\delta\hat{\Psi}_k = -H[\ln|\hat{G}_k| - \ln|G_{k-1}|] \tag{13a}$$

or

$$\delta\hat{\Psi}_k = 1/\hat{f}_0 \frac{1}{1+((kB_0 - \hat{f}_{min})/\hat{f}_0)^2} \tag{13b}$$

In equation (13b), $\delta\hat{\Psi}_k$ is obtained over the segments $[k_1, k_2]$ corresponding to the intervals $[f_1, f_2]$ where $|\hat{G}(f)|$ fades at least 3dB below its running mean, i.e. $k_1 = f_1 / B_0$ and $k_2 = f_2 / B_0$ and $\hat{f}_{min}$ and $\hat{f}_0$ are the estimates of f_{min} and f_0 respectively. The third step consists of estimating the mean μ of $\delta\Psi_k$. This is achieved by obtaining the estimate $\hat{\mu}$ of μ where $\hat{\mu} = \frac{1}{N}\sum_{k=1}^{N}(\triangleleft Z_{0,k} - \triangleleft Z_{0,k-1})$ and $\triangleleft Z_{0,k}$ is the phase of $Z_0(kB_0)$. The fourth step consists of estimating the sign of $\{\delta\Psi_k\}_{k=k_1}^{k_2}$. This is achieved by computing the two values

$$S_+ = \sum_{k=k_1}^{k_2} \min_{\alpha_k \in \{0,\cdots,M-1\}} \{(\triangleleft Z_{0,k} - \triangleleft Z_{0,k-1}) + (\delta\Psi_k + \hat{\mu}) - \alpha_k\}^2 \tag{14a}$$

and

$$S_- = \sum_{k=k_1}^{k_2} \min_{\alpha_k \in \{0,\cdots,M-1\}} \{(\triangleleft Z_{0,k} - \triangleleft Z_{0,k-1}) - (\delta\Psi_k + \hat{\mu}) - \alpha_k\}^2 \tag{14b}$$

If $S_+ < S_-$ then $\delta\hat{\Psi}_k + \hat{\mu}$ from (13a&b) is added to $(\triangleleft Z_{0,k} - \triangleleft Z_{0,k-1})$, otherwise, it is subtracted from $(\triangleleft Z_{0,k} - \triangleleft Z_{0,k-1})$.

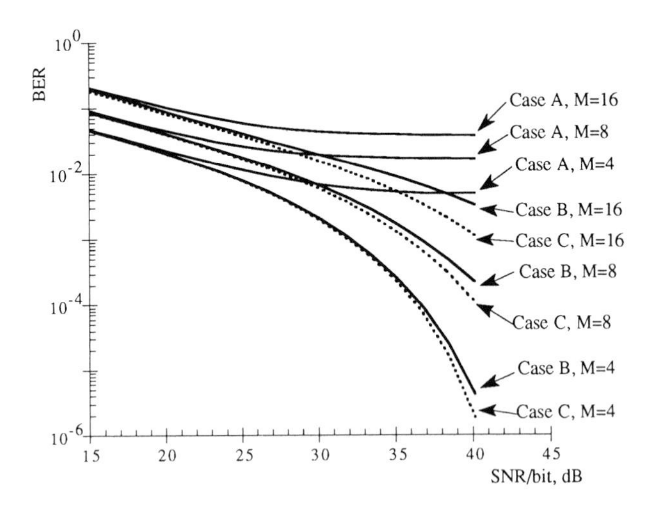

Fig-11 BER curves for M=4, 8 and 16, each for three cases: (A) no phase adjustment, (B) phase adjustment using (13a), and (C) perfect phase adjustment.

Assuming perfect match-filtering and perfect timing recovery at the DMPSK receiver, Fig-11 displays nine BER curves for an OFDM signal transmitted over the selective-fading channel illustrated in Fig-1, with $M = 4,8$ and 16, for each of the following three cases. The first case (Case A) corresponds to no adjustment for the group delay distortion due to the channel. The second case (Case B) corresponds to adjustment for the group delay distortion due to the channel using (13a). The third case (case C) corresponds to perfect adjustment for the group delay due to the channel, i.e. by setting $\delta\Psi_k$ equal to zero. Fig-11 is based on a similar analysis as the one used in [8] for reducing the irreducible BER of DMPSK due to random FM over a flat-fading channel which itself is based on [12]. From Fig-11 one can see that by adjusting for the delay using (13a) one can approach the limiting case where the delay distortion is entirely removed.

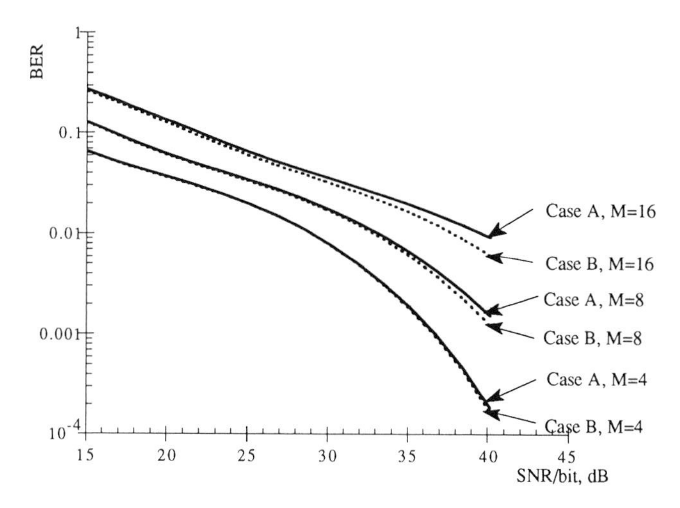

Fig-12 BER curves for M=4, 8 and 16, for two cases: (A) phase adjustment using (13a) and (B) phase adjustment using (13b).

Fig-12 compares the case when the adjustment for the group delay distortion is accomplished using (13a) with the case when the adjustment for the group delay distortion is accomplished using (13b). From Fig-12 one can see that as expected (10) offers a better performance than (1).

6. Conclusion

This paper offered a theoretical justification for using the Hilbert transform to estimate the group delay of a selective-fading channel from its amplitude response. The paper shows that in the neighborhood of a frequency domain null, the system can be either MaxP or MP. This is justified based on the concept that a system whose impulse response is of finite duration and whose frequency response is observed over a finite band $[-B,B]$ can be regarded as a product of a finite number of zeros in the z-domain.

Also, the paper showed that in the neighborhood of a frequency domain null, the frequency response of the frequency-selective fading channel can be well modeled by a single zero in the z-plane. Based on such an observation, the paper replaced the Hilbert-based estimate by a parametric estimate which is simpler to implement and which at high SNR offers a smaller RMSE value. Finally, the paper used both the Hilbert-based estimate and the parametric estimate to successfully reduce the irreducible BER caused by the phase differential due to fading.

References:

[1] P.Hoeher, J. Hagenauer, E. Offer, C. Rapp and H. Schulze, "Performance of an RCPC-Coded OFDM-based Digital Audio Broadcasting (DAB) System," Globecom'91, pp. 2.1.1-2.1.7, Phoenix, Arizona, Dec. 2-5, 1991.

[2] J.F. Helard and B. LeFloch, "Trellis Coded Orthogonal Frequency Division Multiplexing for Digital Video Transmission," Globecom'91, pp. 23.5.1-23.5.7, Phoenix, Arizona, Dec. 2-5, 1991.

[3] B. Hirosaki, S. Hasegawa and A. Sabato, "Advanced Groupband Data Modem using Orthogonally Multiplexed QAM Technique," *IEEE Trans. Commun.*, vol. COM-34, pp. 587-592, June 1986.

[4] L.J. Cimini, Jr, "Analysis and Simulation of a Digital Mobile Channel using Orthogonal Frequency Division Multiplexing,"*IEEE Trans. Commun.*, vol. COM-33, pp. 665-675, July 1985.

[5] E.F. Casas and C. Leung, "OFDM for Data Communications over Mobile Radio FM Channels-Part I: Analysis and Experimental Results," *IEEE Trans. Commun.*, vol. COM-39, pp. 794-807, May 1991

[6] M. Fattouche and H. Zaghloul, "Estimation of Phase Differential of signals Transmitted over a Fading Channel," *Electronics Letters*, vol. 27, no. 20, pp. 1823-24, Sept. 1991.

[7] H. Voelcker, "Toward a Unified Theory of Modulation-Part I: Phase-Envelope Relationships," *Proceedings of the IEEE* , vol. 54, pp. 340-353, March 1966.

[8] M. Fattouche and H. Zaghloul, "Equalization of $\pi/4$ Offset QPSK Transmitted over Flat Fading Channels," ICC'92, pp. 312.2.1-312.2.3, Chicago, Ill., June 14-18, 1992.

[9] M.A. Poletti and R.G. Vaughan, "Reduction of Multipath Fading Effects in Single Variable Modulations," ISSPA'90, pp. 672-676, GoldCoast, Australia, 27-31 August 1990.

[10] H. Zaghloul, G. Morrison and M. Fattouche, "Frequency Response and Path Loss Measurements of the Indoor Channel," *Electron. Lett.*, vol. 27, pp. 1021-1022, June 6 1991.

[11] W.D. Rummler, " A New Selective Fading Model: Application to Propagation Data," *B.S.T.J.*, vol. 58(5), pp. 1037-1071, March-June 1979.

[12] R.F. Pawula, S.O. Rice and J.H. Roberts, "Distribution of the Phase Angle between Two Vectors Perturbed by Gaussian Noise," *IEEE Trans. Commun.*, vol. COM-30, pp. 1828-1841, Aug. 1982.

[13] A.V. Oppenheim and R.W. Shafer, *Discrete-Time Signal Processing*, Prentice-Hall, Englewood Cliffs, 1989.

[14] W.C. Jakes, *Microwave Mobile Communications,* New York: Wiley, 1974.

5

Recent Developments in Applying Neural Nets to Equalization and Interference Rejection*

I. Howitt
Department of Electrical Engineering
and Computer Science
University of California, Davis
Davis, CA 95616
howitt@eecs.ucdavis.edu

J. H. Reed
Mobile and Portable Radio Research Group
Bradley Department of Electrical Engineering
Virginia Tech
Blacksburg, VA 24060
reedjh@vtvm1.cc.vt.edu

V. Vemuri
Department of Applied Science
University of California
Livermore, CA 94550
vemuri@icdc.lln.gov

T.C. Hsia
Department of Electrical Engineering
and Computer Science
University of California, Davis
Davis, CA 95616
hsia@eecs.ucdavis.edu

Abstract

Since the introductory neural net equalization paper by Gibson in 1989, there have been a number of impressive contributions in the area of applying neural nets for equalization and interference rejection. Demonstrated advantages of neural nets over conventional linear filtering and equalization include:

1. *Better compensation of non-linear distortion,*
2. *Superior rejection of noise,*
3. *Better equalization of non-minimal phase channels,*
4. *Better rejection of non-Gaussian interference,*
5. *Capability of rejecting CDMA interference,*
6. *Availability of additional blind equalization algorithms, and*
7. *More robust equalization startup.*

This paper reviews recent developments and research trends in the application of neural nets to equalization and interference rejection.

1.0 Introduction

Neural nets (NNs) have recently become a tool for communication signal processing. In the past few years, NNs have been applied to equalization [1-13, 26], interference rejection

* This work has been sponsored by ARGOSystems, Inc. and the California State MICRO Program.

[11,14, 25, 29], coding [19,20], channel assignment [21,22], and network routing [23,24]. The focus of this literature review is the application of NNs to equalization and interference rejection. In nearly all of the papers reviewed, the NN approach shows better performance than the adaptive linear filtering approach and approximates the performance of optimal receivers.

2.0 Equalization Using NNs

Consider the channel and equalizer model shown in Figure 1. The channel H(z) models multipath distortion. It is apparent from Figure 1 that if the noise is zero, $n_i = 0$, then the adaptive equalizer should have the mapping $G(z) = H^{-1}(z)$. This places limitations on the type of channels from which a *linear transversal equalizer* (LTE) will be able to successfully reconstruct the transmitted sequence. Essentially, the transmitted symbols represented by the input vector to the LTE must be linearly separable in order to obtain complete recovery of the symbols. This is true only if the channel can be represented by a minimum phase system (i.e., if all roots of H(z) lie strictly inside the unit circle) [2,3].

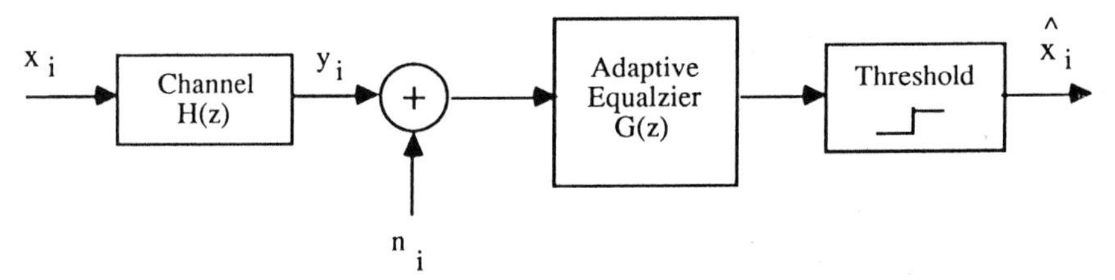

Figure 1. Block Diagram Adaptive Equalizer System

Several examples are given in [2-5] to illustrate the limitations of an LTE and the possible benefits of using a more general nonlinear mapping for the adaptive equalizer other than the simple threshold non-linearity used in conventional equalizers. In the examples, the input signal $x_i \in \{-1,1\}$ is uniformly, independently, and identically distributed. The channel is modeled having a finite impulse response, H(z); and the noise n_i is modeled as an independent zero mean Gaussian distributed sequence. The LTE is a two-tap filter and therefore has two input values y_i and y_{i-1}. Figures 2 and 3 show the channel output points for minimum phase and non-minimum phase system, respectively. The minimum phase system's transfer function is $H(z) = 1 + 0.5z^{-1}$, and the non-minimal phase transfer function is $H(z) = 0.5 + 1.0z^{-1}$.

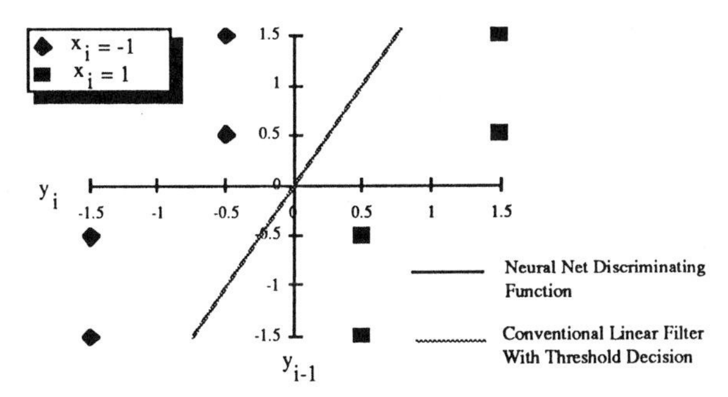

Figure 2. Scatter plot of actual symbols as a function of consecutive output channel samples y_i , y_{i-1}.for a minimum phase channel and no noise.

Two consecutive channel output points $(y_{i,}y_{i-1})$ form eight unique pairs. Two classes can be defined by associating each output with the value of the input signal at time i , i.e., the two classes $c_{-1} = \{(y_i, y_{i-1}) / x_i=-1\}$ and $c_1 = \{(y_i, y_{i-1}) / x_i=1\}$. Figure 2 shows that the two classes can be separated by a line. In geometric terms, an LTE 's decision boundary forms a hyperplane in its input vector space. For the minimum phase example, an LTE with the correct weights and a simple threshold device uniquely distinguishes the two classes assuming.

In Figure 3, however, the two classes cannot be separated by a single hyperplane. In addition, if noise is considered, then the optimum decision boundaries for an equalizer of a given order M can be determined using a Bayes classifier with prior knowledge of the underlying statistics of the signal and noise [2,3,4]. The resulting decision boundary can be shown to produce the least possible *bit error rate* (BER) [2].

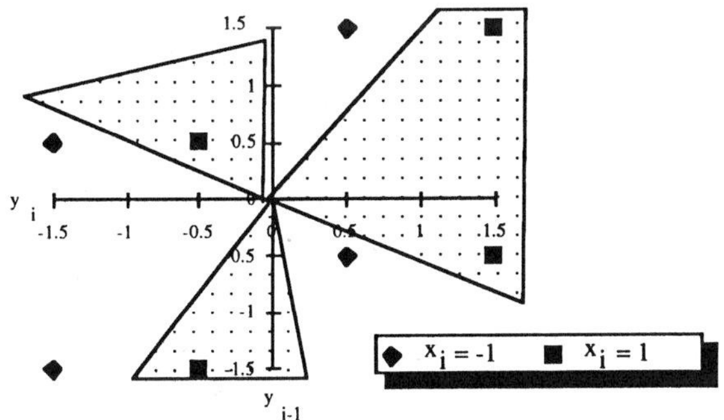

Figure 3. Scatter plot for a non-minimal phase channel and the neural net discriminating function.

The optimal decision boundaries depicted in Figure 2 are determined by finding the set of points [18]

$$\left\{(y_j,y_{j-1})\in\Re^2 \,/\, p_{c_1}(y_j,y_{j-1})=p_{c_{-1}}(y_j,y_{j-1})\right\}.$$

Here p_{c_1} and $p_{c_{-1}}$ are the probability density functions for the two classes. Since it is known apriori that the input signals are uniformly distributed and the noise is Gaussian, then the equation above reduces to

$$\left\{(y_j,y_{j-1}) \in \Re^2 \,/\, \sum_{x_i\in c_1} p(x_i)p_{c_1}(y_j,y_{j-1}/x_i)-\sum_{x_k\in c_{-1}} p(x_k)p_{c_1}(y_k,y_{k-1}/x_k)=0\right\}.$$

Since x is i.i.d. $p(x_i)=p(x_k)\vee x_i\in c_1, x_k\in c_{-1}$, then,

$$\left\{(y_j,y_{j-1})\in\Re^2 \,/\, \sum_{x_i\in c_1} p_{c_1}(y_i,y_{i-1}/x_i)-\sum_{x_k\in c_{-1}} p_{c_1}(y_k,y_{k-1}/x_k)=0\right\}.$$

Here $p(x_i)$ is the density function for the input signal and $p_{c_1}(y_j,y_{j-1}/x_i)$ and $p_{c_{-1}}(y_k,y_{k-1}/x_k)$ are the conditional density functions for the channel output given the input signal. As can be seen from Figure 2, the optimal decision paths are non-linear even when the noise power is small. Therefore, even when the channel can be modeled as a minimum phase system, the LTE cannot produce a decision boundary that is optimal when noise is present.

Another problem encountered with the LTE is that of noise enhancement [2,4]. This essentially limits the effectiveness of increasing the order of an LTE, since as the order of the equalizer increases, the total power of the noise of the equalizer input also increases, thus providing further support for examining the use of low order nonlinear filters.

Implementation and testing of nonlinear adaptive equalizers using a feed forward NN with *backpropagation* (BP) have been carried out by a number of researchers [2,3,5,6,8-10]. The general implementation scheme is a straightforward extension of the LTE as shown in Figure 4. The figure includes an extension to the basic transversal equalizer that involves feedback of the detected signal, hence the name decision feedback. Reference [5] examines the use of NNs for this case.

It is shown in [2,3] that for the minimum phase example illustrated in Figure 2, the BP trained NN almost identically forms the optimal decision boundary based on the Bayesian classifier. In addition, simulation results show that transversal the NN equalizer provides at least an order of magnitude improvement in BER over the LTE when the SNR is greater than 20 dB. The NN provides a consistent 3 to 4 dB improvement in SNR for a given BER when compared to the LTE. When the NN is extended to using decision feedback, an additional improvement in BER is obtained compared to both the NN transversal equalizer and the linear decision feedback equalizer [5].

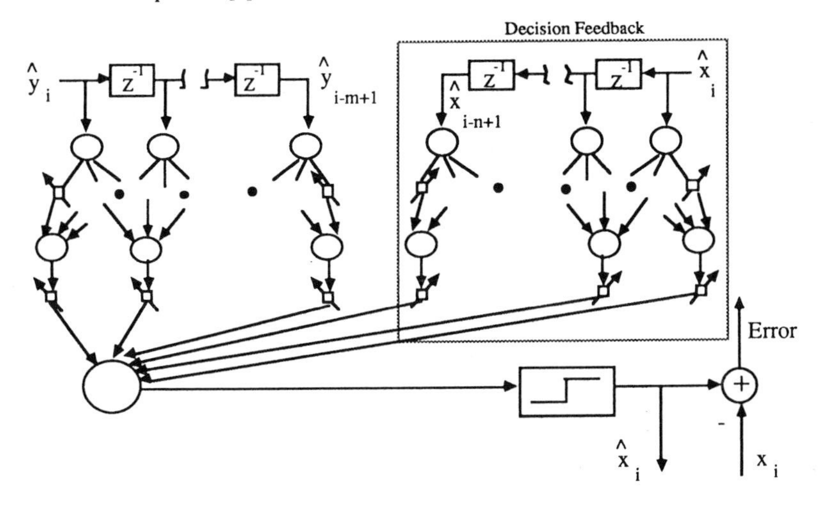

Figure 4. Feed Forward NN Adaptive Equalizer with Optional Decision Feedback

In order to handle QAM signals the feed forward network needs to be modified along with the BP algorithm to handle complex signals. Examples of these straightforward extensions are given in [8-10]. For the most part, the NN and its extensions provide an improvement of 3 to 5 dB MSE over that of the corresponding LTE.

The application of neural net equalization of non-linear channels is explored in [6]. In this study, the channel model is divided into three components: linear dispersive, clamping non-linearity, and additive noise. The results show approximately two orders of magnitude improvement in the BER at a SNR of 30 dB when comparing the NN equalizer to the LTE.

A major drawback of BP trained NN is the number of iterations required to converge. In addition, in order to obtain the performances given above, large training sets are required for the NNs. In a number of applications (speech processing, character recognition, etc.) the feed forward NN is trained off line and then the trained networks are embedded into the system.

3.0 Interference Rejection Using NNs

Demonstrations of the capability of neural nets to reject interference are contained in [11,14,25,19]. The most promising work to date is presented in [25] by Chen. Chen shows that an adaptive radial basis function neural net equalizer can implement the optimal Bayesian symbol-decision equalizer using a two-stage learning algorithm. The first stage is a supervised or decision-directed clustering algorithm which learns the centers of the desired signal states, and the second stage is a variation of an unsupervised k-means clustering algorithm for modeling the effect of the interference. The results obtained by Chen are quite impressive. In one example, the neural net provides an effective reduction in SINR by 7 dB over the transversal equalizer for a BER of 10^{-4}. The algorithm converges remarkably fast even compared to traditional equalization algorithms. The results also indicate that the algorithm might be very effective for a negative SIR provided the SNR is high.

A decision feedback multilayer perceptron design using a fast adaptive algorithm for training has been shown to be useful for rejecting interference for a mobile communication channel subject to Rayleigh fading, noise, co-channel interference and multipath [11]. The fast adaptive training algorithm is based on a combination of the steepest descent method and the conjugate gradient method. The step size (or learning rate) is adaptively adjusted at each iteration by using a line search algorithm. In addition, heuristics techniques are applied to selectively update the weights. The results reported showed an improvement of at least an order of magnitude in the BER for all SNRs tested. In some cases the improvement in BER over the conventional decision feedback equalizer is on the order of 100 .

Direct sequence spread-spectrum modulation used in *code division multiple access* (CDMA) creates interference among co-users of the spectrum. In particular, extreme interference is experienced by co-users when the received signal power level of any particular user greatly exceeds the average received power level of all other users. This problem is called the near-far problem and has been addressed theoretically in [27] using a demodulator optimized for CDMA interference. This optimal solution is computationally expensive, growing exponentially with the number of active channel users (this occurs for both synchronous and asynchronous operations) [28].

A neural network has been proposed and evaluated using binary PSK direct sequence signals with k active users and a spreading sequence of length N [14] in an effort to find a simplified approximation to the optimal solution. The neural network is a multi-layer perceptron trained with backpropagation. A slight modification to the BP algorithm is used in some cases where the additional information about the corrupting interference signal is used to restrict the nodes updated in the hidden layer. The multi-layer perceptron is trained to form a decision boundary for synchronous transmission (single user and multi-user demodulation) that is nearly identical to the optimal receiver boundary. Tests results are provided using k=3 users, N=3 length code, and SIRs with respect to each co-user of -6 dB and 0 dB. The resulting BER vs. SNR curve shows the NN demodulator tracks the optimal receiver's performance within 2 dB over a wide range of the SNR. These test cases include synchronous channels, where the information bits from the different sources arrive at the receiver at the same time, and from asynchronous channels, where the relative time delays are assumed to be constant but having unknown values throughout the training and testing. BER versus SNR curves show the NN demodulator provides orders of magnitude improvement in the BER for SNRs greater than 8 dB and a SIR of -6 dB. For a more realistic spreading sequence, k=2 and N=31, the NN demodulator provides orders of magnitude better BER than the match filter for SNRs greater than 8 dB.

4.0 Conclusion

The technical literature shows NNs to be substantially better in mitigating the effects of noise, linear and non-linear distortion, and interference than traditional equalization techniques. Our experience in verifying some of these results has shown that much experimental trial and error is necessary to realize the full performance potential of the NNs. For instance, it is necessary to experimentally determine the number of layers and nodes, initialization procedure, adaptation parameters, etc. In addition, a gradient descent algorithm is typically used to train the NN equalizer, and as a result, there is no guarantee of reaching an optimal solution. Even if an optimal solution is reached it is most likely a locally optimal solution. Thus training may require restarting the NN multiple times to find the best solution. In most instances, the convergence rate of the NN equalization techniques is very slow. These are problems that must be overcome before NN equalizers can be incorporated into actual systems, especially for systems that contend with a dynamic channel. The challenging future research work is to determine robust and practical NN structures for equalization and interference rejection applications.

5.0 References

[1] Gibson, G.J., S. Siu, and C.F.N. Cowan, "Application of multilayer perceptrons as adaptive channel equalizers," *Proceedings of the IEEE International Conference on Acoustics, Speech, and Signal Processing*, Glasgow, pp. 1183-1186, May 1989.

[2] G. J. Gibson, S. Siu, S. Chen, C. F. N. Cowan, and P. M. Grant, "The application of nonlinear architectures to adaptive channel equalisation," *IEEE International Conference on Communications*, vol. 2, pp. 649-653, 1990.

[3] G. Gibson J, S. Siu, and C. F. N. Cowan, "Multilayer perceptron structures applied to adaptive equalisers for data communications," *International Conference on Acoustics, Speech, and Signal Processing*, pp. 1183-1186, 1991.

[4] S. Chen, G. J. Gibson, and G. F. N. Cowan, "Adaptive channel equalisation using a polynomial-perceptron structure," *IEE Proceedings*, vol. 137 part 1, no. 5, pp. 257-264, Oct. 1990.

[5] S. Siu, G. J. Gibson, and C. F. N. Cowan, "Decision feedback equalisation using neural network structures and performance comparison with standard architecture," *IEE Proceedings*, vol. 137 part 1, no. 4, pp. 221-225, Aug. 1990.

[6] Q. Zhang, "Adaptive equalization using the backpropagation algorithm," *IEEE Transactions on Circuits and Systems*, vol. 37, no. 6, pp. 848-849, June. 1990.

[7] K. Raivio, O. Simula, and J. Henriksson, "Improving design feedback equaliser performance using neural networks," *Electronics Letters*, vol. 27, no. 23, pp. 2151-2153, 7 Nov. 1991.

[8] N. Benvenuto, M. Marchesi, F. Piazza, and A. Uncini, "Non linear satellite radio links equalized using blind neural networks," *International Conference on Acoustics, Speech, and Signal Processing*, pp. 1521-1524, 1991.

[9] K. Giridhar and J. J. Shynk, "A perceptron-based adaptive equalizer for PAM signals," , pp. 1581-1584, 1992.

[10] M. Peng, C. L. Nikias, and J. G. Proakis, "Adaptive equalization for PAM and QAM signals with neural networks," *The Twenty-Fifth Asilomar Conference on Signals, Systems & Computers*, pp. 496-500, 1991.

[11] Z. Xiang and B. Guangguo, "Fractionally spaced decision feedback multilayer perceptron for adaptive MQAM digital mobile radio reception," *International Conference on Communications '92*, vol. 3, pp. 1262-1266, 1992.

[12] Z. Xiang and G. Bi, "New lattice polynomial perceptron based M-QAM digital communication reception systems," *Electronics Letters*, vol. 28, no. 19, pp. 1839-1841, 10 Sept. 1992.

[13] G. de Veciana and A. Zakhor, "Neural net based continuous phase modulation receivers," *IEEE International Confernce on Communications*, vol. 2, pp. 419-423, 1990.

[14] B. Aazhang, Bernd-P. Paris, and G. C. Orsak, "Neural networks for multiuser detection in code-division multiple-access communications," *IEEE Transactions on Communications*, vol. 40, no. 7, pp. 1212-1222, July. 1992.

[15] D. E. Rumelhart and J. L. McClelland, *Parallel distributed processing: explorations in the microstructure of cognition*. Cambridge: MIT Press, 1986.

[16] R. P. Lippmann, "An introduction to computing with neural nets," *IEEE ASSP Magazine*, vol. 4, 1987.

[17] T. Kohonen, "The "neural" phonetic typewriter," *Computer*, vol. 21, no. 3, pp. 11-22, March. 1988.

[18] J. T. Tou and R. C. Gonzalez, *Pattern recognition principles*. Addison-Wesley, 1974.

[19] C. Jeffries, "High order neural models for error correcting code," *Proceedings form the SPIE*, pp. 510-517, 1990.

[20] M.E. Santamaria, M.A. Lagunas and M. Cabrera, "Neural nets filters: integrated coding and signaling communication systems," *Melecon'89*, pp. 532-535, 1989.

[21] K. Nakano, et. al., "Channel assignment in cellular mobile communication systems using neural networks," *Singapore Int. Conf. on Communication Systems*, pp. 531-534, Nov. 1990.

[22] D. Kunz, "Channel assignment for cellular radio using neural networks," *IEEE Trans. on Vehicular Technology*, vol. 40, no. 1, pp. 188-193, Feb 1991.

[23] A. Hiramatsu, "ATM communications network control by neural network," *IJCNN 89, Washington D.C.*, pp I/259-266, 1989.

[24] J.E. Jensen, M.A. Eshara, and S.C. Barach, "Neural network controller for adaptive routing in survivable communications networks," *IJCNN 90, San Diego, CA*, pp. II/29-36, 1990.

[25] S. Chen and B. Mulgrew, “Overcoming co-channel interference using an adaptive radial basis function equalizer,” *Signal Processing*, Vol. 28, pp. 91-107, 1992.

[26] Chen S., B. Mulgrew, and S. McLaughlin, “Adaptive Bayesian decision feedback equalizer based on a radial basis function network,” *IEEE International Conference on Communications*, pp. 1267-1271, 1992.

[27] Verdu, S., “Minimum probability of error for asynchronous Gaussian MA channels,” *IEEE Transactions on Information Theory*, Vol IT-32, pp. 85-96, Jan. 1986.

[28] Verdu, S., "Computational complexity of optimal multiuser detection," *Algorithmica*, Vol 4, no. 3, pp. 303-312, 1989.

[29] Aazhang, A. , "Neural networks for multi-user detection in code-division multiple access communications," *IEEE Transactions on Communications*, Vol. 40, No. 7, pp. 1212-1232, July 1992.

6

Direct Sequence Spread Spectrum Interference Rejection Using Vector Space Projection Techniques

John F. Doherty*
330 Coover Hall
Department of Electrical Engineering and Computer Engineering
Iowa State University
Ames, Iowa 50011

Abstract

A new method of rejecting narrowband interference in direct-sequence spread spectrum communications is presented. A typical approach is to reject the interference using a filter with large attenuation at the interference frequencies prior to despreading. The interference rejection method presented in this paper incorporates vector space projection techniques to suppress the correlated interference. Several signal characteristics are formulated which lead to constraint surfaces in the vector space of possible solutions. These constraint surfaces describe interference rejection solutions which introduce minimal distortion to the chip sequence and simultaneously remove the interference. The constraint surfaces essentially correspond to chip sequence estimates which, after interference rejection, conform to known characteristics of the transmitted chip sequence. The formulation of the surfaces relies on prior knowledge about the chip sequence correlation and spectral properties.

1. Introduction

The presence of relatively narrowband interference in direct sequence spread spectrum (DSSS) systems is often an unavoidable problem. The interference may be intentional, as in military communications, or it may be unintentional, as in a spectral overlay system. In any case, the receiver achieves higher signal to noise ratios at the decision device input if a interference rejection filter is used prior to despreading [2]. The rejection filter is usually adaptive and relies on the

*This work was supported by a Harpole-Pentair Faculty Fellowship.

pseudo-white properties of the chip sequence[1]. However, there is some benefit from using the *a priori* information available about the pseudo-noise sequence [3]. This information can subsequently be related to the bit modulated chip sequence. The known properties of the chip sequence can be used to formulate linear constraint surfaces, which can yield an alternative adaptation process for the interference rejection.

This paper introduces a technique that combines linear and nonlinear processing to remove the interference from the chip sequence prior to despreading. The interference rejection is accomplished without the need for a rejection filter. This nonlinear technique requires some prior knowledge about the chip sequence in both the time and frequency domains.

2. Projection Based Method

The direct sequence used to construct the chip sequence has deterministic properties that may be used to form constraint surfaces which correspond to better solutions for the chip sequence. The discrete-time signal model at the input to the decision device is given by

$$\widetilde{a}_n = \operatorname{sgn}\left[\sum_{k=1}^{K} p(k)\widehat{c}_n(k)\right] \tag{1}$$

where $\operatorname{sgn}[\bullet]$ is the signum function, $\widetilde{a}_n$ is the nth bit estimate, p_k is the direct sequence, and $\widehat{c}_n(k)$ is the chip sequence estimate for the nth bit interval. There are certain properties of the chip sequence that we will use to improve our estimate $\widetilde{a}_n$.

1. **Property:** The magnitude of the discrete Fourier transform of $c_n(k)$ is given by

$$|C_n(m)|^2 = K \qquad \forall m \neq 1 \tag{2}$$

   and $|C_n(1)| = 1$.

2. **Property:** The energy of $c_n(k)$ is given by

$$\sum_{k=1}^{K} c_n^2(k) = K \tag{3}$$

   where it is assumed that we have binary signaling.

3. **Property:** The chip sequence has bounded amplitude

$$|c_n(k)| = 1 \qquad \forall k \tag{4}$$

4. **Property:** The average value of the chip sequence is given by

$$\bar{c}_n = \sum_{k=1}^{K} c_n(k) = \mp 1 \tag{5}$$

   Note that (5) implies that $|C_n(1)| = 1$.

5. **Property:** The cross correlation of the chip sequence with the direct sequence is given by

$$\sum_{k=1}^{K} p(k) c_n(k) = \pm K \tag{6}$$

6. **Property:** The cross correlation of the chip sequence with a rotated version of the direct sequence is given by

$$\sum_{k=1}^{K} p_m(k) c_n(k) = \mp 1 \qquad m = 1, 2, \ldots, K-1 \tag{7}$$

   where $p_m(k)$ is the direct sequence cyclically rotated m positions.

The goal of the rejection process will be to produce an estimate of the chip sequence that satisfies all these properties, based on observing the chip sequence corrupted by correlated interference and uncorrelated noise. The interference and noise will not, in general, satisfy any of the properties listed above. This allows the rejection process to discriminate between the chip sequence and the extraneous interference.

2.1. Iterative Chip Sequence Estimation using Nonlinear Projection Updates

2.1.1. Orthogonal projections onto linear surfaces

Each of the above listed properties can be forced upon any chip sequence estimate that we produce. Some of these properties can be achieved with multiple orthogonal projections onto a set of linear constraint surfaces.

$$\mathbf{c}_n^{j+1} = \mathbf{c}_n^j + \mu \left[d - \mathbf{x}^T \mathbf{c}_n^j \right] \frac{\mathbf{x}}{||\mathbf{x}||^2} \tag{8}$$

The equation (8) is readily recognized as the normalized LMS update equation.

2.1.2. Orthogonal projections onto nonlinear surfaces.

The chip sequence can be made to conform to the properties indicated in (2),(3), and (4) by using orthogonal projections onto nonlinear surfaces. The properties associated with projecting onto these particular surfaces are described in detail in [5]. We will only provide the mechanism by which the rejection processor achieves these objectives. The orthogonal projection associated with the spectral boundedness property of (2) is given by

$$C_m^{j+1}(n) = \begin{cases} C_m^{j+1}(n) & |C_m^{j+1}(n)|^2 \leq K \\ \sqrt{\frac{K}{|C_m^{j+1}(n)|}} C_m^{j+1}(n) & |C_m^{j+1}(n)|^2 > K \end{cases} \tag{9}$$

The energy property of (3) can be achieved using

$$c_n^{j+1}(k) = \sqrt{\frac{K}{\sum_k \left(c_n^{j+1}(k) \right)^2}} \; c_n^{j+1}(k) \tag{10}$$

The amplitude bound of (4) can be implemented using the soft-limiting operation

$$c_n^{j+1}(k) = \begin{cases} c_n^{j+1}(k) & |c_n^{j+1}(k)| \leq 1 \\ \operatorname{sgn}\left[c_n^{j+1}(k) \right] & |c_n^{j+1}(k)| > 1 \end{cases} \tag{11}$$

2.2. Noise reduction properties

The linear and nonlinear operations previously described have noise reduction capability.

2.2.1. Linear projections

The update equation in (8) has the equivalent form

$$\mathbf{c}_n^{j+1} = \left[\mathbf{I} - \mu \frac{\mathbf{x}_k \mathbf{x}_k^T}{||\mathbf{x}_k||^2} \right] \mathbf{c}_n^j + \mu d_k \frac{\mathbf{x}_k}{||\mathbf{x}_k||^2} \tag{12}$$

where we are updating using the kth constraint property. This can be simplified in notation by using the substitutions

$$\mathbf{P}_k = \left[\mathbf{I} - \mu\left(\mathbf{x}_k\mathbf{x}_k^T\right) / \left(\|\mathbf{x}_k\|^2\right)\right] \tag{13}$$

and $\mathbf{v} = \mu d_k \mathbf{x}_k / \|\mathbf{x}_k\|^2$. We note that the matrix $\mathbf{P}_k$ is a projection operator which projects onto the kth linear surface, passing though the origin, described by $\mathbf{x}_k$. When $\mu = 1$, this is known as an affine projection. The vector $\mathbf{v}_k$ provides the translation necessary to achieve the correct amplitude d_k. The combination of all the linear projections results in

$$\mathbf{c}_n^r = \prod_{l=1}^{r} \mathbf{P}_l \mathbf{c}_n^0 + \sum_{m=1}^{r-1} \left(\prod_{l=m+1}^{r} \mathbf{P}_l \right) \mathbf{v}_m + \mathbf{v}_r \tag{14}$$

where each integer increment of the index r is equivalent to projecting once onto each of the constraint surfaces. Define $\mathbf{A} = \prod_{l=1}^{r} \mathbf{P}_l$ and $\mathbf{b} = \sum_{m=1}^{r-1} \left(\prod_{l=m+1}^{r} \mathbf{P}_l\right) \mathbf{v}_m + \mathbf{v}_r$, then multiple projections onto all the constraint surfaces produces

$$\hat{\mathbf{c}}_n = \mathbf{A}^M \mathbf{c}_n^0 + \sum_{k=0}^{M-1} \mathbf{A}^k \mathbf{b} \tag{15}$$

Here it is noted that the initial chip sequence estimate is the summation of the actual chip sequence, $\mathbf{c}_n$, and the correlated and white interference, $\mathbf{w}_n$. Substituting for $\mathbf{c}_n^0$ in (15) yields an estimate of the form

$$\hat{\mathbf{c}}_n = \left[\mathbf{A}^M \mathbf{c}_n + \sum_{k=0}^{M-1} \mathbf{A}^k \mathbf{b} \right] + \mathbf{A}^M \mathbf{w}_n \tag{16}$$

The bracketed term in (16) is identically equal to $\mathbf{c}_n$, since it represents a projection of $\mathbf{c}_n$ onto those surfaces that describe $\mathbf{c}_n$. The remaining term in (16) is the residual noise of the estimation process. The nth symbol estimate resulting from (16) is

$$\hat{a}_n^{pro} = \mathbf{p}^T \hat{\mathbf{c}}_n = a_n + \mathbf{p}^T \mathbf{A}^M \mathbf{w}_n \tag{17}$$

The output of the nominal despreader, without interference rejection, is

$$\hat{a}_n = a_n + \mathbf{p}^T \mathbf{w}_n \tag{18}$$

So that is remains to be shown that

$$\mathbf{p}^T \mathbf{A}^M \mathbf{w}_n < \mathbf{p}^T \mathbf{w}_n \tag{19}$$

in which case the projection estimator provides interference suppression. Let the matrix $\mathbf{A}$ have the eigen-decomposition

$$\mathbf{A} = \mathbf{U}\mathbf{\Lambda}\mathbf{U}^T \tag{20}$$

where $\mathbf{U}$ is a unitary matrix and $\mathbf{\Lambda}$ is diagonal. Furthermore, let $\mathbf{p}' = \mathbf{U}^T\mathbf{p}$ and $\mathbf{w}_n' = \mathbf{U}^T\mathbf{w}_n$. Using these substitutions, we obtain

$$\mathbf{p}^T\mathbf{A}^M\mathbf{w}_n = \mathbf{p}'^T\mathbf{\Lambda}^M\mathbf{w}_n' \tag{21}$$

Thus, if we can show that all of the eigenvalues of $\mathbf{A}$ satisfy $0 < \lambda_k < 1$, then the inequality in (19) holds.

The eigenvalues of $\mathbf{A}$ satisfy the relationship[4]

$$\prod_{m=1}^{M} \min\left[\text{eig}\left(\mathbf{P}_m\right)\right] \le \lambda_k \le \prod_{m=1}^{M} \max\left[\text{eig}\left(\mathbf{P}_m\right)\right] \qquad \forall k \tag{22}$$

where each projection matrix in (13), $\mathbf{P}_m$, has $K-1$ eigenvalues equal to unity, and one eigenvalue equal to $1-\mu$. The right hand side of the inequality (22) achieves equality only when all the matrices, $\mathbf{P}_m$, share the same principal axis system, which is not true for the constraint surfaces considered here. Thus, the eigenvalues of $\mathbf{A}$ satisfy the inequality

$$0 \le (1-\mu)^M \le \lambda_k < (1)^M \tag{23}$$

2.2.2. Amplitude bound

The amplitude bounding operation can be analyzed by considering the resulting reduction in the estimate error at its output. If the following relationship holds, then the noise has been reduced.

$$\sum_{k=1}^{K}\left(c_n(k) - [c_n(k)+w(k)]'\right)^2 < \sum_{k=1}^{K} w^2(k) \tag{24}$$

where $[\bullet]'$ represents the result of the soft-limiter. Consider two ranges of the indices $\aleph_1$ and $\aleph_2$, such that $\aleph_1 \cup \aleph_2 = \aleph = \{k \mid k = 1, 2, \ldots, K\}$ represents the entire set of indices in (24). We will define the sets as

$$\aleph_1 = \{k \mid |c_n(k) + w(k)| \le 1\} \tag{25}$$

$$\aleph_2 = \{k \mid |c_n(k) + w(k)| > 1\} \tag{26}$$

The summations in (24) can be decomposed into these two ranges, which yields

$$\sum^{\aleph_2} (c_n(k) - \mathrm{sgn}\,(c_n(k) + w(k)))^2 < \sum^{\aleph_2} w^2(k) \tag{27}$$

where

$$\mathrm{sgn}\,(c_n(k) + w(k)) = \begin{cases} c_n(k) & \text{if } |w(k)| \leq |c_n(k)| \quad \text{Region:}\aleph_3 \\ -c_n(k) & \text{if } |w(k)| > |c_n(k)| \quad \text{Region:}\aleph_4 \end{cases} \tag{28}$$

where $\aleph_3 \cup \aleph_4 = \aleph_2$. Using the dichotomy of the indices in(28), the inequality in (27) can be expressed as

$$4\sum^{\aleph_4} c_n^2(k) < \sum^{\aleph_3} w^2(k) + \sum^{\aleph_4} w^2(k) \tag{29}$$

The left hand side of (29) is just $4\aleph_4^*$, where $\bullet^*$ denotes the number of elements in that set. The right hand side may be bounded by replacing $w(k)$ with its lower bound in each range, which produces a lower bound of $\delta = \sum^{\aleph_3} 0^2 + \sum^{\aleph_4} 2^2 = 4\aleph_4^*$. Thus, the inequality of (29) holds, which means that the original inequality of (24) is true, and the amplitude bound always reduces the noise.

3. Simulation Results

The projection method will be numerically tested using the discrete time received signal model

$$r_n(k) = c_n(k) + \jmath_n(k) + \nu_n(k) \tag{30}$$

where $\jmath_n(k)$ is the correlated interference and $\nu_n(k)$ is the thermal wideband noise.

The update equation (8) is modified due to the uncertainty associated with the polarity of the information bit. The projection update used is

$$\mathbf{c}_n^{j+1} = \mathbf{c}_n^j + \mu \left[\mathrm{sgn}\left[\mathbf{x}^T\mathbf{c}_n^j\right] d - \mathbf{x}^T\mathbf{c}_n^j\right] \frac{\mathbf{x}}{||\mathbf{x}||^2} \tag{31}$$

The direct sequence length is set to 511 and two (2) iterations are used. That is, each surface

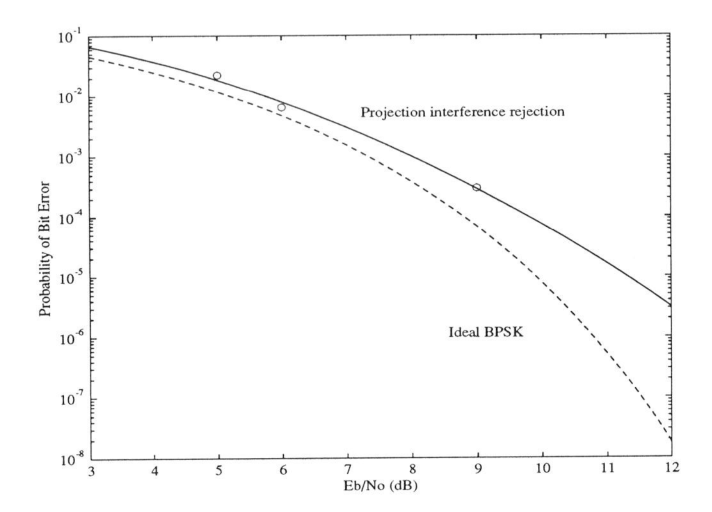

Figure 1: Experimental results using SIR=-20dB.

is projected onto twice. Only the first two rotations of (7) are used. The constant $\mu = 1$ is used.

The correlated interference occupies 10% of the signal bandwidth centered at $f = 0.25$ in normalized frequency. The results of the rejection process are shown in Figure 1 for a SIR of -20dB. The results for an SIR of -30dB are shown in Figure 2.

4. Conclusion

A new projection based interference rejection method was presented which suppresses the correlated interference in a direct sequence spread spectrum system. The interference rejection is accomplished without the use of a filter, in its usual form. The rejection takes place by projecting onto constraint surfaces that conform to the chip sequence. In general, the interference does not conform to these constraint surfaces, and is therefore removed. The projection method was demonstrated using computer simulation, where its efficacy for removing the interference was demonstrated.

References

[1] L. Milstein and R. Iltis, "Signal processing for interference rejection in spread spectrum communications," *IEEE Acoustics, Speech, and Signal Processing Magazine*, vol. 3, pp. 18–31, April

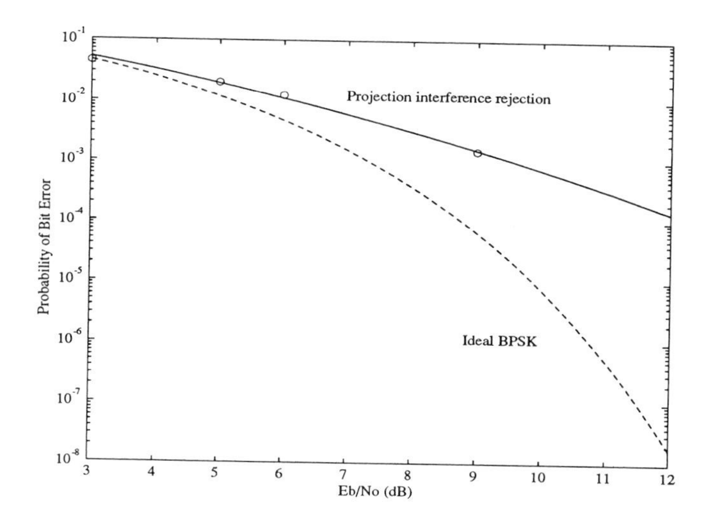

Figure 2: Experimental results using SIR=-30dB.

1986.

[2] J. Ketchum and J. Proakis, "Adaptive algorithms for estimating and suppressing narrowband interference in PN spread spectrum systems," *IEEE Transactions on Communications*, vol. COM-30, pp. 913–923, May 1982.

[3] J. Doherty, "Linearly constrained direct sequence spread spectrum interference rejection," *IEEE Transactions on Communications*, to appear.

[4] T. Chow,"Matrices and Linear Algebra", Handbook of Applied Mathematics,Van Nostrand Reinhold, New York, 1983.

[5] D. Youla,"Generalized Image Restoration by the Method of Alternating Orthogonal Projections", *IEEE Transactions on Circuits and Systems*, vol. 25, 1978.

7

Solving the Near-Far Problem: Exploitation of Spatial and Spectral Diversity in Wireless Personal Communication Networks

Brian G. Agee
Radix Technologies, Inc.
329 North Bernardo Avenue
Mountain View, California 94043

Abstract

A general approach is presented for overcoming the near-far power management problem in wireless communication networks, by exploiting the spatial or spectral diversity inherent to the communication network. It is shown that the stability and efficiency of near-far power management strategies used in CDMA, TDMA, or FDMA communication networks are greatly enhanced by exploiting the spatial diversity of the communication network. It is also shown that the same improvements in stability and efficiency can be obtained by exploiting the spectral diversity of CDMA networks. In particular, it is shown that use of an M-element multiport antenna array at the base station of any communication network can increase the frequency reuse of the network by a factor of M and greatly broaden the range of input SINRs required for adequate demodulation of the subscriber signals in the network. A similar result is obtained for CDMA networks employing an M-chip modulation on symbol (MOS) *DSSS spreading formats where the code sequence repeats once per message symbol, even if a single antenna is used at the base stations in the network. These results are supported via computer simulations for an FDMA network employing a 3-element antenna array to separate 3 AMPS-type FM signals received at the same frequency, and for a CDMA communication network employing a single-antenna optimal linear despreader to separate 32 users with 64-chip MOS-DSSS modulation formats.*

1 Introduction

The *near-far problem* (reception of signals over greatly disparate propagation ranges) poses a serious challenge to the design of any wireless personal communication network. Conventional solutions to this problem manage the transmit power of each user in the network to provide a usable SNR for all of the users after their recovery from the communications channel, for example, after the direct-sequence despread operations at the base stations of a CDMA network. Implementation of such techniques requires close supervision of the communication channel to determine necessary changes in the user power levels, as well as continuous feedback to the network users to ensure fast reaction to channel variations. Moreover, these requirements can introduce severe instability into the communication network, due to inherent delays in the power management link, mobility of the system users, and dynamic multipath distortion in the wireless communication channel.

This paper demonstrates that this problem can be overcome by exploiting the *spatial diversity* inherent to the geographical distribution of the users in the communication network, or the *spectral diversity* inherent to the modulation format employed by typical communication networks. It is shown that multiport blind array adaptation techniques can be used in the presence of either form of diversity to separate and demodulate the multiple signal waveforms and/or message sequences received by the network, where the array input signal is derived from a set of spatially distributed antennas (spatial diversity exploitation) or from an appropriate time or frequency channelization operation (spectral diversity exploitation). It is also shown that diversity exploitation can allow separation of far more signals than conventional demodulation techniques, at far greater disparities in receive power level. In particular, it is shown that these techniques can separate signals with *arbitrary* disparities in receive power, as long as the signals are received at a power level that is sufficiently greater than the *nondiverse* signals (e.g., the background interference) in the received signal environment.

2 Background: Conventional Power Management Strategy

The conventional power management strategy is motivated by considering the environment where K_{user} time-coincident subscriber signals are simultaneously received on a single frequency channel at the base station in a cell of a wireless communication network. The data waveform $x(t)$ provided by this receiver can then be modelled by

$$x(t) \quad = \quad i(t) + \sum_{k=1}^{K_{\text{user}}} g_k e^{j2\pi f_k(t-\tau_k)} s_k(t-\tau_k) \tag{1}$$

after downconversion to complex baseband representation, where $\{s_k(t)\}_{k=1}^{K_{\text{user}}}$ are the transmitted subscriber signals and $\{g_k, \tau_k, f_k\}_{k=1}^{K_{\text{user}}}$ are the complex scalings, delays, and Doppler shifts induced on each signal over the communications channel, and where $i(t)$ is the remaining noise and interference received over the communication channel. The subscriber signals are all assumed to be signals of interest to the base station, i.e., the goal of the base station is to successfully demodulate all K_{user} subscriber signals. Conversely, the interference signal $i(t)$ is assumed to be composed of signals *not* of interest to the receiver, and can include background noise, subscriber signals transmitting on the same frequency channel to base stations in different clusters of the communication network (adjacent-channel interference), and sidebands of subscriber signals transmitting on different frequency channels to base stations in the same cluster of the communication network (adjacent-cluster interference).

The *near-far problem* arises in this scenario due to the disparate propagation ranges between the subscribers base stations in the communication network. This can result in power disparities of 20-to-30 dB for every factor-of-ten difference in distance between the subscribers and the base station, for example, if two end nodes in a wireless local area network are located 3 ft and 30 ft from the base station in the network. This problem is of greatest importance in CDMA systems where $K_{\text{user}} > 1$ and the subscriber signals are separated using only the spreading gain of the CDMA modulation format. However, this problem can also occur in conventional FDMA systems, due to reception of strong (near) signals on channels adjacent to weaker (far) signals of interest, such that the strong signal sidebands "poke through" into the weak signal channel. In these systems, *power management* strategies must be employed to overcome these power disparities and ensure that all of the subscribers can be demodulated with adequate quality.

Power management is accomplished in CDMA communication systems by adjusting the power of each subscriber signal to satisfy the dual constraints

$$\gamma_k \leq \gamma_{\max} \qquad \text{and} \qquad \hat{\gamma}_k \geq \frac{\hat{\gamma}_{\text{dmd}}}{M_{\text{dmd}}} \tag{2}$$

where γ_k and $\hat{\gamma}_k$ are the signal-to-interference-and-noise ratios (SINRs) of the kth subscriber signal measured with respect the background interference $i(t)$ and total interference seen by that signal, respectively,

$$\gamma_k \quad \triangleq \quad \frac{|g_k|^2 R_{s_k s_k}}{R_{ii}} \tag{3}$$

$$\hat{\gamma}_k \;\triangleq\; \frac{|g_k|^2 R_{s_k s_k}}{R_{ii} + \sum_{\ell \neq k} |g_\ell|^2 R_{s_\ell s_\ell}} \;=\; \frac{\gamma_k}{1 + \sum_{\ell \neq k} \gamma_\ell}, \tag{4}$$

and where $\hat{\gamma}_{\text{dmd}}$ is the minimum SINR required at the demodulator input (despreader output) to reliably demodulate the subscriber signals, M_{dmd} is the processing gain of the CDMA modulation format (typically equal to the number of chips per message symbol in conventional CDMA systems), and $\gamma_{\max}$ is a maximum allowable subscriber-free SINR constraint added to prevent excessive interference with the other base stations in the network. Substituting (2) into (4) yields equivalent inequality

$$\gamma_{\min} \;\leq \gamma_k \leq\; \gamma_{\max}, \qquad \gamma_{\min} \;\triangleq\; \frac{\hat{\gamma}_{\text{rcv}}}{1 - (K_{\text{user}} - 1)\,\hat{\gamma}_{\text{rcv}}}, \tag{5}$$

which exists if and only if K_{user} satisfies

$$K_{\text{user}} \;\leq\; 1 + \left(\frac{M_{\text{dmd}}}{\hat{\gamma}_{\text{dmd}}} - \frac{1}{\gamma_{\max}}\right) \;\triangleq\; \hat{K}_{\max}. \tag{6}$$

The lower inequality in (5) are achieved if and only if $\gamma_k \equiv \gamma_{\min}$. Moreover, this lower inequality rises to

$$\gamma_{\min}(L) \;=\; \frac{(1 + L\gamma_{\max})\,\hat{\gamma}_{\text{dmd}}}{M_{\text{dmd}} \;-\; (K_{\text{user}} - L - 1)\,\hat{\gamma}_{\text{dmd}}} \tag{7}$$

if L subscriber signals are received at the maximum allowable subscriber-free SINR, and is achieved if and only if all $K_{\text{user}} - L$ signals have subscriber-free SINRs equal to $\gamma_{\min}(L)$. As a consequence, the maximum spread between subscriber-free SINRs is given by

$$\frac{\gamma_{\max}}{\gamma_{\min}(L)} \;=\; 1 + \frac{\gamma_{\max}}{1 + L\gamma_{\max}}\left(\hat{K}_{\max} \;-\; K_{\text{user}}\right) \tag{8}$$

in this reception scenario, with the largest spread for *any* scenario occuring when a single subscriber is received with SINR $\gamma_k = \gamma_{\max}$ and the remaining subscribers are received with SINRs $\gamma_k \equiv \gamma_{\min}(1)$. This number is typically small, e.g., less than 6 dB if $K_{\text{user}} \geq \hat{K}_{\max} - 3$.

Equation (6) has a direct bearing on the *frequency reuse efficiency* achievable using conventional CDMA networks, defined here as the fraction of message bandwidth that can be received at a given cell over the total bandwidth of the network. The frequency reuse efficiency for a CDMA network employing the power management strategy described above is

$$\frac{K_{\text{user}}}{M_{\text{dmd}}} \;\approx\; \frac{1}{\hat{\gamma}_{\text{dmd}}} \tag{9}$$

i.e., the CDMA network is less efficient than a conventional FDMA system for $\hat{\gamma}_{\text{dmd}} > 7$ (8.5 dB). For example, if the CDMA network is implemented using a 64-chip spreading code and requires a 10 dB despread SINR for adequate demodulation of the message signals, then a conventional single-sensor CDMA network cannot separate more than 7 signals. The CDMA modulation format is highly inefficient in this example, requiring a $\times 64$ increase in message bandwidth to carry 7 subscribers.

In addition, as (8) shows, this reuse efficiency cannot be obtained without extremely careful management of the power levels of all of the subscriber transmitters. For the network described above, the received subscriber SINRs must be within 3 dB of each other if $\gamma_{\max}$ is large and the network is operating at capacity. Moreover, this SINR range can change abruptly as subscribers enter and leave the communication network. As (9) shows, addition of link margin to improve this sensitivity comes at great cost in network capacity – a 3 dB addition of link margin, for example, reduces $\hat{K}_{\max}$ by a factor of two.

3 Spatial Diversity Exploitation

Spatial diversity can be exploited for any networking approach and modulation format, by employing a *multiport adaptive antenna array* to separate the time-coincident subscriber signals prior to the demodulation operation. This processor is motivated by considering the environment where K_{user} subscriber signals are received at the base station in a cell of a wireless communication network using an array of M spatially separated antennas. If the inverse bandwidth of the receiver is small with respect to the electrical distance between the array elements and any multipath is due to near-field reflectors in the vicinity of the array and/or subscribers, then the received data waveform provided by this array can be expressed as an $M \times 1$ vector waveform $\mathbf{x}(t)$ modelled by

$$\begin{aligned} \mathbf{x}(t) &= \mathbf{i}(t) + \sum_{k=1}^{K_{\text{user}}} \mathbf{a}_k e^{j2\pi f_k(t-\tau_k)} s_k(t-\tau_k) && (10) \\ &= \mathbf{i}(t) + \mathbf{A}\tilde{\mathbf{s}}(t) && (11) \end{aligned}$$

$$\mathbf{A} \stackrel{\Delta}{=} [\mathbf{a}_1 \cdots \mathbf{a}_{K_{\text{user}}}]$$

where $\{\mathbf{a}_k\}_{k=1}^{K_{\text{user}}}$ are a set of $M \times 1$ (slowly time vaying) complex *aperture vectors* that multiply the subscriber signals, and where $\tilde{\mathbf{s}}(t) = \left[e^{j2\pi f_k(t-\tau_k)} s_k(t-\tau_k)\right]$ is the $K_{\text{user}} \times 1$ vector representation of the received, delayed, and Doppler-shifted subscriber signals.

If the subscriber signals are temporally uncorrelated, then the kth subscriber signal can be extracted from $\mathbf{x}(t)$ using the linear estimator $y_k(t) = \mathbf{w}_k^H \mathbf{x}(t)$, where $(\cdot)$ denotes the conjugate (Hermitian) transpose operation and $\mathbf{w}_k$ is set to the linear combiner that extracts $s_k(t)$ with maximum SINR. This solution is given by

$$\begin{aligned} \mathbf{w}_k &\propto \mathbf{R}_{\mathbf{i}_k\mathbf{i}_k}^{-1}\mathbf{a}_k \\ &\propto \mathbf{R}_{\mathbf{xx}}^{-1}\mathbf{a}_k, && (12) \end{aligned}$$

where $\mathbf{R}_{\mathbf{xx}} = \left\langle \mathbf{x}(t)\mathbf{x}^H(t) \right\rangle_\infty$ and $\mathbf{R}_{\mathbf{i}_k\mathbf{i}_k} = \mathbf{R}_{\mathbf{xx}} - \mathbf{a}_k\mathbf{a}_k^H$ are the *limit* (infinite time-average) autocorrelation matrices of the received data waveforms $\mathbf{x}(t)$ and the interference seen by the kth subscriber signal, respectively, and where $\langle\cdot\rangle_\infty$ denotes infinite time averaging and the signal powers $R_{s_k s_k}$ are subsumed in $\mathbf{a}_k$. The linearly combined antenna feeds form an effective antenna pattern with a beam in the direction of $s_k(t)$ and nulls in the directions of the other spatially coherent signals in the environment (including the interfering subscriber signals $\{s_\ell(t)\}_{\ell\neq k}$), allowing the processor to extract $s_k(t)$ from arbitrary levels of interference.

The resultant multiport processor structure is shown in Figure 1. The processor consists of an array of spatially separated antennas feeding a bank of receivers that downconvert the array to complex baseband representation. The $M \times 1$ complex baseband signal $\mathbf{x}(t)$ formed by this processor is then multiplied by an $M \times K_{\text{user}}$ weight matrix $\mathbf{W} = [\mathbf{w}_1 \cdots \mathbf{w}_{K_{\text{user}}}]$ to form the $K_{\text{user}} \times 1$ estimated signal vector $\mathbf{y}(t) = \mathbf{W}^H\mathbf{x}(t-\Delta)$. The data delay Δ is typically set to a large enough value to allow the processor to detect and compensate for changes in the environment, for example, as subscriber signals appear and disappear in the communication channel.

The maximum-SINR weight vectors can be obtained in a variety of manners. If the aperture vectors $\{\mathbf{a}_k\}_{k=1}^{K_{\text{user}}}$ are known at the receiver, the optimal weights can be determined directly from (12) using a finite-time estimate of $\mathbf{R}_{\mathbf{xx}}$. Similarly, if the communication format allows the inclusion of known training signals in the subscriber waveform, the optimal weights can be computed using a nonblind least-squares or LMS adaptation algorithm. In addition, a number of *blind* adaptation algorithms can be used to determine the optimal weights without the use of known training signals, by exploiting the known properties of the modulation format employed in the network. Examples of such algorithms include the *self-coherence restoral (SCORE)* algorithms, which exploit the known

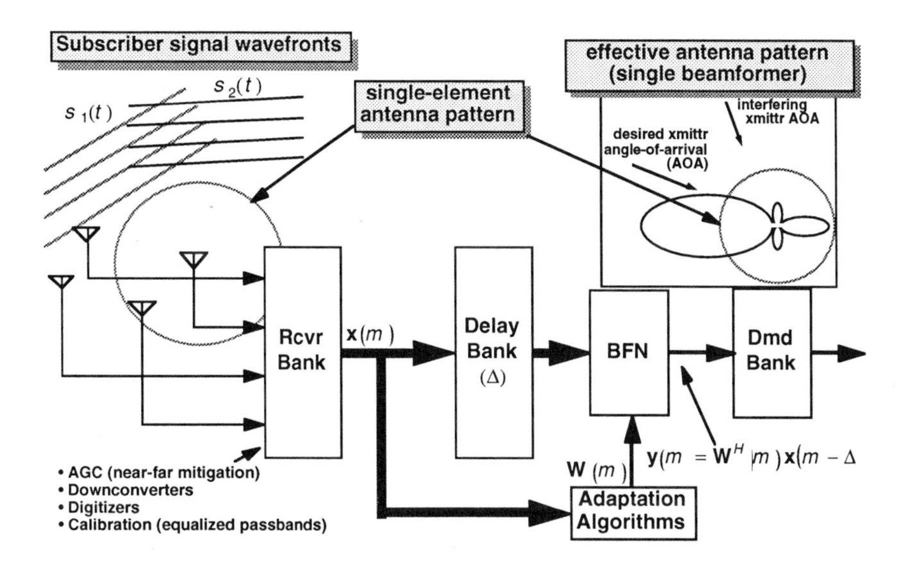

Figure 1: Multiport Adaptive Antenna Array Processor

baud-rate or phase symmetry of bauded (PSK, PCM QAM, CPFSK, GMSK, etc.) and constant phase (BPSK, MSK, DSB-AM, etc.) waveforms [3, 4]; the *multitarget constant modulus algorithm (MT-CMA)* [5], which exploits the low modulus variation of FM, CPFSK, GMSK, and PCM QAM waveforms; and *anticipatory* algorithms that exploit the limited time and/or frequency support of burst (e.g., TDMA), narrowband (e.g., FDMA) and spread spectrum (e.g., CDMA) waveforms [6, 7, 8]. The best choice of adaptation algorithm is dependent on a number of factors, including the modulation format employed by the network, the nonstationarity of the communication channel (e.g., due to mobility of the subscribers or the multiple access strategy employed by the network), and the cost and complexity of the base station employing the array.

The maximal SINR obtained by the weight vector given in (12) is equal to

$$\begin{aligned} \hat{\gamma}_k &= \mathbf{a}_k^H \mathbf{R}_{\mathbf{i}_k \mathbf{i}_k}^{-1} \mathbf{a}_k \\ &\rightarrow (1 - \tilde{\rho}_k)\, \tilde{\gamma}_k \end{aligned} \tag{13}$$

where $\bar{\mathbf{A}}_k = [\cdots \mathbf{a}_{k-1}\, \mathbf{a}_{k+1} \cdots]$ is the matrix of aperture vectors with $\mathbf{a}_k$ removed, and where $\tilde{\gamma}_k$ and $\tilde{\rho}_k$ are the *subscriber-free maximum-attainable SINR* and *in-cell interference SINR loss factor* for the kth subscriber signal,

$$\tilde{\gamma}_k \triangleq \mathbf{a}_k^H \mathbf{R}_{\mathbf{ii}}^{-1} \mathbf{a}_k \tag{14}$$

$$\tilde{\rho}_k \triangleq \frac{\mathbf{a}_k^H \mathbf{R}_{\mathbf{ii}}^{-1} \bar{\mathbf{A}}_k \left(\mathbf{I} + \bar{\mathbf{A}}_k^H \mathbf{R}_{\mathbf{ii}}^{-1} \bar{\mathbf{A}}_k\right)^{-1} \bar{\mathbf{A}}_k^H \mathbf{R}_{\mathbf{ii}}^{-1} \mathbf{a}_k}{\mathbf{a}_k^H \mathbf{R}_{\mathbf{ii}}^{-1} \mathbf{a}_k} \tag{15}$$

$$< \frac{\mathbf{a}_k^H \mathbf{R}_{\mathbf{ii}}^{-1} \bar{\mathbf{A}}_k \left(\bar{\mathbf{A}}_k^H \mathbf{R}_{\mathbf{ii}}^{-1} \bar{\mathbf{A}}_k\right)^{-1} \bar{\mathbf{A}}_k^H \mathbf{R}_{\mathbf{ii}}^{-1} \mathbf{a}_k}{\mathbf{a}_k^H \mathbf{R}_{\mathbf{ii}}^{-1} \mathbf{a}_k}. \tag{16}$$

The inequality in (16) is close if the subscriber-free maximum-attainable SINRs are strong ($\tilde{\gamma}_\ell \gg 1$ for $\ell \neq k$). The loss factor $\tilde{\rho}_k$ ranges between 0 and 1, and is maximized if $\mathbf{a}_k$ is linearly dependent on the other subscriber aperture vectors ($\mathbf{a}_k = \bar{\mathbf{A}}_k \mathbf{g}_k$). In addition, $\tilde{\rho}_k$ is not dependent on the *power* of any of the subscriber signals if those signals are strong, but is instead dependent only on the *generalized spatial correlation coefficient* between the subscriber aperture vectors,

$$\rho_{k\ell} \triangleq \frac{\mathbf{a}_k^H \mathbf{R}_{\text{ii}}^{-1} \mathbf{a}_\ell}{\sqrt{\left(\mathbf{a}_k^H \mathbf{R}_{\text{ii}}^{-1} \mathbf{a}_k\right)\left(\mathbf{a}_\ell^H \mathbf{R}_{\text{ii}}^{-1} \mathbf{a}_\ell\right)}}, \tag{17}$$

which is typically small if the subscribers are emanating from widely separated geographical locations. Consequently, $\hat{\gamma}_k$ should not be affected by in-cell near-far interference if the subscribers are spatially separated and have sufficiently strong power levels.

The maximum attainable SINR reduces to a simple average value if the interference waveform $\mathbf{i}(t)$ is dominated by large numbers of subscriber signals in adjacent cells and clusters of the network and the subscriber aperture vectors are arbitrary functions of the azimuth and elevation (θ, φ) of the subscriber signal wavefronts impinging on the array. In this case, the interference autocorrelation matrix can be approximated by

$$\mathbf{R}_{\text{ii}} = R_{ii} \left\langle \mathbf{a}(\theta,\varphi)\, \mathbf{a}^H(\theta,\varphi) \right\rangle_{(\theta,\varphi)}$$

where $\langle \cdot \rangle_{(\theta,\varphi)}$ denotes *spatial* averaging over azimuth and elevation, given by

$$\langle \eta(\theta,\varphi) \rangle_{(\theta,\varphi)} \triangleq \frac{1}{4\pi} \int_{-\pi}^{\pi} \int_{-\frac{\pi}{2}}^{\frac{\pi}{2}} \eta(\theta,\varphi)\, f_{(\Theta,\Phi)}(\theta,\varphi) \cos(\varphi)\, d\varphi\, d\theta$$

for arbitrary function $\eta(\theta,\varphi)$ and spatial density $f_{(\Theta,\Phi)}(\theta,\varphi)$. If the in-cell subscribers and adjacent-cell interferers are distributed with the same density in azimuth and elevation, then the *average* maximum-attainable SINR is given by

$$\begin{aligned}
\langle \tilde{\gamma}_k \rangle_{(\theta,\varphi)} &= \operatorname{Tr}\left\{ \mathbf{R}_{\text{ii}}^{-1} \left\langle \mathbf{a}_k(\theta,\varphi)\, \mathbf{a}_k^H(\theta,\varphi) \right\rangle_{(\theta,\varphi)} \right\} \\
&= \operatorname{Tr}\{\mathbf{I}\} \frac{R_{s_k s_k}}{R_{ii}} \\
&= M \gamma_k,
\end{aligned} \tag{18}$$

where γ_k is the subscriber-free input SINR for the kth subscriber-signal. A similar result can be obtained for the averaged maximum-attainable SINR $\hat{\gamma}_k$, yielding

$$\langle \hat{\gamma}_k \rangle_{(\theta,\varphi)} > (M - K_{\text{user}} + 1)\, \gamma_k, \tag{19}$$

where the inequality in (19) is close if $\{\gamma_\ell\}_{\ell \neq k}$ are large. These average values can be achieved in practice by assigning spatially close subscribers (which can have high spatial correlation and therefore may violate (19)) to separate channels in the networking format (e.g., to different frequency channels in a conventional FDMA network).

If (19) holds, then the power management criteria given in (2) can be satisfied by the constraint

$$\frac{\hat{\gamma}_{\text{rcv}}}{M - K_{\text{user}} + 1} \leq \gamma_k \leq \gamma_{\max} \tag{20}$$

using an M-element optimal antenna array, where $\hat{\gamma}_{\text{rcv}}$ is the minimum array output SINR required for adequate demodulation of the subscriber signals, and where the inequality exists if and only if K_{user} satisfies

$$K_{\text{user}} \leq M + \left(1 - \frac{\hat{\gamma}_{\text{rcv}}}{\gamma_{\max}}\right) \triangleq \tilde{K}_{\max}. \tag{21}$$

The array gain factor M is analogous to the matched filter processing gain M_{dmd} exploited in CDMA modulation formats, except that the array gain can be obtained for arbitrary signal modulations and is based only on the number of elements in the antenna array.

This result is considerably less stringent than the power management conditions given in (5)-(8) for conventional CDMA networks. In particular, each received subscriber-free SINR γ_k is allowed to cover a much broader range of values, and is not dependent on the SINRs of the other subscribers received by the base station. In addition, the M-element antenna array can support at least M subscribers on each channel of the networking format if $\hat{\gamma}_{\text{rcv}} \leq \gamma_{\max}$, yielding a $\times M$ increase in frequency reuse efficiency at at no increase in cost per user.

Figure 2 illustrates this frequency-reuse capability for a three-element antenna array operating against a set of real field-test data containing three frequency and time coincident FM signals ($M = K_{\text{user}} = 3$). The antenna array employs the MT-CMA described in [5] to extract the FM signals from a single 25 kHz frequency channel (consistent with an AMPS voice channel), using 16 msec segments of data collected once every 10 seconds from the data stream provided by the array. The reception platform is mobile, such that the FM signals are received with angular separations varying between 5° and 30° and power separations varying between 0 dB to 30 dB over the course of the experiment. In addition, the "FM 2" subscriber signal undergoes an outage two minutes into the experiment, such that its maximum attainable SINR drops below -10 dB. The algorithm simultaneously tracked all of three of the FM signals at nearly the maximum SINR attainable by the array over every data collect, failing only in the outage case where the maximum SINR dropped below the demodulation threshold. In addition, the algorithm tracked the weaker "FM 2" subscriber signal right into and out of the outage segment, demonstrating fast recovery from adverse reception geometries. Furthermore, this performance is accomplished without knowledge of the direction of arrival or content of the subscriber signals. This result demonstrates the improved efficiency (channel capacity) and stability (relaxed power management conditions) achievable through spatial diversity exploitation, as well as the broad applicability of the approach.

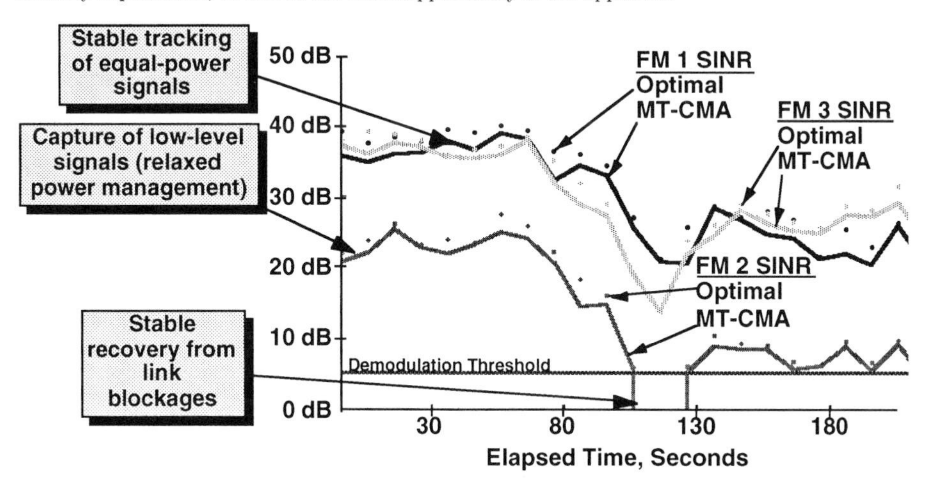

Figure 2: MT-CMA Performance Against Overlapping AMPS-Like Subscriber Signals

4 Spectral Diversity Exploitation

Spectral diversity is easily exploited in CDMA networks employing *modulation-on-symbol* direct-sequence spread spectrum (MOS-DSSS) modulation formats where the direct-sequence code repeats once per message signal. In this case, a simple multiport linear-combiner structure can be used to directly extract all of the message sequences from the wideband data channel using a single-antenna receiver, with output SINRs that are much larger than the output SINRs given in (4) for the conventional CDMA network [3].

This structure is motivated by considering the effect of time-channelizing a sampled data signal with rectangular chip shaping and an MOS-DSSS structure. The sampled signal is modelled by

$$s(n) = c\left((n)_{M_{\text{spread}}}\right) d\left(\left\lfloor \frac{n}{M_{\text{spread}}} \right\rfloor\right), \tag{22}$$

after sampling at the signal chip-rate, where $d(n)$ is the message (prespread) sequence, $\{c(n)\}_{n=0}^{M_{\text{spread}}-1}$ is the pseudorandom spreading sequence, and M_{spread} is the code period and spreading factor (equal to M_{dmd} in this example), and where $(\cdot)_M$ and $\lfloor\cdot\rfloor$ are the modulo-M and greatest-integer operations, respectively. This signal can be transformed to a format that allows direct demodulation of the message symbol sequence, by organizing $s(m)$ into $M_{\text{spread}} \times 1$ contiguous and nonoverlapping data blocks. The resultant vector signal sequence is modelled by

$$\begin{aligned} \mathbf{s}(m) &\triangleq \begin{bmatrix} s\left(nM_{\text{spread}}\right) \\ \vdots \\ s\left(nM_{\text{spread}} + M_{\text{spread}} - 1\right) \end{bmatrix} = \begin{bmatrix} c(0) \\ \vdots \\ c\left(M_{\text{spread}} - 1\right) \end{bmatrix} d(m) \\ &= \mathbf{c}d(m), \end{aligned} \tag{23}$$

where $\mathbf{c}$ is the vector representation of the code sequence. Similarly, any delayed and Doppler shifted version of $s(n)$ can be modelled by

$$\tilde{s}(n) = ge^{j2\pi f_0 n}\, s(n-\Delta), \qquad 0 \le \Delta < M_{\text{spread}} \tag{24}$$

$$\Rightarrow \quad \tilde{\mathbf{s}}(m) = \begin{bmatrix} 0 & \tilde{c}(0) \\ \vdots & \vdots \\ 0 & \tilde{c}\left(M_{\text{spread}} - \Delta - 1\right) \\ \tilde{c}\left(M_{\text{spread}} - \Delta\right) & 0 \\ \vdots & \vdots \\ \tilde{c}\left(M_{\text{spread}} - 1\right) & 0 \end{bmatrix} \begin{bmatrix} \tilde{d}(m) \\ \tilde{d}(m-1) \end{bmatrix}, \tag{25}$$

$$= \tilde{\mathbf{C}}(\Delta)\,\tilde{\mathbf{d}}(m) \tag{26}$$

after time channelization, where $0 \le \Delta < M_{\text{spread}}$, and where $\tilde{c}(n) = e^{j2\pi(n-\Delta)f_0}c(n)$ and $\tilde{d}(m) = \tilde{d}(m) = e^{j2\pi M_{\text{spread}} f_0 m} d(m)$ are the Doppler-shifted code and message sequences that conserve the MOS-DSSS structure given in (22).

Using this model, the single-antenna received data signal $x(t)$ given in (1) can be transformed to an $M_{\text{spread}} \times 1$ vector sequence $\mathbf{x}_k(m)$ modelled by

$$\mathbf{x}_k(m) = \mathbf{i}_k(m) + g_k \mathbf{c}_k d_k(m), \qquad \mathbf{i}_k(m) = \tilde{\mathbf{i}}_k(m) + \sum_{\ell \ne k} g_\ell \tilde{\mathbf{C}}_\ell\left(\Delta_{k\ell}\right)\tilde{\mathbf{d}}_\ell(m) \tag{27}$$

after sampling at the network chip-rate, synchronization (carrier and timing recovery) to the kth subscriber signal, and time-channelization at the network message rate, where $\Delta_{k\ell} = \tau_\ell - \tau_k$ is the differential delay between the signals and $\tilde{\mathbf{i}}_k(m)$ is the sampled, synchronized, and time-channelized background interference. Equation (27) bears a strong resemblance to the signal generated by a narrowband antenna array receiving $2K_{\text{user}} - 1$ spatially-coherent signals in the presence of background interference. In this case the scaled code matrices $g_k \mathbf{c}_k$ and $\left\{g_\ell \tilde{\mathbf{C}}_\ell(\Delta_{k\ell})\right\}_{\ell \neq k}$ and message sequences take on the same function as the signal and interference code matrices and signal waveforms. Consequently, the *message sequence* $d_k(m)$ for each subscriber signal can be extracted from $\mathbf{x}_k(m)$ with high SINR using the linear estimator $y_k(m) = \mathbf{w}_k^H \mathbf{x}_k(m)$, where the *optimal* processor weight vector $\mathbf{w}_k$ that maximizes the SINR of the output message stream is given by

$$\mathbf{w}_k \ \propto \ \mathbf{R}_{\mathbf{x}_k\mathbf{x}_k}^{-1} \mathbf{c}_k, \tag{28}$$

and where $\mathbf{R}_{\mathbf{x}_k\mathbf{x}_k}$ is the *limit* (infinite time-average) autocorrelation matrix of the channelized data sequence $\mathbf{x}_k(m)$.

The block diagram for this processor is shown in Figure 3. The processor structure is identical to the conventional matched-filter processor, except that the processor weights $\{w_k(m)\} \leftrightarrow \{\mathbf{w}_k\}$ are given by (28) rather than $\{c_k(n)\} \leftrightarrow \{\mathbf{c}_k\}$. However, the optimal weights provide much better performance in the presence of multiple MOS-DSSS waveforms, by directing strong *nulls* on the other subscriber signals in the communication network. Moreover, this nulling capability is provided using a single antenna and receiver at the base station.

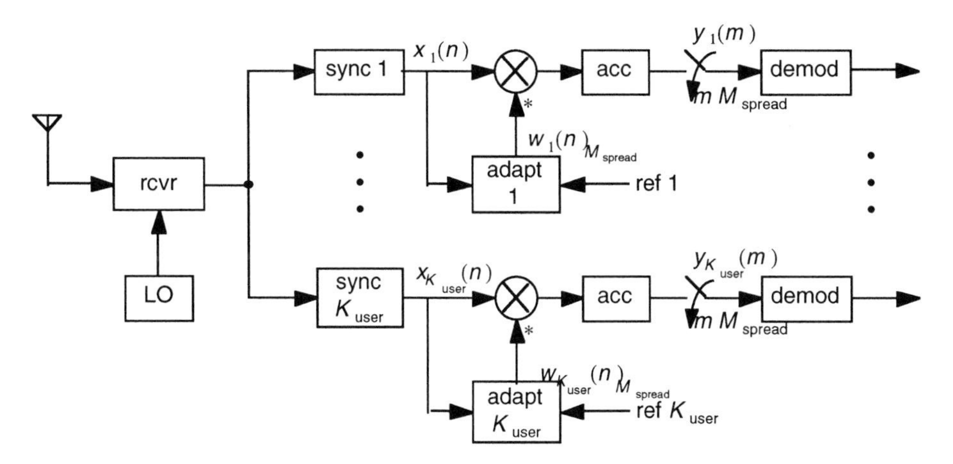

Figure 3: Optimal Linear CDMA Receiver Structure

The maximum-SINR combiner weights can be obtained using *exactly* the same algorithms discussed in Section 3. In particular, the weights can be determined directly from (28) if the code sequences $\{\mathbf{c}_k\}_{k=1}^{K_{\text{user}}}$ are known at the receiver, or using nonblind adaptation algorithms if known preamble, pilot, or intermittently-applied training sequence are added to the message waveforms $\{d_k(m)\}_{k=1}^{K_{\text{user}}}$. In addition, blind adaptation algorithms can be employed to determine the optimal weights without the use of known training signals or code sequences, by exploiting the known properties of the message sequences, for example, constant modulus if the network employs BPSK or QPSK message sequences. As with the array adaptation algorithms, the best approach is dependent on a number of factors, including the exact properties of the DSSS modulation format employed by the network, the nonstationarity of the communication channel (e.g., due to mobility of the subscribers or the multiple access strategy employed by the network), and the cost and complexity of the base stations in the network.

The extraction performance of this processor can be taken directly from the analysis performed in Section 3. In particular, the maximal SINR obtained by the processor for the kth message sequence is given by

$$\hat{\gamma}_k = (1 - \tilde{\rho}_k)\,\tilde{\gamma}_k \tag{29}$$

where $\tilde{\gamma}_k$ and $\tilde{\rho}_k$ are the maximum-attainable subscriber-free message SINR and in-cell interference loss factors for the kth message sequence, respectively,

$$\tilde{\gamma}_k \triangleq \mathbf{c}_k^H \mathbf{R}_{\bar{\mathbf{i}}\bar{\mathbf{i}}}^{-1} \mathbf{c}_k \, |g_k|^2 \tag{30}$$

$$\tilde{\rho}_k \triangleq \frac{\mathbf{c}_k^H \mathbf{R}_{\bar{\mathbf{i}}\bar{\mathbf{i}}}^{-1} \bar{\mathbf{C}}_k \left(\mathbf{I} + \bar{\mathbf{C}}_k^H \mathbf{R}_{\bar{\mathbf{i}}\bar{\mathbf{i}}}^{-1} \bar{\mathbf{C}}_k\right)^{-1} \bar{\mathbf{C}}_k^H \mathbf{R}_{\bar{\mathbf{i}}\bar{\mathbf{i}}}^{-1} \mathbf{c}_k}{\mathbf{c}_k^H \mathbf{R}_{\bar{\mathbf{i}}\bar{\mathbf{i}}}^{-1} \mathbf{c}_k} \tag{31}$$

$$< \frac{\mathbf{c}_k^H \mathbf{R}_{\bar{\mathbf{i}}\bar{\mathbf{i}}}^{-1} \bar{\mathbf{C}}_k \left(\bar{\mathbf{C}}_k^H \mathbf{R}_{\bar{\mathbf{i}}\bar{\mathbf{i}}}^{-1} \bar{\mathbf{C}}_k\right)^{-1} \bar{\mathbf{C}}_k^H \mathbf{R}_{\bar{\mathbf{i}}\bar{\mathbf{i}}}^{-1} \mathbf{c}_k}{\mathbf{c}_k^H \mathbf{R}_{\bar{\mathbf{i}}\bar{\mathbf{i}}}^{-1} \mathbf{c}_k}, . \tag{32}$$

and where $\bar{\mathbf{C}}_k = \left[\cdots \; g_{k-1}\tilde{\mathbf{C}}_{k-1} \; g_{k+1}\tilde{\mathbf{C}}_{k+1} \; \cdots\right]$ is the $M_{\text{spread}} \times (2K_{\text{user}})$ array of extended interference code vectors and the inequality in (32) is close if the subscriber-free maximum-attainable SINRs $\{\tilde{\gamma}_\ell\}_{\ell \neq k}$ are strong. As before, $\tilde{\rho}_k$ is maximized if $\mathbf{c}_k$ is linearly dependent on the interference code matrices $\left\{\tilde{\mathbf{C}}_\ell\right\}_{\ell \neq k}$, and is not dependent on the power of the subscriber signals if those signals are strong. In addition, if the signals in the network employ temporally independent codes and the interference is dominated by the CDMA subscriber signals in adjacent cells and clusters of the network, then $\tilde{\gamma}_k$ and $\hat{\gamma}_k$ are given on average by

$$\langle \tilde{\gamma}_k \rangle_{\mathbf{c}} = M_{\text{spread}} \gamma_k \tag{33}$$

$$\langle \hat{\gamma}_k \rangle_{\mathbf{c}} = (M_{\text{spread}} - 2K_{\text{user}} + 2)\, \gamma_k, \tag{34}$$

where γ_k is the subscriber-free received SINR for the kth subscriber signal and $\langle \cdot \rangle_{\mathbf{c}}$ denotes averaging over the set of available code sequences. Consequently, $\hat{\gamma}_k$ is nearly unaffected by in-cell near-far interference if the code sequences have typical correlation for independent sequences.

Using (34), the power management criteria given in (2) is satisfied by the constraint

$$\gamma_{\min} \leq \gamma_k \leq \gamma_{\max}, \qquad \gamma_{\min} \triangleq \frac{\hat{\gamma}_{\text{dmd}}}{M_{\text{spread}} - 2K_{\text{user}} + 2} \tag{35}$$

using the optimal linear CDMA processor, where the inequality exists if and only if K_{user} satisfies

$$K_{\text{user}} \leq 1 + \frac{1}{2}\left(M - \frac{\hat{\gamma}_{\text{dmd}}}{\gamma_{\max}}\right) \triangleq \tilde{K}_{\max}. \tag{36}$$

Furthermore, the range between the minimum and maximum received signal SINRs is given by

$$\frac{\gamma_{\max}}{\gamma_{\min}} = 1 + 2\left(\tilde{K}_{\max} - K_{\text{user}}\right)\frac{\gamma_{\max}}{\hat{\gamma}_{\text{dmd}}}, \tag{37}$$

in the presence of arbitrary numbers of interfering subscribers received with arbitrary power levels. This result is considerably less stringent than the power management conditions given in (6)-(8) for conventional CDMA networks. In particular, the optimal linear processor can support as many as $\frac{M_{\text{spread}}}{2}$ subscribers on each channel of the networking format if $\hat{\gamma}_{\text{dmd}} \leq 2\gamma_{\max}$, yielding a $\times\frac{\hat{\gamma}_{\text{dmd}}}{2}$ increase in frequency reuse efficiency over the matched filter processor at *no* increase in system cost. In addition, this performance can be sustained over a much broader range of subscriber-free SINRs, without imposing a complicated dependency between the SINRs.

This observation is illustrated by Figure 4, which plots the locus of allowable SINRs $\{\gamma_k\}_{k=1}^{2}$ for a CDMA network containing two users ($K_{\text{user}} = 2$) and employing a DSSS modulation format with an 18 dB spreading gain ($M_{\text{dmd}} = M_{\text{spread}} = 64$), 10 dB required SINR for demodulation ($\hat{\gamma}_{\text{dmd}} = 10$ dB), and 12 dB maximum allowable receive SINR ($\gamma_{\max} = 12$ dB). The spreading codes used by the two subscribers are additionally assumed to have a cross-correlation of $\frac{1}{M_{\text{spread}}}$ (-18 dB) with identical timing phases (worst case $\{\tilde{\rho}_k\}_{k=1}^{2}$). The optimal linear processor can tolerate almost 19.9 dB of independent variation in each of the received signal SINRs in this environment, shrinking to a minimum value of 5 dB for $K_{\text{user}} = 32$. In contrast, the matched filter processor can only tolerate nor more than 7.8 dB of (highly-dependent) variation in the received signal SINRs, shrinking to a minimum value 1.2 dB for $K_{\text{user}} = 7$. This corresponds to a 12 dB improvement in allowable SINR variation, and a $\times 4.4$ increase in frequency reuse efficiency.

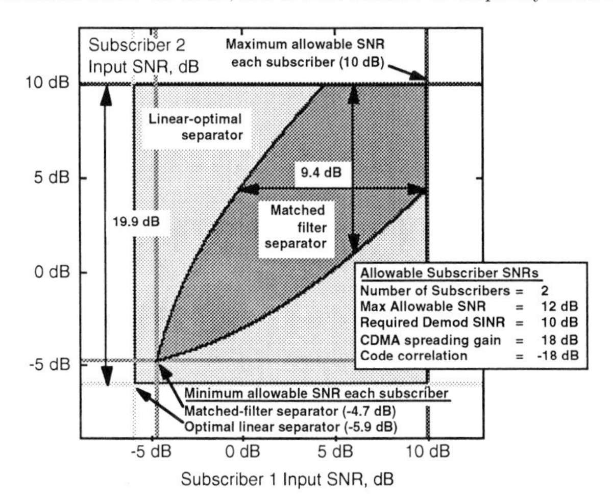

Figure 4: Allowable Subscriber-Free Input SINRs, Matched-Filter and Optimal Linear CDMA System, $K_{\text{user}} = 2$, $M_{\text{dmd}} = 64$, $\hat{\gamma}_{\text{dmd}} = 10$ dB, and $\gamma_{\max} = 12$ dB

The relative performance of these processors is further illustrated in Figure 5, showing the demodulator input SINRs provided at the one port of the matched-filter and optimal linear CDMA receiver over 400 independent Monte Carlo simulations, in a fully loaded environment where $M_{\text{spread}} = 64$, $K_{\text{user}} = 32$, and $\hat{\gamma}_{\text{dmd}} = 10$ dB. In each test, the subscriber signals are modulated using a randomly-derived QPSK code sequence, and managed to provide an SINR ranging between 7 dB and 10 dB at the base station, i.e., within 3 dB of the minimum SINR allowed under this reception scenario. As expected, the matched filter processor fails completely in this environment. In contrast, the optimal linear processor succeeds in every test in this environment, providing an output SINR that is well above the 10 dB nominal SINR required for successful demodulation of the message sequence.

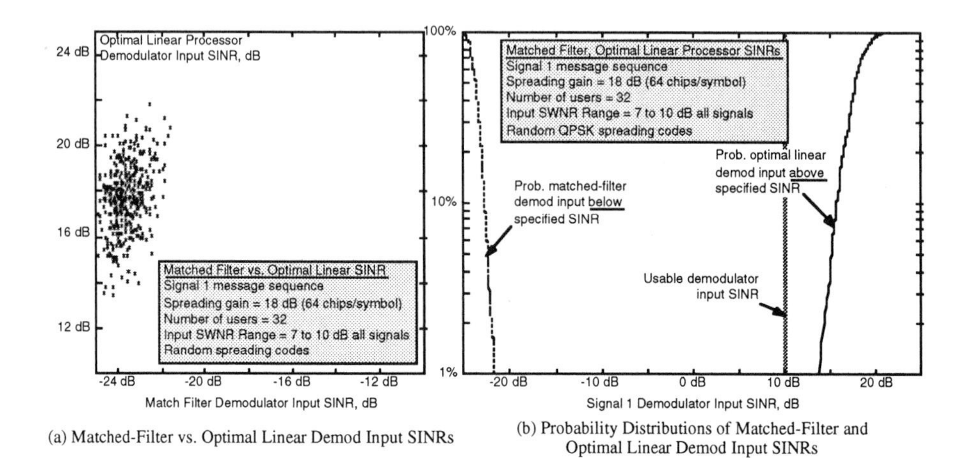

(a) Matched-Filter vs. Optimal Linear Demod Input SINRs

(b) Probability Distributions of Matched-Filter and Optimal Linear Demod Input SINRs

Figure 5: Matched-Filter and Optimal Linear Demod Input SINRs, $K_{\text{user}} = 32$, $M_{\text{dmd}} = 64$, $\hat{\gamma}_{\text{dmd}} = 10$ dB, Random Spreading Codes, 3 dB Variation in Input SINR

References

[1] W. A. Gardner, "On the Meanings and Uses of Coherence in Signal Processing," Tech. Rep. No. SIPL-89-2, Dept. of EECS, University of California, Davis, December 1988

[2] W. A. Gardner, "Exploitation of Spectral Correlation as Inherent Frequency Diversity for Signal Correction in Digital Communication Systems," Tech. Rep. No. SIPL-89-5, Department of EECS, University of California, Davis, 1989

[3] B. Agee, "The Property Restoral Approach to Blind Adaptive Signal Extraction," Ph.D. Dissertation, Department of EECS, University of California, Davis, June 1989

[4] B. G. Agee, S. V. Schell, W. A. Gardner, "Spectral Self-Coherence Restoral: A New Approach to Blind Adaptive Signal Extraction Using Antenna Arrays," *Proc. IEEE*, vol. 78, no. 4, pp. 753-767, April 1990

[5] B. Agee, "Blind Separation and Capture of Communication Signals Using a Multitarget Constant Modulus Beamformer," in *Proc. 1989 IEEE Military Comm. Conf.*, 1989

[6] K. Bakhru, D. J. Torrieri, "The Maximin Algorithm for Adaptive Arrays and Frequency-Hopping Communications," *IEEE Trans. Ant. and Prop.*, vol. AP-32, Sept. 1984

[7] D. Dlugos, R. Scholtz, "Acquisition of Spread Spectrum Signals by an Adaptive Array," *IEEE Trans. ASSP*, vol. 37, no. 8, pp. 1253- 1270, Aug. 1989

[8] B. Agee, "Fast Acquisition of Burst and Transient Signals Using a Predictive Adaptive Beamformer," in *Proc. 1989 IEEE Military Comm. Conf.*, 1989

8

A Narrowband PCS Advanced Messaging System

Walt Roehr

Telecommunication Networks Consulting

11317 South Shore Road, Reston VA 22090

Rade Petrovic

Center for Telecommunications

University of Mississippi, University MS 38677

Dennis Cameron

MTEL Technologies

P.O. Box 2469, Jackson MS 39225

Abstract

The Nationwide Wireless Network (NWN) System economically provides wireless electronic mail type services to small, low powered transceivers by combining efficient modulation techniques (described in the companion paper Multicarrier Permutation Modulation for Narrowband PCS); time division multiplexing of system-wide and zonal broadcasts; a dense deployment of reverse channel receivers to allow for low-power transceivers while providing location determination; and a high functionality Network Operations Center. The synergistic effectiveness of these system components was recognized by the FCC, when in its August 1992 Notice of Proposed Rule Making and Tentative Decision, it tentatively concluded that the NWN merits a pioneer's preference in the 900 MHz narrowband PCS service.

1 Introduction

The Nationwide Wireless Network (NWN) design focuses on economically providing two-way wide area store-and-forward messaging to small user devices (Portable Messaging Units or PMU) that can operate for weeks on a few penlight batteries. Attaining this goal, while insuring that the allocated RF spectrum is efficiently used, necessitated adoption of novel approaches for modulation, multiplexing, and control. Each of these areas is treated in the following sections of this paper. However, the utility of the NWN exceeds the summation of each of these individual innovations -- there is true synergy among all the components of the NWN. This area is explored in the section titled "System Synergy". Finally, a comparison of the service provided by the focused NWN is compared with what might be expected from an omnibus Universal Personal Communications system.

Overview of the NWN

NWN is a centrally controlled system -- all traffic is gathered at the Network Operations Center (NOC), as shown in Figure 1. Forward Channel (NOC toward PMU) traffic is allocated capacity and uplinked over a VSAT one-way broadcast system to the Base Transmitters. Approximately 500 Base Transmitters are required nationwide. All of these Base Transmitters operate on the same frequency and are synchronized so that they can, when necessary, operate as a single nationwide simulcast system. However, in order to allow reuse of channel capacity, the bulk of the information is transmitted on a zonal basis -- a subset of Base Transmitters located in one area transmit to local PMUs while other subsets of transmitters, located in other areas, are transmitting to PMUs in their vicinity. The only traffic passed on a nationwide basis is that destined for PMUs located in border areas between zones and short probe messages for PMUs that are "lost". PMUs receiving messages are given Reverse Channel (PMU to NOC) time assignments. Since it was expected that a second frequency channel would not be available for the Reverse Channel, NWN Time Division Duplexes a single RF channel (Ping-Pong Access). (The FCC has indicated that it is considering allocating additional frequencies for 2-way working; if this comes to pass it will enhance the efficiency of the NWN.) Reverse traffic may be acknowledgements of Forward messages, Reverse messages, or requests for Reverse Channel

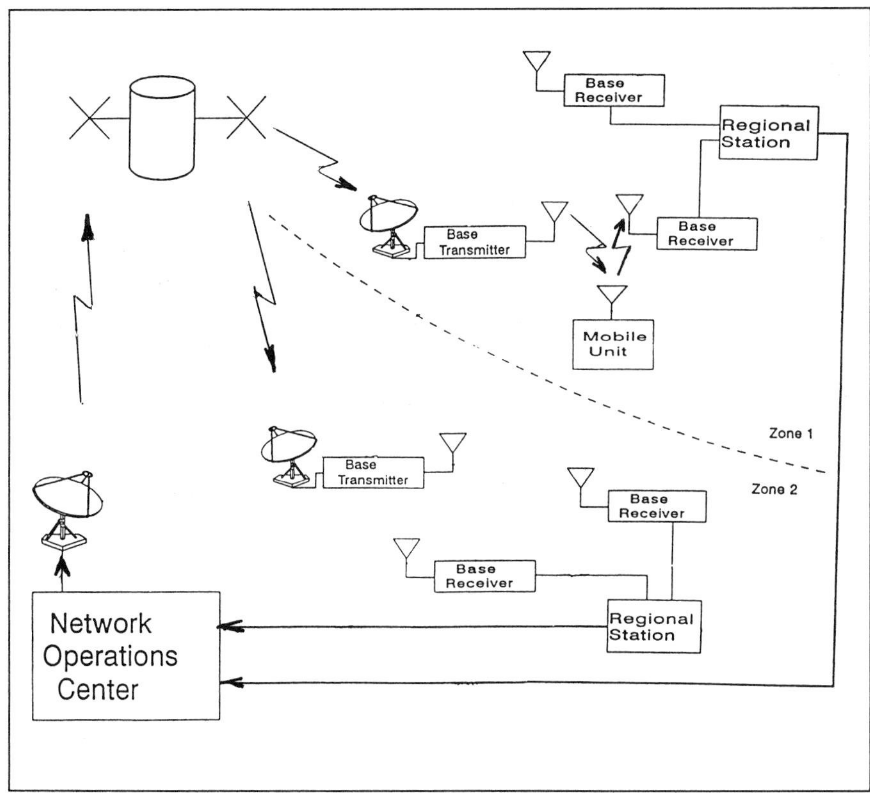

Figure 1: High Level NWN System Diagram

capacity that are transmitted in a time slice dedicated to contention access. Several thousand Base Receivers are deployed across the nation to insure that there is a receiver close to the low powered PMU transmitters. Spacial re-use of this dense grid of Base Receivers substantially increases the capacity of the Reverse Channel.

2 Modulation

The foundation of the NWN is Multi-Carrier Modulation technology that supports bit rates up to 24 kilobits per second in a simulcast environment. In a simulcast system a number of synchronized transmitters, operating on the same frequency, broadcast the same information into overlapping coverage areas. While this introduces the possibility of destructive interference, it also extends the coverage area and decreases the impact of shadowing (log-normal fading).

Simulcast economically provides superior coverage of extended regions -- high powered transmitters and high elevation antennas make it feasible to blanket areas that extend hundreds (or even thousands) of miles. The phase cancellation form of destructive interference (equal magnitude signals arriving 180 degrees out of phase) is relatively benign -- equi-signal areas are small and minor differences in carrier frequency are sufficient to insure that the phase opposition condition does not persist at any given location. A modest error correction capability can easily clean-up the occasional error this degradation introduces. However, misalignment of the modulation waveforms from individual transmitters is a malignant interference -- it is equivalent to a delay dispersion and in a digital system it causes intersymbol interference.

The dispersion in a modern paging system is typically 200 microseconds. By carefully synchronizing the modulation waveforms at the transmitters the dispersion can only be reduced to approximately 50 microseconds. Two approaches are then possible: the symbol duration can be made long, relative to the dispersion, or adaptive equalization (e.g., transversal filters) can be employed. Adaptive equalization is not compatible with low cost and (particularly) low power consumption. Therefore, NWN accepted the constraint of using long duration symbols. In particular, NWN proceeded with Multicarrier Permutation Modulation (described in the companion paper Multicarrier Permutation Modulation for Narrowband PCS) that delivers 6 bits per symbol and operates at 4 kilobaud, for a channel bit rate of 24,000 bits per second.

Since all messages are acknowledged, addition of error detection and retransmission (ARQ) capabilities are a trivial extension. However, this leads to significant capacity savings. Most of the PMUs enjoy very high fade margins and low error rates. It is only when a PMU is operating near the edges of the coverage area that a heavy duty Forward Error Control

(FEC) code is justified. In the Forward Channel the full FEC is applied to header and control information, including the notification of arriving traffic. If the PMU successfully receives the notification but fails to receive the lightly protected message there is a strong indication that FEC is justified for the re-transmission. Analysis has shown that the savings are significant even if 10% of the lightly protected messages are errored and the full FEC was completely effective. Both of these assumption are conservative.

3 Multiplexing

While a 24 kilobits per second simulcast system is very impressive (the highest speeds to which current commercial paging systems aspire are 2 to 6 kilobits), the traffic loads expected in a nationwide narrowband PCS are 40 to 100 times greater. Efficient means for sharing and re-using this channel capacity are needed. In this section "multiplexing" is used in its broadest sense, to include any use of a resource (e.g., rf spectrum, frame time on the rf channel) for multiple purposes. There are three major classes of multiplexing in the NWN:

Time Division Duplexing (TDD) of the forward and reverse channels;

Time Division Multiplexing of the Forward Channel to provide for zonal and system-wide traffic;

Spacial Multiplexing of Base Receivers used for the reverse channel.

All of this multiplexing is most evident in the cycle structure, as shown in Figure 2. At the highest level there is multiplexing of the forward and reverse intervals for both system-wide and zonal usage. The lengths of all the intervals and batches are continuously adjusted by the NOC -- there are no fixed duration fields that consume resources when they are not fully utilized.

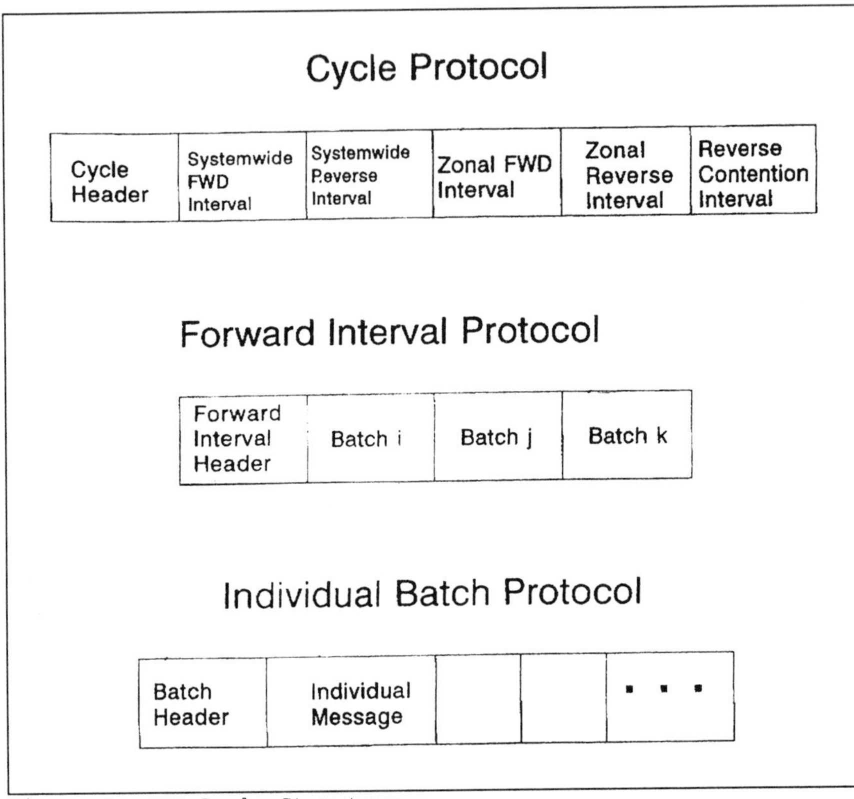

Figure 2: NWN Cycle Structure

The NOC strives to minimize the traffic carried in the system-wide portions of the cycle because the system loading due to that traffic is effectively multiplied by the number of zones in the system. (Time allocated to the system-wide portions can not be re-used; zonal time can be re-used in every zone.) Zone boundaries are placed in areas with low PMU density. This minimizes the traffic to areas that must be served during the system-wide intervals. If traffic to a boundary area builds to unacceptable levels the boundary can be moved (by changing the assignment of a Base Transmitter). In the limit, Dynamic Zone Boundaries, which move at every cycle, can be used to completely eliminate the need to transmit to a PMU while it is in a boundary area -- on a subsequent cycle the PMU will be well within a re-drawn boundary. The penalty is the heavy control load.

Each Forward Interval is subdivided into 64 batches and a given PMU only listens to its assigned batch. Furthermore, each batch begins with a roll call of PMUs that are to receive traffic in that batch. Therefore, a PMU need only be "awake" for a short interval to determine if there is traffic for it. The remainder of the cycle the PMU goes into a power saving "sleep" state, with only timer circuitry continuing to draw power. With a nominal cycle length of 30 seconds, an idle PMU need only be awake a small fraction of one percent of the time. These extended sleep periods are the prime mechanism for extending the battery life of the PMU.

The spacial multiplexing of Base Receivers within zones increases the effective throughput of the Reverse Channel by a factor of more than ten. Market studies predict that the total Reverse Channel load will be comparable to the forward load. The spacial multiplexing allows simple PMU transmitters, with burst speeds in the 4 to 10 kilobit per second range, to handle this load in less than 10 percent of the cycle time.

4 Control

High bit rate forward channel technology and efficient multiplexing are not sufficient to provide the capacity a mature NWN will require -- even heavily multiplexed 24 kilobit per second channels would saturate if means were not found for efficiently allocating the resources these techniques provide. A fundamental design principle was to avoid, wherever possible, the fixed assignment of resources. Critical to efficient management of this flexibility was a commitment to centralized control of all these allocations -- assuring the necessary degree of state synchronization in a distributed control system would have been a high risk development.

Two areas of flexibility, adjustment of cycle component durations and zone boundaries, have been described in the Multiplex section, above. A third example of the flexibility built into the NWN is the dynamic control of auto-registration. NWN uses customer location information, obtained via the adaptive registration process, to vector messages to a single zone, and thus conserve spectrum. Adaption allows the gathering of this information to be "tuned" to each customer's usage pattern. Location information is obtained whenever the portable terminal transmits. Acknowledgement of forward messages, requests for reverse channel capacity, and transmission of reverse messages all update the location database. Autonomous registration, wherein the portable terminal notes from the cycle header that it is located in a new zone and autonomously emits a REG message during the contention portions of a cycle, is another means of updating the location database. Adaptive registration allows turning autonomous registration ON or OFF, on a customer-by-customer basis, depending upon that customer's recent message reception history.

An autonomous registration is a waste of spectrum if the customer does not receive a message before it next transmits either another REG message or a request for reverse channel capacity, either of which will update the location data base. On the other hand, an autonomous registration is a more efficient use of spectrum than a nation-wide search for a customer that crossed a zone boundary. By keeping track of these two events:

(1) REG, followed by another portable initiated use of the reverse channel before there is a forward message;

(2) nation-wide search with discovery in another zone;

for each portable terminal, the central control can set the Adaptive Registration status of the terminal (by updating a flag in the terminals non-volatile memory) to maximize the long-term average spectral efficiency.

Note that in neither of the Adaptive Registration states (REG ON nor REG OFF) are wasteful polling or timer driven algorithms used to obtain location information. Those algorithms, which are not driven by the portable noting a zone change, must be wasting spectrum on redundant location transmissions if they are to have any hope of having current zone location information.

Further tuning of the adaption algorithm, down to noting the peculiarities of specific portable terminals (e.g., terminal 456-7431 gets lots of messages in different zones on Monday and Friday but is idle the rest of the week) allows moving from a long term average optimization to an optimization for the day of the (week) (month) (year), or an optimization that is driven by location (whenever terminal 235-7856 gets outside California it never gets messages). Development of these profile driven algorithms will be an on-going effort.

5 System Synergy

The combination of locating users, combining zoning and system wide coverage, and efficiently administering these resources on a centralized basis, are all needed if NWN is to yield its promise. Particularly critical to the design is the commitment to two-way operation. The Reverse Channel provides location information and location information is essential for facilitating zoning of the Forward Channel and spacial multiplexing in the Reverse Channel. Furthermore, acknowledgements allow the use of ARQ and decrease the FEC burden. Consideration was briefly given to offering a one-way paging service at a lower price. However, the cost of serving these receive only terminals in the system-wide interval far outweighed any savings that came from eliminating the Reverse Channel.

6 Comparison with Universal Communication Solutions

NWN's heritage is paging rather than mobile telephony -- the focus of the NWN is on providing economical wide-area messaging coverage to small user terminals that operate for up to a month on a few disposable penlight cells. Extended "sleep" periods and a protocol that focuses on efficiency rather than delay are key attributes. The success of this focused approach challenges the industry's fixation with Universal Personal Communication. While there is no doubt that a full function UPC system could provide messaging services, there is significant doubt as to UPC's ability to provide a service that is as sparing in its use of funds, power, and volume.

7 Conclusion

A focused approach to a well defined PCS mission (store-and-forward two-way delivery of messages) has yielded an elegant solution. At various points during the development of the NWN there has been consideration of attempting applications that are of a more real-time nature, e.g., interactive terminal support or the delivery of traffic streams. Very quickly significant inefficiencies arose -- power consumption increased, throughput degraded, costs rose. Returning to the original mission, with an emphasis on efficient modulation, flexible multiplexing, and a powerful centralized control of the dynamic allocation of system resources once again put the NWN back on track.

9

Multicarrier Permutation Modulation for Narrowband PCS

Rade Petrovic
Center for Telecommunications
University of Mississippi, University, MS 38677
Walt Roehr
Telecommunication Networks Consulting
11317 South Shore Road, Reston, VA 22090
Dennis Cameron
MTEL Technologies
P.O. Box 2469, Jackson, MS 39225

Abstract: In multicarrier permutation modulation a symbol is generated by transmitting, simultaneously, n out of m possible distinct subcarriers ($1<n<m$). This technique is proposed for the forward channel in a 930 MHz band narrowband PCS. In a 50 kHz channel m = 8 subcarriers are spaced 5 kHz appart with n = 4 of them turned on. The signaling rate is 4 kbaud and bit rate 24 kbit/s. Experimental transmitters consist of four 4-FSK subtransmitters, with outputs combined through hybrid circuits. The experimental receiver is composed of an RF down-converter, an A/D converter, and a PC performing DFT and other signal processing. The transmitters have a very compact spectrum which closely follows the emission mask defined by the FCC. Field tests showed that the receiver has high sensitivity, and operates successfully in simulcast overlap areas, stationary or moving, in fast fades, and other hostile conditions. The results presented here illustrate the probability of error versus signal strength in fading environments, and the influence of frequency deviation on the probability of error.

1. Introduction

The Nationwide Wireless Network (NWN) is designed to provide wireless electronic mail type services to small, low powered devices. It was proposed by MTEL to the FCC [1] and, later, NWN was granted a tentative pioneer's preference in the field of Narrowband Personal Communication Services (NPCS) [2]. Its architecture, features, and a comparison to alternative networks are discussed in the accompanying paper "A Narrowband PCS

Advanced Messaging System". One of the innovations included in the NWN proposal is Multicarrier Permutation Modulation (MPM), which is the subject of this paper.

The choice of a the MPM technique was based on the requirement that the forward channel (from the network to portable devices) should operate in the simulcast mode, with a number of transmitters simultaneously transmitting the same information in mutually overlapping areas. This technique provides extended coverage area, improved building penetration, and reduced effects of shading and multipath fading. On the other hand it restricts maximum symbol rates due to the effects of differential propagation delays. Also, it is considered to be incompatible with coherent detection techniques because of the complexity of tracking signals from multiple transmitters in the simulcast overlap. Further, extended equalization time is incompatible with respect to the battery saving technique, which powers the receiver only during short, prescheduled intervals.

The above constraints led us to propose a variant of multicarrier modulation techniques [3], whose general features include good immunity to impulse noise and fast fades, and no need for equalization. The proposed variant can be classified as Permutation Modulation, [4] which has features of constant energy for each symbol, efficient spectrum utilization and a low symbol energy to noise density ratio requirement [5].

2. Description of the Proposed MPM

In the FCC ruling, based on the MTEL proposal, it was determined that the channel spacing in the 930 MHz band should be 50 kHz. The emission mask boundaries are obtained by shifting boundaries of the current part 22 mask ± 15 kHz away from the center frequencies (see dashed lines in Figs. 1 and 2). In this manner a 40 kHz pass-band is defined in the middle of the channel wirh 5 kHz guard bands on each side for the skirts of the spectrum.

In order to fully utilize the allocated spectrum, and provide fast fall-off of the spectrum in the guard band, we propose eight subcarriers spaced 5 kHz apart. Further, we propose that during each symbol interval a combination of four carriers is ON while the other four are OFF. The total number of distinct symbols is $4C8 = 8!/[(8-4)!4!] = 70$, which gives $\log_2 70 = 6.13$ bits per symbol.

We propose that 64 out of 70 symbols are used for data transmission, which corresponds to 6 bits per symbol, while the remaining 6 symbols are reserved for overhead functions (synchronization, control, etc.).

The symbol rate is chosen to be 4 kbaud, which is adequate for simulcast systems [6]. During the 250 μs symbol interval the parameters are maintained stable for 200 μs, while symbol transition occurs during remaining 50 μs. The 200 μs window corresponds to the frequency resolution of 5 kHz, i.e. to the spacing of the subcarriers.

The total bit rate per channel is 4 kbaud x 6 bit/symbol = 24 kbit/s, which is much higher than other simulcast systems (POCSAG goes up to 2.4 kbit/s with 2-FSK and ERMES has bit rate of 6.25 kbit/s with 4-FSK). Also the spectrum efficiency is higher, with 0.48 bit/s/Hz for NWN versus 0.25 bit/s/Hz for ERMES and up to 0.096 bit/s/Hz for POCSAG.

3. Experimental Transmitter Design

Two experimental transmitters, each with 20 W output power, were constructed by Glenayre, Quincy, IL. Each transmitter has four subtransmitters capable of 4-FSK over a subset of the 8 frequencies. In particular, the frequencies assigned to each subtransmitter were {1,2,3,4},{1,2,4,6},{3,4,7,8} and {5,6,7,8}, where digits 1,2,..,8 refer to frequencies in ascending order. In this setup only 66 out of 70 symbols can be generated. This is so because for each subtransmitter there is one 4 frequencies combination in which it can not participate. For example, the first subtransmitter can not participate in generating a symbol having frequencies {5,6,7,8} ON, and the other three subtransmitters cannot make four subcarriers. Therefore, for all 70 symbols we would need 5-FSK subtransmitters, or some other transmitter design.

Fig. 1 shows the spectrum of a signal having frequencies {4,6,7,8} ON continously. It is used to determine the "power of unmodulated carrier", the term used in the FCC spectrum mask description, which can be interpreted as the sum of powers of all subcarriers.

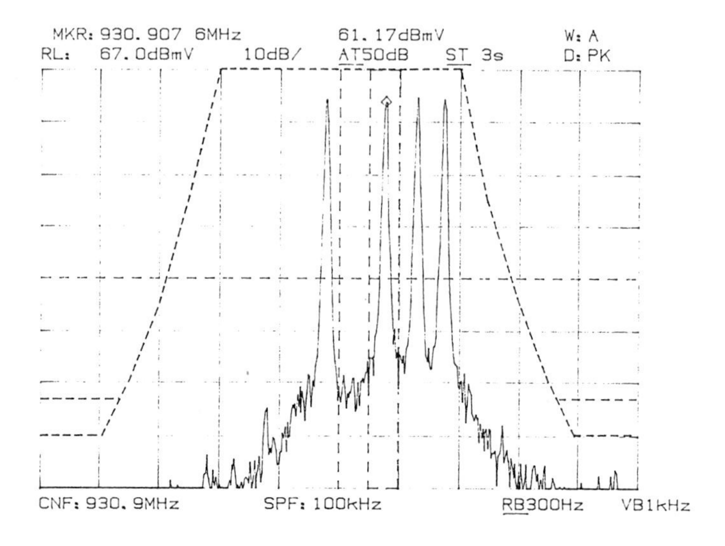

Fig. 1. Spectrum of the signal with constant frequencies

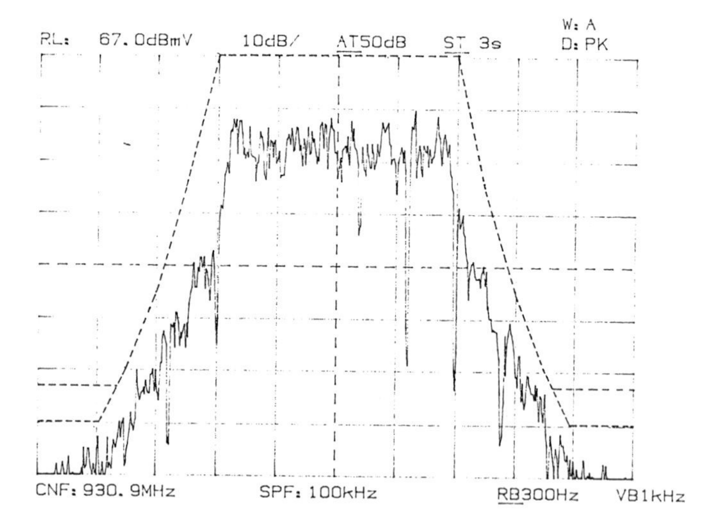

Fig. 2. Spectrum of the signal with random bit pattern

This power determines the referent level for the spectrum mask, and it is represented by a horizontal dashed line at the top of the diagram.

Fig. 2 shows the spectrum of a signal carrying pseudo random data from a 2^{13} - 1 bit long period generator, which is used for BER count. Obviously, the signal spectrum fits under the required spectrum mask. Also, it follows closely the boundaries defined by FCC, which indicates high utilization of the allocated frequency band.

One of the issues that we were seriously concerned was the effect of frequency selective fading. The first test results were alarming. The spectrum for the four carriers measured in the field test showed significant differential fading, as illustrated in Fig. 3. In order to be sure that it is not the effect of fast fading during the sweep of the spectrum analyzer, we performed spectrum analysis using an FFT on a 200 μs window of signal. The results, for five independent windows, are presented in Fig. 4, where consecutive FFT bins coincide with subcarrier frequencies. Again the differential fading is apparent.

Fortunately, after extensive laboratory and field tests we concluded that these phenomena occurred due to the leakage radiation from the subtransmitters, not due to the frequency selectivity. In short, by careful shielding of subtransmitters and combiners we managed to almost eliminate any selective fading.

4. Experimental Receiver Design

The basic block diagram of the experimental receiver is given in Fig 5. The antenna is connected through an attenuator to a RF section which contains a two step down-convertor built from off-shelf components. The local oscillators are external signal generators. In the RF section the signal is down-converted so that the lowest of the eight frequencies is at 15 kHz while the highest is at 50 kHz. Such a signal is sampled by a 16 bit A/D converter at a 120 kHz sampling rate. The 16 bit/sample precision is used in order to avoid AGC in the RF section. The sampling rate corresponds to 30 samples per symbol interval, and it is chosen based on an optimized symbol detection algorithm.

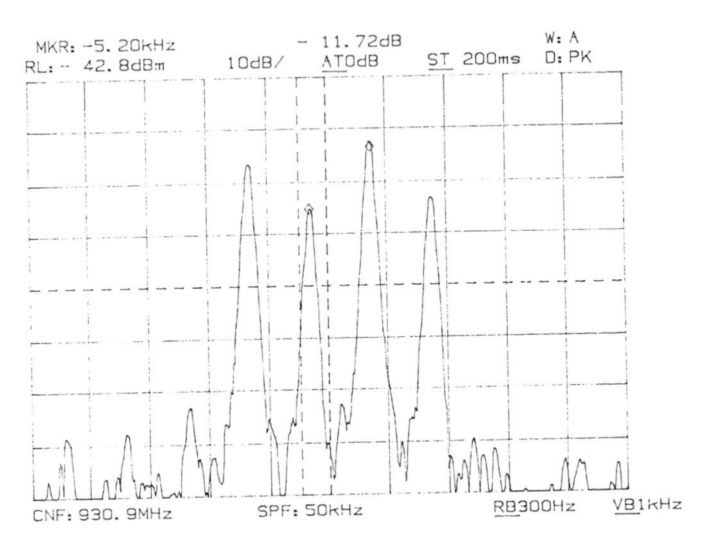

Fig. 3. Spectrum of the signal with constant frequencies in a selective fading environment

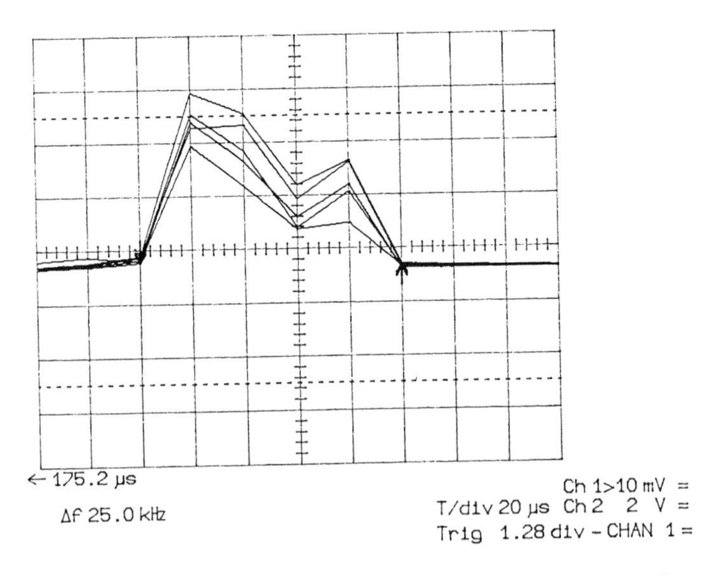

Fig. 4. Spectrum of the signal with constant frequencies as obtained by FFT processing. Resolution is 5 kHz/div, and frequency bins coincide with subcarrier frequencies

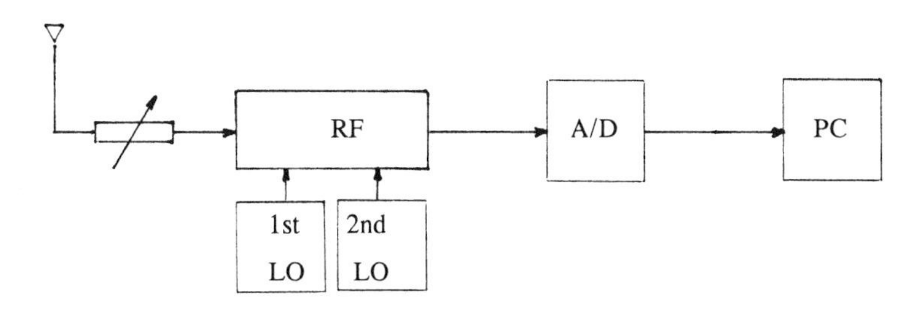

Fig. 5. Block diagram of the experimental receiver.

Symbol detection and other processing (bit and frame synchronization, bit error rate count, status display etc.) is done by an IBM compatible PC with a 486 processor. Symbol detection is based on a 24 point DFT followed by bin comparison and decoding by a look-up table. The 24 samples correspond to a 200 μs window, which equals the interval of stable symbol parameters. Generally speaking, any number of samples per window larger than 16 can be used for symbol detection, and choice made is based on minimum processing time. It was found that 8 tones can be detected by a 24 sample DFT with as few as 14 integer multiplications. The assembly routine performing DFT, magnitude square calculation, sorting, and symbol extraction from a look-up table, can be done with approximately 1800 clock cycles, i.e. in 36 μs in our 50 MHz PC. This means that only 14.4% of the symbol time is needed for symbol decoding, while the rest can be used for other processing.

In laboratory tests the signal was simulated by an arbitrary waveform synthesizer (a 16 symbol frame was synthesized). It was used to amplitude modulate a signal generator so that the upper side-band fits the 50 kHz pass band. The receiver was able to decode such a signal without errors for hours.

5. Field Tests

Field tests performed in Oxford, Mississippi, used two transmitters separated approximately 7 miles, so that a strong overlap signal area is created (signal field intensity of approximately 40 dBμV/m). Tests included stationary and moving receivers, strong and weak signal areas, nearly equal signal areas as well as a single transmitter dominated areas, and variation in system parameters such as jitter and frequency errors. Generally speaking the modulation technique proved to provide good sensitivity, and robust operation in a hostile environment. As an illustration, the following results are presented.

Fig. 6 shows result of bit error rate tests performed statically in a building one mile from the nearest transmitter, where the field intensity measurements indicated 42 dBμV/m field strength in average. Tests were run with messages 5400 bits long, preceded by the preamble and frame delimiter header. The attenuation in the antenna path is varied as indicated on the abscissa, and each test run was 2 minutes, except for low bit error rates, when the test was extended in order to accumulate statistically significant error count. The variations in bit error counts are attributed to the fading effect. From the diagram we conclude that one order of magnitude improvement in the probability of errors can be obtained if the signal strength is increased by 4 dB, as indicated by the slope of the middle dashed line. Upper and bottom lines indicate the slowest and the fastest rise of BER, respectively, as obtained in the test. For 0 dB attenuation the test was run for 24 hours yilding 9.8 10^{-7} bit-error-rate.

Fig. 7 shows bit-error-rate variations with frequency deviation of the demodulated subcarriers from their nominal position (15 Hz - 50 kHz), which is expected as a result of instabilities in receiver crystal oscillators. This variation is achieved by tuning the frequency of the second local oscillator. In this test the procedure was the same as in previous test, and the attenuator was adjusted for 15 dB attenuation. This figure shows that the bit-error-rate would rise an order of magnitude when the frequency deviation increases by approximately 300 to 400 Hz. This rate of change is somewhat steeper if the signal is stronger and less steep for a weaker signal (which corresponds to lower and higher bottom of the wedge shaped curve, respectively, while the edges are only slightly changed). It can be estimated that approximately 1 dB of penalty in receiver sensitivity is paid for each 100 Hz of frequency offset. This indicates that some form of frequency stabilization should be implemented in receivers. Current investigations are focussed on digital filtering methods (set during preamble) for frequency offset detection.

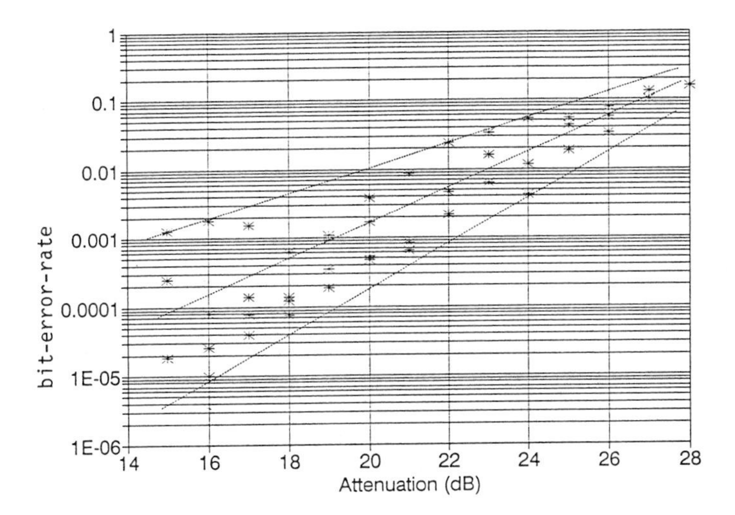

Fig. 6. Bit-error-rate versus attenuation in antenna path in fading environment

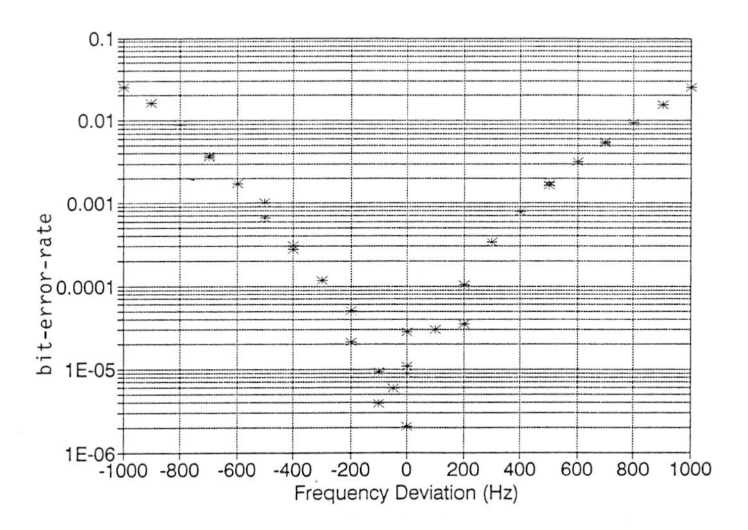

Fig. 7. Bit-error-rate versus frequency deviation of the second local oscillator in the receiver

6. Conclusion

The transmitter design, with independent subtransmitters combined through hybrid circuits, has very low intermodulation products, but has significant losses in combiners (9 dB approximately). Also, careful shielding is necessary to avoid selective fading due to the leakage radiation. An alternative design, with synthesized symbols and a linear amplifier is currently under testing.

The receiver design is developed around a DSP in a PC. Efficient algorithms for the DFT were developed and successfully tested. Clock extraction is also achieved through a DSP. Clock extraction is equivalent to choosing a set of 24 samples from the continuous sample stream, which fits the interval of stable symbol parameters. Current work is directed toward DSP frequency offset detection, and implementation of the algorithms in hardware.

Field testing has proved the technical feasibility of the proposed modulation technique. It has also revealed its robustness in simulcast areas and in an environment of small towns, and forested hills. Future tests are planned for a metropolitan environment.

REFERENCES

[1] Mobile Telecommunications Technologies Corporation, "Petition for Rulemaking to Allocate 150 kHz in the 930-931 MHz Band and to Establish Rules and Policies for Nationwide Wireless network (NWN) Service", Before the FCC, Nov. 12, 1991.

[2] FCC, "New Personal Communications Services Proposed (Gen. Docket No. 99-314, ET Docket 92-100)", FCC news, July 17, 1992.

[3] J.A.C Birnhan, "Multicarrier Modulation for Data Transmission: An Idea Whose Time Has Come", IEEE Communications Magazine, May 1990, pp 5-14.

[4] D. Slepian, "Permutation Modulation". [proceeding of the IEEE, March 1965. pp 228-237.

[5] E. Brookner, "The Performance of FSK Permutation Modulation in Fading Channels on Their Comparison Based on a General method for the Comparison of M-ary Modulations," IEEE Trans. on Comm. Tech., Vol. COM-17. Mp/ 6. Dec. 1969, pp 616-639.

[6] PacTel Paging Telesis Technologies Laboratory, "Advanced Architecture Paging Experimental License", Presentation to the Telocator High Speed Committee, May 29, 1992.

10

Video Compression for Wireless Communications

Teresa H. Meng†, Ely K. Tsern‡, Andy C. Hung‡, Sheila S. Hemami‡, and Benjamin M. Gordon†
†Computer Systems Laboratory ‡Information Systems Laboratory
Department of Electrical Engineering
Stanford University

Abstract: *This research centers on providing digital video on demand to portable receivers through wireless communications. The three main technological issues for wireless video communication are compression efficiency, error recovery, and low-power implementation. The algorithmic goal is to develop compression algorithms that maintain consistent visual quality for image and video signals transmitted over a noisy channel. These algorithms must therefore provide efficient compression and provisions for recovery in situations of severely degraded transmission. The hardware goal is to demonstrate low-power decoder modules that implement the decompression algorithms with recovery capability for lost information.*

1.0 Compression in a Wireless Environment

As the demand for wireless personal communications continues to grow, the need arises for video compression that not only provides good compression and fidelity, but also offers low-complexity, low-power implementation that permits battery-operated portable designs. Most of the compression hardware available on the market implements standards algorithms which ensure proper protocol. However, these standards algorithms, primarily the H.261, JPEG, MPEG, and most recently MPEG-2, assume 1) a robust transmission channel where errors are minimized, 2) a transmission bandwidth that is a-priori fixed, and exhibit 3) a design strategy in which power consumption is not an issue. They were not designed for rapidly fluctuating channels and high packet loss, as are experienced in a wireless transmission link. Furthermore, these standards algorithms do not consider reconstruction and recovery techniques which are necessary to guarantee visual quality at the portable receiver. We address the unique issues of reliability and portability in designing the algorithms and hardware for video transmission in a wireless environment.

Providing universal access to a video database using wireless communications calls for revolutions in intelligent networks, efficient channel utilization, and error-resilient source coding. Our work concentrates on the design techniques for delivering entertainment-quality video with relaxed assumptions on channel or network reliability. Major challenges are development of fault-tolerant compression algorithms for low-power implementation, real-time reconstruction of lost video signals due to random packet loss, and an encoding platform that adapts to varying transmission bandwidth. In this paper we present two projects related to wireless video compression. The first project addresses the algorithm development and hardware design of implementing a low-power portable decoder for decoding compressed digital video transmitted through a dedicated channel (infrared or RF). The design goal is to deliver reasonably good quality video under severe bursty channel errors without using error correcting codes. The second project concerns real-time reconstruction of video signals lost in packetized transmission, an approach which can serve as baseline processing to facilitate automatic rate control in a congested network.

2.0 Low-Power Error-Resilient Video Compression*

This section describes a video compression scheme that performs pyramid lattice vector quantization (PVQ) of subband coefficients and a VLSI architecture for the PVQ decoder. This algorithm not only

---

This research has been sponsored by DARPA with additional support from JSEP and the NSF.

* This work has been submitted to *1993 IEEE Workshop on VLSI Signal Processing*[1].

provides good compression/fidelity performance, but also results in a low complexity, low power VLSI implementation. Furthermore, this algorithm demonstrates excellent resiliency under severe channel distortion without the use of error correction hardware. Low-complexity encoding and decoding schemes allow for real-time video coding and eliminate the need for large memories. The algorithm and its performance are described in detail, and the chip architecture of the PVQ decoder is presented.

2.1 Compression Algorithm Requirements

When transmitting digital video over wireless channels, the compression algorithm must meet two performance criteria: good rate-distortion performance and error-resiliency to channel bit errors. To obtain the best image compression, many systems utilize variable-rate coding (such as entropy-coding, run-length coding) together with error-correction coding to combat channel errors. Variable-rate coding provides better compression than fixed-rate coding, but requires greater hardware complexity and additional buffering and control. Variable-rate coding is also highly susceptible to bit errors, often resulting in catastrophic errors and requiring complex resynchronization control. Error correction coding can be very effective in reducing bit errors, but loses its effectiveness under severe loss and bursty situations, while adding additional hardware complexity and 40% to 60% overhead to the overall bit rate. With indoor and mobile channels that experience low distortions most of the time, but suffer intermittent severe channel degradation occasionally (e.g. indoor infrared links), error correction codes add unnecessary bandwidth overhead under low distortion conditions and are ineffective in handling severe channel degradation.

We present a video compression algorithm using pyramid lattice vector quantization (PVQ) of subband coefficients. Two key aspects of this algorithm are 1) the use of small vector dimension (N=4) to improve error resiliency and image quality, and 2) an error-resilient PVQ codeword indexing scheme, which improves PSNR up to 2 dB under severe bit errors. This algorithm is a fixed-rate code that offers competitive rate-distortion performance with variable-rate coding and error correction and provides excellent error resiliency, even under bursty, high bit-error rate situations, without the use of error correction codes. The fixed-rate nature of this algorithm not only improves error resiliency, but also greatly reduces hardware complexity.

2.2 Background: Subband Coding and Lattice VQ

In this subsection, we give a brief overview of subband coding and lattice vector quantization. A more complete description of subband coding can be found in [2],[3]. Additional references for PVQ are [4],[5].

2.2.1 Subband Coding

As shown in Figure 1, subband coding divides each video frame into various subbands by passing the image through a series of 2-D low-pass and high-pass filters. Each "level" divides the image into four subbands. We can "hierarchically" decompose the image by further subdividing the lowest band to four more bands. We refer pixel values in each subband as subband coefficients.

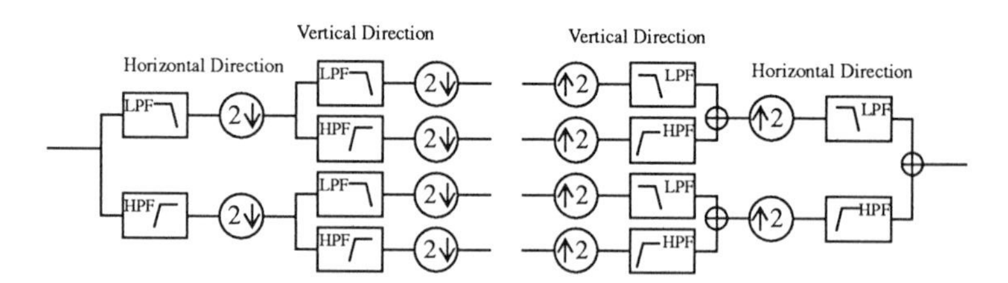

Figure 1 One level of 2-D subband filtering

The filtering process decorrelates the image information and compresses the image energy into the lower frequency bands, a natural result of the fact that most image information lies in low spatial frequencies. The higher spatial frequency information, such as edges and small details, lies in the higher subbands, which have less energy. We achieve compression by quantizing each of the subbands at different bit rates according to the amount of energy (variance) in each subband and the relative visual important of the subbands. A common approach to quantizing subband coefficients is the use of scalar uniform quantization followed by entropy coding. This approach, although nearly optimal in rate-distortion performance for high rates, results in a variable-rate code, which is error-sensitive and requires significant buffering hardware. We instead use pyramid lattice vector (PVQ) quantization, which achieves comparable performance with a fixed-rate code.

2.2.2 Pyramid Lattice Vector Quantization

Fisher first introduced PVQ as a fast method of encoding Laplacian-distributed i.i.d. vectors and later applied this technique to encoding high frequency DCT coefficients, which can be approximated by the Laplacian distribution [6],[7]. PVQ involves grouping data into N-dimensional vectors, mapping these vectors on to an N-dimensional pyramid structure (Figure 2 below), and finding the nearest lattice point on this pyramid. The lattice points on the pyramid form the PVQ codebook, and each lattice point is assigned a binary codeword index. Since the pyramid structure is regular, encoding and decoding can be performed with simple computations and minimal memory look-ups. This eliminates the need for codebook storage and training found in LBG-based VQ designs and allows very large codebooks without being restricted by physical memory size. Furthermore, the computational complexity for both encoding and decoding depends linearly on the vector size, not exponentially as with the encoding required for most LBG-based VQ schemes. This allows for real-time encoding and decoding, making PVQ a practical, effective method of quantizing image data at video rates.

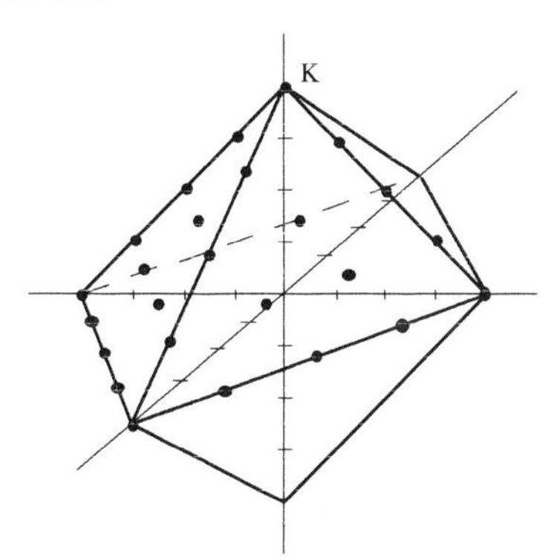

Figure 2 Integer lattice points on a 3-D pyramid of "radius" K=4.

The pyramid shape can be derived using the asymptotic equipartition principle (AEP) from information theory. If vector data are i.i.d. and have Laplacian distribution, the AEP implies that for large vector dimension, N, almost all vectors will fall in an equiprobable N-dimensional pyramid surface [4].

However, subband coefficients of real images, although roughly Laplacian in nature, are far from being independent and identically distributed. As a result, most real image vectors land far from the assumed pyramid surface. Antonini and Barlaud addressed this problem by including all the lattice points "inside" and on the pyramid surface into the codebook and using a single scaling factor to map all vectors into the "volume" inside the pyramid [8]. However, our analysis has shown that Fisher's original product-code PVQ method of mapping vectors on to the pyramid surface offers better performance for subband coefficients.

2.3 Compression Algorithm Implementation

2.3.1 Selection of Vector Dimension

Selection of the vector dimension size is an important factor in the performance and implementation of PVQ. Although the formal theory suggests that larger dimensions achieve better performance, we found that smaller dimensions, e.g. N=4, actually perform slightly better in coding real image data, as shown in Figure 6. More importantly, however, we found that using a smaller vector dimension vastly improves error resiliency. A smaller vector dimension localizes the effect of a single bit error; corrupted vectors with larger dimensions, especially those from the lower frequency subbands, can distort a much larger section of the image, causing significantly worse performance under noisy channel conditions. This performance will be detailed in section 2.4.

Another important benefit of using small vector dimension is a significant reduction in hardware and computation. The PVQ encoding and decoding algorithm requires table look-ups of pre-computed factorial calculations. The size of this table is 2*(N-2)*(K-2), where N is the vector dimension and K is the pyramid radius. The required memory size in bits is (size of table)*(vector bit width) = 2(N-2)*(K-2)*(bits/pel)*N. With a larger vector dimension, a larger radius K is required to maintain the same bit rate, as shown in Table 1. Thus, the amount of memory needed to store the table significantly increases with larger vector dimension. As an example, for N=4 and a bit rate of 5.0 bits/pel (K=73), the memory size is 2*(4-2)*(73-2)*(5)(4) = 5680 bits. For N=16 at the same bit rate (K=123), and the corresponding memory size is 2*(16-2)*(123-2)*(5)(16) = 271,040 bits, a factor of 48 times more memory.

Table 1 Radius K and codebook size as a function of the bit rate and vector dimension.

Bit rate (bits/pel)	Vector dimension, N	Radius, K	Codebook size
3.0	4	11	3.60800e+03
3.0	8	16	1.51583e+07
3.0	16	27	2.12749e+14
5.0	4	73	1.03777e+06
5.0	8	80	1.06988e+14
5.0	16	123	1.13912e+24

Smaller vector size also results in smaller codeword indices and correspondingly smaller datapath widths. The required datapath width is directly proportional to the vector dimension. For instance, for a bit rate of 5.0 bits/pel, with N=4, the datapath is 20 bits wide; with N=16, the datapath would increase to 80 bits wide.

Finally, since the PVQ decoding computation time is roughly proportional to N*K, one can also obtain significant computation savings using smaller vector dimensions.

2.3.2 An Error-Resilient PVQ Index Scheme

In [4], Fisher proposed an enumeration technique for assigning codeword indices to lattice points in the PVQ codebook. We propose a new indexing method which significantly improves channel robustness without increasing the bit rate and has roughly the same encoding/decoding complexity. Every vector can be described by three characteristics: 1) the *pattern*, which describes the positions of the non-zero vector elements, 2) the *shape*, which describes the magnitude of the non-zero vector elements, and 3) the *sign* pattern, which describes the signs of the non-zero vector elements. Our index scheme codes each of these characteristics separately, then combines them in an implicit product code. Separating out these characteristics in a product code helps limit the effect of single bit errors.

To enumerate the sign pattern, we notice that there is a one-to-one correspondence between the number of sign bits and the number of non-zero vector elements. Let the number of non-zero elements be s. Then the number of ways to assign signs to s non-zero elements is 2^s. Thus, the sign index word can be concatenated together into a string of s bits long.

To code the pattern, we notice that the total number of possible patterns equals $P(N, s) = \binom{N}{s}$, which describes selecting s positions from a vector length N. We assign an index value to each possible pattern using the following recursion, based on whether a non-zero digit is in the first element of the vector, as

$$P(N, s) = P(N-1, s) + P(N-1, s-1) \tag{1}$$

Finally, the total number of possible shape vectors, given s non-zero vector elements whose magnitudes must sum to K, is $S(s, K) = \binom{K-1}{s-1}$. Like we did for the pattern, we assign an index value to each possible shape using a recursive expression for $S(s,K)$:

$$S(s, K) = \sum_{i=1}^{K} S(s-1, K-i) \tag{2}$$

The total number of points on the pyramid is obtained by taking the product of all possible sign combinations, patterns, and shapes and summing the product over all possible values of s:

$$X(N, K) = \sum_{s=1}^{m} 2^s P(N, s) S(s, K) \tag{3}$$

where m is the $min(N,K)$.

To form a codeword index for a vector, we first enumerate the pattern, multiply that by the number of possible shapes, and then add the enumerated shape. Finally, we shift in the sign bits into the least significant bits. The decoding algorithm requires at most (NK + 5) memory fetches, (NK + 8) compares, (NK + 5) subtracts, and 1 divide.

The enumeration is robust to error because the sign bits form an implicit binary product code; a single bit error in the sign bits only causes a sign change in a single pixel without affecting other pixels. Furthermore, the pattern and shape indices are combined so that the pattern information rests in the most significant bits of the product code. Thus, single bit errors that occur in the less significant bits merely affect the shape or sign; only bit errors that occur in the most significant bits corrupt both the pattern and shape.

2.3.3 Description of the Subband/PVQ Algorithm

This subsection describes the overall algorithm. Note that this technique is entirely intraframe; i.e., each video frame is coded separately and independently.

- Each video frame is hierarchically subband decomposed into 13 subbands, as shown in Figure 3. This corresponds to "four levels" of subband decomposition. We use a 9-tap QMF filter, as specified in [9].
- A bit rate (bits/pel) is assigned for every subband. This bit allocation, shown in Figure 3, is based on the amount of compression desired, the energy in each band, and subjective weightings based on visual importance.
- Given its high entropy and non-Laplacian distribution, the lowest subband (upper left corner) is uniformly scalar quantized (PCM) to 8 bits. Redundancy bits are added to the top three most significant bits of each pel to protect against bit errors in these most visually sensitive bits.

- Of the remaining high frequency subbands, each subband is subdivided into vectors of 4 pels each. Each four-dimensional vector is then lattice quantized: the vector is scaled onto the proper pyramid surface, and closest lattice point on the pyramid is found. Using the error-resilient indexing scheme, described in section 2.3.2, the representative index for the encoded lattice point is computed and sent over the channel, along with the vector's scaling factor. We quantize this scaling factor using a non-uniform logarithmic compandor, which matches the Laplacian nature of this scaling factor.
- The decoder translates each codeword index to a code vector. The decoded vector is then scaled, according to the received scaling factor, and sent to a subband filter bank to reconstruct the image. Note that the subband bit allocation is fixed, as is the order in which subband coefficients are quantized. Since this information is known by the decoder before hand, it does not require transmission.

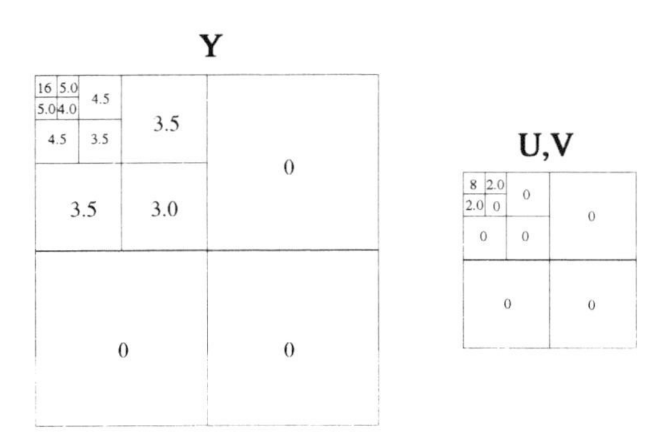

Figure 3 Subband bit allocation for YUV.

2.4 Performance

We evaluate our compression scheme based on its rate-distortion performance and its error resilient properties. To provide a relative measure of performance, we compare our fixed-rate, intraframe technique to a standardized variable-rate method, JPEG, which is an intraframe, DCT-based algorithm utilizing run-length and Huffman coding to achieve a high-degree of compression. To make the comparison fair, we also add error-correction coding to JPEG, which is a standard error protection technique and critical to reasonable performance under noisy channel conditions. We selected a (45,73) "difference-set" error correction code and a threshold decoding scheme, because it offers good protection (Hamming distance =10) and has a very simple decoder hardware implementation. This (45,73) code adds about 60% overhead to the overall transmission rate [10].

Figure 4 graphically shows the performance of our algorithm under noisy channel conditions for the Lena image. Looking at the JPEG plus error correction curve, notice that the error correction code does an excellent job of eliminating bit errors, but starts to fail at around BER = 10^{-2}. Once the BER increases past this point, catastrophic errors occur, causing severe block loss and rapid drop in PSNR performance. Our fixed-rate scheme, however, maintains a gradual decline in fidelity, even under the worst BER conditions. This gradual degradation performance makes the subband/PVQ scheme well suited for situations where image quality must be maintained under deep channel fades and bursty bit loss.

Figure 5 shows coded images under such channel conditions, where the subband/PVQ image can still be recognizable with reasonable quality under BER=10^{-1}. Notice that the bit errors cause ringing artifacts in the subband/PVQ images, which locally distort the image. In the JPEG images, bit errors become very noticeable when synchronization bits are corrupted, causing total block loss in some cases. Here, the resynchronization interval is set to four DCT blocks to improve its performance.

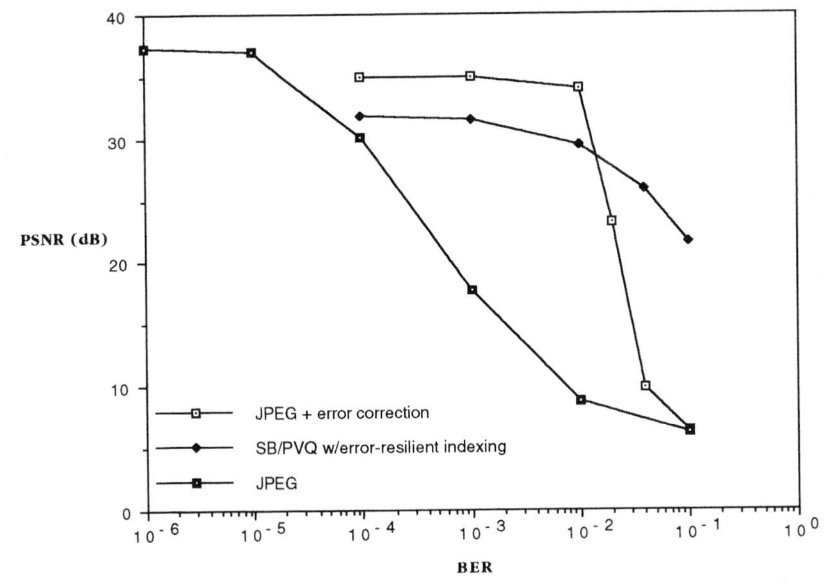

Figure 4 PSNR vs. BER for subband/PVQ and JPEG (Lena image). Compression = 8 to 1.

When viewing the compressed video (which will be shown in the presentation), the random loss of blocks is more dramatic, causing a flickering that makes it difficult to view details of the video. With the subband/PVQ video, the bit errors cause increased blurriness and wavering artifacts, but most of the image detail remains and is distinguishable.

We attribute the resiliency of our algorithm to several factors:

- Fixed-rate nature of algorithm prevents error propagation and eliminates the need for resynchronization bits.
- The effect of every bit error is localized. Selection of small vector dimension (N=4) for PVQ helps localize the effect of single bit errors. Notice in Figure 6 that when a 16-dimensional vector is used, the PSNR drops much faster under high BER. In the most visually important subband, the low DC band, PCM coding is used, which limits bit errors to a small region and requires that bit flips occur in more significant bits of a pel to have a noticeable effect.
- Our indexing scheme improves image quality by 2 dB under high BER conditions (see Figure 6).

Notice that this performance is achieved without the use of additional error correction hardware.

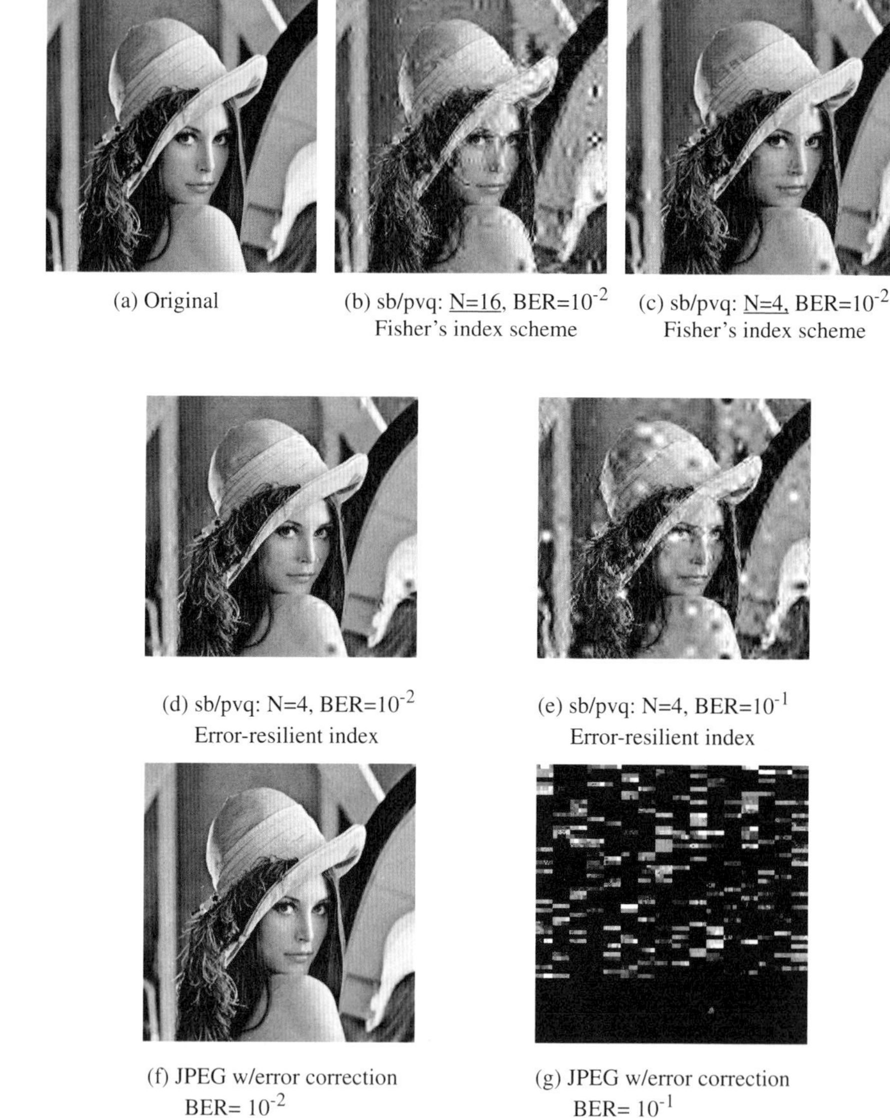

Figure 5 Compressed Lena image under bit errors. Compression ratio = 8:1. (b) and (c) compare error performance with different vector dimensions under Fisher's index scheme. (d) through (g) compare our subband/PVQ scheme with JPEG/error correction code. In (c) and (d), note the improvement using the error-resilient index scheme.

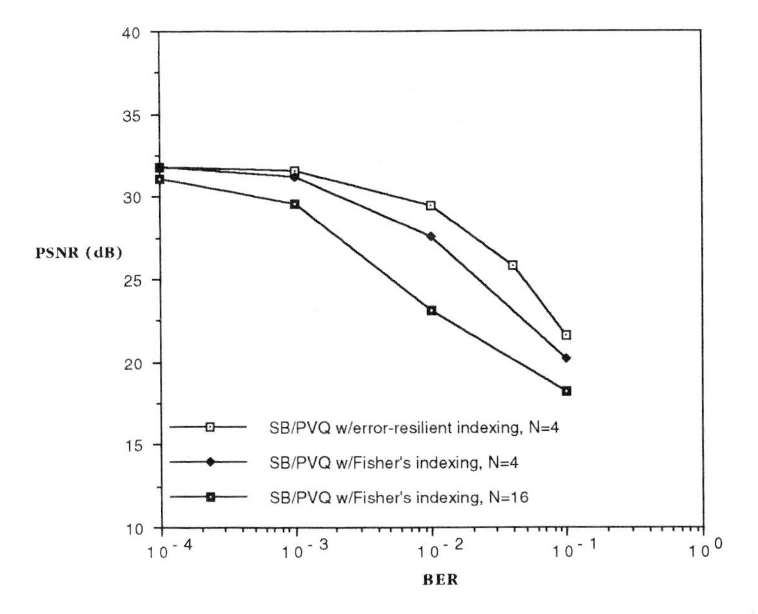

Figure 6 PSNR vs. BER for subband/PVQ with different index schemes and vector dimensions N.

2.5 VLSI Architecture for Pyramid VQ Decoder

The hardware design trade-offs, discussed in the subsections above, permit a relatively low-complexity decoder implementation with minimal buffering and memory requirements. When coupled with a subband decoder chip that does line-by-line processing, this design also eliminates the need for frame buffers, which consume a significant portion of the overall power budget.

The decoding algorithm described in section 2.3 can be implemented by three processing elements (PE's): 1) the index separator, which divides the incoming codeword into sign, pattern, and shape indices, 2) the vector generator, which generates a 4-dimensional vector based on the decoded indices, and 3) a 12-bit pipelined multiplier, which performs the final vector scaling. To increase throughput, these 3 PE's form a three-stage pipeline, where three vectors are being decoded simultaneously. What makes this pipeline interesting is that the latency of the second stage is non-deterministic and depends on the "radius" of the vector and the vector elements. To account for this uncertainty, the processor uses a simple handshaking scheme, in which all PE's acknowledge completion before the pipeline proceeds to the next cycle.

The index separator includes a 20-bit wide datapath performing compares, subtracts, and integer divide with remainder, as shown in Figure 7 (a). The 24-bit shift register separates the sign index from the codeword. The division unit separates out the pattern and shape indices. Local registers and a local bus allow for local data transfers and processing, and final results are placed in global registers, where the vector generator can access them on the next cycle. Figure 7 (b) shows the vector generator design, which also includes a simple 20-bit wide datapath to perform compares and subtracts. The shape generator is a simple counter circuit which keeps track of the vector sum, radius K, and sets each vector element value.

The critical path of this design is the vector generator processor. Its computation time is dominated by memory fetches; thus, the critical path of the entire processor depends on the speed of the memory accesses. We selected a 6-T SRAM design that has an access time of about 38 ns (2K x 64 bit) at 1.5 V [11], the target supply voltage for this chip. Keeping all the memory on-chip will keep access times and power consumption low. The memory will primarily store the pre-computed factorial calculations and greatly reduces the amount of computations in the decoding process. As a result, the decoder chip will operate at a power dissipation on the order of tens of *m*Watts for real-time 30 frames/sec video decoding.

A final comment should be made on the codebook size and memory requirements. At 5.0 bits/pel with a 4-D vector, the effective codebook size is about 10^6. Table look-up decoding requires three orders of magnitude more memory than that used in our design, and consumes much greater power. The PVQ scheme trades memory for computation, thus reducing the overall power consumption and physical decoder size.

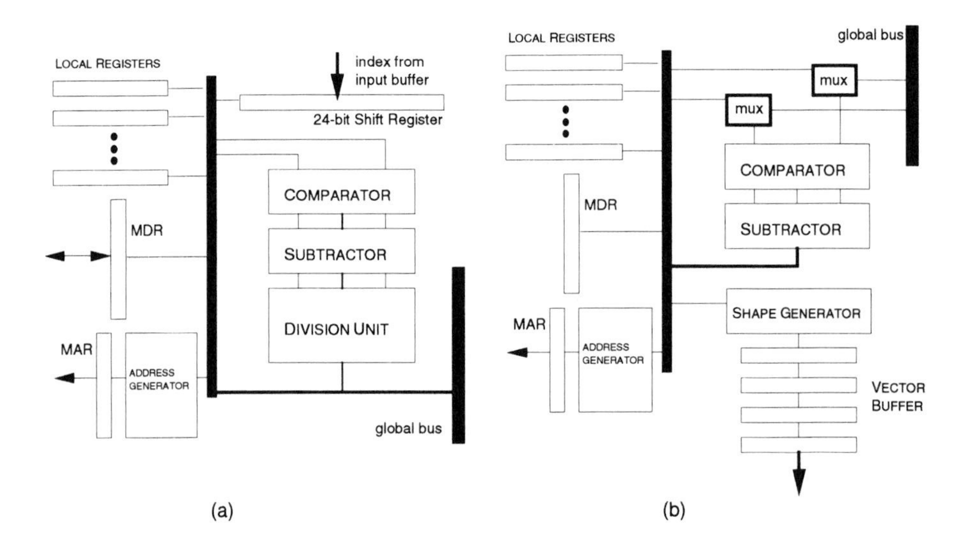

Figure 7 (a) Index separator block diagram and (b) Vector generator block diagram.

2.6 Summary

We propose a subband/PVQ compression algorithm that exhibits desired characteristics for wireless video transmission. By selecting a small vector dimension, using an error-resilient indexing scheme, and keeping the entire scheme fixed-rate, we simultaneously achieved excellent error resiliency and kept hardware complexity low. Without error correction, the PVQ scheme outperformed the JPEG/error correction scheme by up to 15 dB under severe channel error conditions. The proposed PVQ decoding algorithm maps into a simple architecture that can easily be implemented on a single chip and provides real-time decoding at video rates.

3.0 Spatial and Temporal Reconstruction for Packetized Video†

In this section we describe another important topic in wireless video compression: reconstruction of lost video signals due to packer loss. Transmission of packet video over a radio-frequency (RF) link presents challenges that do not arise when a physical link such as a fiber optic cable or twisted pair is used. A "guaranteed channel" is often assumed when using a physical link, in which higher priority information experiences a much lower packet loss rate than lower priority data. In an RF link, all data is equally subject to loss. Because of this difference, layered coding for fault-tolerance is not effective in safeguarding high-priority data from loss. The decoder must therefore compensate for loss of all parts of the transmitted video data. This section presents a reconstruction scheme for block-based coded video, capable of recovering from loss of the spatial intraframe and the temporal motion vector information. The power consumption of this reconstruction algorithm is on the order of tens of *m*Watts for real time operation. First, the video coding algorithm is described and packetization assumptions made in developing the reconstruction techniques are explained. The spatial and temporal reconstruction techniques are then described and their computational complexities presented. Finally, the performance of the complete algorithm is discussed.

3.1 Video Coding Algorithm and Packetization Assumptions

The assumed video coding algorithm is similar to that proposed in the MPEG-1 standard [13]. Frames are coded as one of two types, intraframe and interframe. Every sixth frame is coded as an intraframe, in which both the luminance and the chrominance are coded using a discrete cosine transform (DCT) on 8x8 pixel blocks. The five frames between any two intraframes are coded as interframes. The luminance component of each interframe is motion-compensated on a half-pixel grid using bi-directional prediction on 8x8 blocks, and the same motion vectors, appropriately scaled, are used for the chrominance. The search window for motion estimation is of size ±8 pixels from the previously predicted frame (telescoping). Difference images for both the luminance and the chrominance components are generated from the motion compensated frames, and then vector quantized using a vector size of 4x4 pixels. Thus three types of information are transmitted: (1) spatial intraframe, transform-coded blocks (luminance and chrominance), (2) temporal interframe, motion vectors and prediction directions, and (3) spatial interframe, vector quantized difference images (luminance and chrominance). While specific block sizes and intraframe intervals ("frame intervals") are given here, the algorithm is independent of these specifics.

The video coding algorithm provides a natural segmentation of data for packetization. Each block within an intraframe is transformed, and the resulting coefficients can be grouped either as entire blocks, in which all coefficients for a transform block are transmitted together, or as coefficients, in which specific coefficients from a given number of transform blocks are transmitted together. An intraframe data segment refers to either of these groupings. Interframes are also segmented into blocks, to each of which corresponds one motion vector (or two, in the case of interpolative prediction) and prediction direction. The motion vector(s) and corresponding direction for a block comprise a temporal information data segment and are transmitted together, so that both must be reconstructed together. Finally, the vectors used to quantize difference image information are blocks of a given size, and each vector index is a difference information data segment.

3.2 Spatial Reconstruction

3.2.1 Intraframes

Lost intraframe information consists of either partial or complete blocks of transform coefficients, depending on the packetization, in both the luminance and the chrominance components. Transform coefficients are reconstructed using the technique presented in [14], which exploits the correlation between coefficients in adjacent blocks. This technique is reviewed here. In the following discussion, specific

† This work was previously presented at the *Fifth International Workshop on Packet Video*, March 22-23, 1993, Berlin, Germany [12].

matrix and vector dimensions are given for ease of understanding. However, this technique is applicable to any block size. Local image characteristics are maintained by using transform coefficients from neighboring blocks to interpolate lost coefficients. Up to four weights are computed, one for each adjacent block, and each lost coefficient is interpolated as a linear combination of the coefficients at the same position from the surrounding blocks. To ensure that the reconstructed block is smoothly connected to its neighbors, the weights are computed subject to a smoothing criterion such that the squared difference is minimized between border pixels of the reconstructed block and border pixels of the adjacent blocks.

The reconstructed coefficient block $\hat{C}_Z$ (of dimension $N \times N$) can be written as

$$\hat{C}_Z = C_Z + w_T \overline{C_T} + w_B \overline{C_B} + w_L \overline{C_L} + w_R \overline{C_R} \tag{4}$$

where C_Z is the matrix of correctly received coefficients for the block and $\overline{C_X}$ is the matrix containing the complemented set of coefficients from adjacent block X ($X = T, B, L, R$, as depicted in Figure 8); that is, the set of coefficients to be interpolated. The coefficients allotted to these matrices are illustrated in Figure 9. If an adjacent block is unavailable (for example, on the edges of the image, or if adjacent blocks have also been lost), then its corresponding weight is set to zero. Note that the linearity of the DCT permits rewriting (4) using pixels instead of coefficients as

$$\hat{P}_Z = P_Z + w_T \overline{P_T} + w_B \overline{P_B} + w_L \overline{P_L} + w_R \overline{P_R}. \tag{5}$$

The reconstructed block can be formed directly using (5), avoiding an inverse transform.

To ensure that the reconstructed block connects smoothly to the available adjacent blocks, a smoothing criterion is imposed, similar to that used in [15]. This criterion quantifies the pixel differences across the boundaries between the reconstructed block and the available adjacent blocks, and is referred to as the total squared edge error:

$$\varepsilon^2 = \varepsilon_T^2 + \varepsilon_B^2 + \varepsilon_L^2 + \varepsilon_R^2 \tag{6}$$

where the subscripts refer to the adjacent blocks. If an adjacent block is unavailable, the corresponding squared edge error is set to zero. Each squared edge error is written in terms of pixels as

$$\varepsilon_T^2 = (\hat{p}_{Zt} - p_{Tb})'(\hat{p}_{Zt} - p_{Tb}) \tag{7}$$

$$\varepsilon_B^2 = (\hat{p}_{Zb} - p_{Bt})'(\hat{p}_{Zb} - p_{Bt}) \tag{8}$$

$$\varepsilon_L^2 = (\hat{p}_{Zl} - p_{Lr})'(\hat{p}_{Zl} - p_{Lr}) \tag{9}$$

$$\varepsilon_R^2 = (\hat{p}_{Zr} - p_{Rl})'(\hat{p}_{Zr} - p_{Rl}) \tag{10}$$

where the upper case subscript refers to the adjacent block to which the pixels belong and the lower case subscript refers to the column vector of edge pixels as shown in Figure 10.

Figure 8 Labelling of adjacent blocks (lost block in center).

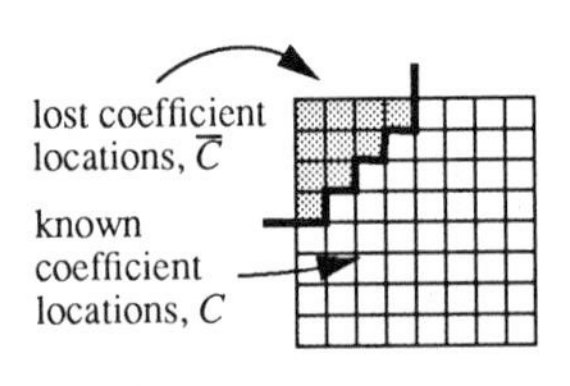

Figure 9 Coefficients comprising the matrices C and $\overline{C}$.

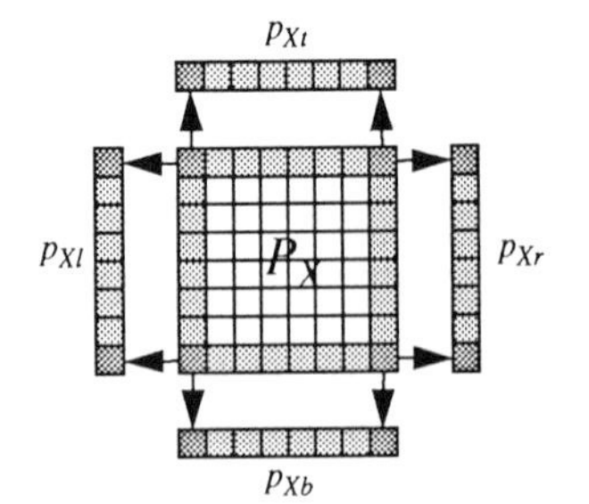

Figure 10 Formation of pixel column vectors for computation of edge errors where $X = T, B, L, R$.

Minimizing the total squared edge error with respect to the 4 weights w_T, w_B, w_L, and w_R, is simply a linear least squares problem and can be easily solved.

The computational complexity for reconstructing an 8x8 pixel block using 4 adjacent blocks is given in Table 2. If the intraframes are coded using a discrete cosine transform (DCT), the total number of operations per pixel is less than 2.2 times the number of operations required for simply decoding a block using a fast, recursive DCT, and less than 1.2 times the number of operations required for decoding using a non-recursive DCT.

Table 2 Computational Complexity of the Intraframe Reconstruction Technique

Floating Pt. Operations	add/subtract	multiply	divide
Form equations	434	448	
Solve equations	52	42	11
Create block	192	256	
Total, per block	678	746	11
Total, per pel	11	12	0.2

3.2.2 Vector Quantized Difference Images

Interframe difference information, which is vector quantized, is also spatial information and individual vector indices can be lost during transmission. Loss of interframe difference information is simply ignored, as it is primarily high frequency information and less important to the overall reconstruction quality than the other two types of information.

3.3 Temporal Reconstruction

Lost temporal information requires the decoder to generate an estimate of both the prediction direction from which the block was predicted and the motion vector for the block. The temporal information reconstruction relies on a smooth movement assumption; that is, if a block in the image is moving, then blocks around it will also be moving in the same general direction with high probability. Thus a block's lost prediction direction is taken as the prediction direction of the majority of the known adjacent blocks (up to four). The lost motion vector is computed as the simple mean of the known motion vectors of adjacent blocks. The complexity of the temporal reconstruction is negligible compared to that of the spatial reconstruction, requiring at most 8 multiplies, 8 additions, and 2 divisions per lost motion vector, and a majority vote of at most 4 directions per lost prediction direction.

Both spatial reconstruction and temporal reconstruction can be easily implemented in real time at a frame rate of 30 frames per second using a commercially available DSP processor, which provides a simple and effective reconstruction scheme for video signals compressed using standards algorithms and transmitted in a packet-switched network. Our design criterion, however, is a low-power decoder module that uses reconstruction as a rate-control strategy, in that the encoder (transmitter), regulates transmission bandwidth by intelligently throwing away packets while the decoder (portable receiver), reconstructs lost packets without knowing whether the packets were deliberately thrown away by the transmitter or lost in transmission. Therefore the reconstruction computation is an integrated part of the decoder module, and a DSP processor solution would consume too much power. Preliminary hardware design reveals that less than 30 *m*Watts is needed for reconstructing up to 10% loss of all data.

3.4 Performance Evaluation of Reconstructed Video

The video reconstruction algorithm was evaluated using three sequences, the familiar "table tennis" and "flower garden" sequences, and a third named "air." The third sequence, from a basketball game, was chosen because it contains more and faster movement than the other two sequences, and thus provides a more demanding test of the technique. The frame size was 352x240, in 4:1:1 YUV format, at a frame rate of 30 frames/second. The video coding used the following parameters: 8x8 DCT block size for intraframes with MPEG quantization matrices, intraframe intervals of 6 and 10 frames, 8x8 and 16x16 motion estimation (ME) block sizes, and 4x4 difference image vector size. Each sequence experienced a 1% random loss on an interleaved stream of data. Reconstruction of each type of information was performed separately at first, assuming that the other two types were perfectly received, to assess the errors caused by reconstruction. A 1% loss of all data was then simulated to evaluate the complete reconstruction technique.

In the following discussions, emphasis is placed on the performance of the algorithm on the luminance component. Luminance errors are much more visible than chrominance errors, and when the reconstructed video is viewed at full frame rate, chrominance errors are virtually unnoticeable, while luminance problems can be quite visible. Performance is described subjectively, rather than using SNR, because SNR comparisons do not match visual results.

3.4.1 Spatial Reconstruction

The intraframe reconstruction technique minimizes blockiness in intraframes by creating a block that connects smoothly to its neighboring blocks. Horizontal, near-horizontal, vertical, and near-vertical edges and lines continue smoothly through reconstructed blocks, and gradual luminance gradients are maintained. In these cases, the structure of the reconstructed block is similar to that of blocks adjacent to it, and as such can be well represented by a linear combination of the adjacent blocks. Because all transform coefficients are used in reconstruction, random high frequency patterns are preserved and the technique performs well on textures, such as the wall in the "table tennis" sequence and the flowers in the "flower garden" sequence. Reconstruction of diagonal edges and lines is fair to poor. The quality of the reconstructed frame is therefore content-dependent.

Because intraframes are used in the decoding process to create interframes, visible errors in the reconstructed intraframes can propagate over several frames. The extent of error propagation is independent of the frame interval and motion estimation block size, but is content-dependent, as expected. The "table tennis" sequence demonstrated minimal errors at the simulated 1% loss rate, exhibiting less than 10 errant 8x8 blocks in a 150 frame sequence. The visible errors were caused by reconstructed blocks along the diagonal edges of the table, which propagated over several frames. The visual quality was judged excellent. Likewise, the busy "air" sequence exhibited minimal errors and quality was also excellent. The "flower garden" sequence, however, exhibited "flicker" on some diagonal roof edges and highly detailed areas of the house. The details are of approximately the same size as the transform block size (8x8 pixels), and therefore cannot be reconstructed well from adjacent blocks. At a frame interval of 6, the reconstructed intraframe information was the cause of the virtually all of the visible errors in "table tennis" and "flower garden," due to error propagation.

The three sequences exhibited robustness to omission of the lost vector quantized difference image information. At both frame intervals, no errors were visible in either the "table tennis" or the "air" sequences. Slight flicker on roof edges in several frames of the "flower garden" sequence was visible at a frame intervals of 10, and no errors were visible at a frame interval of 6. Compared to intraframe and motion vector reconstruction, omission of this information caused the fewest errors.

3.4.2 Temporal Reconstruction

To measure the performance of the temporal reconstruction, all coded motion vectors and prediction directions in the 150 frame sequences were first reconstructed using the proposed technique (assuming that all adjacent blocks were available). Results are presented in Table 3 for both 8x8 and 16x16 ME block sizes. The reconstructed motion vector mean error is calculated as the mean length of the error vector, which is given by the difference between the reconstructed motion vector and the true motion vector.

At the smaller ME block size of 8x8, the percentage of correctly predicted directions only decreases slightly with increasing frame intervals but is a function of the amount of the movement in the sequence. The "air" sequence contains faster and more varied motion than the other two sequences, and as a result approximately 13% fewer of the "air" sequence prediction directions are correctly reconstructed compared to the "table tennis" and "flower garden" sequences. However, the accuracy of the reconstructed motion vectors decreases more significantly with increasing frame intervals. As the intraframes become farther apart in time, the amount of object movement between intraframes increases and blocks that were originally motion compensated together (i.e., blocks that were moving in the same direction at the same rate) may drift apart farther away from the intraframes. The accuracy of the reconstructed motion vectors is also a function of the amount of movement in the sequence, as the "air" sequence also had the largest error vectors. However, the sizes of the error vectors provide a relative, rather than absolute, measure of visible error in the sequence, as described below.

At the larger ME block size of 16x16, larger differences in the percentages of correctly predicted directions are evident as the frame interval increases, as are larger error vectors. Larger blocks can contain more independent movement, especially when the object size is comparable to the block size, as in the "air" sequence, where a 16x16 block can encompass the basketball or a large portion of a player. As such, the smooth movement assumption does not hold as often as in the case of the 8x8 ME block size, and more and larger errors ensue.

Table 3 Motion Vector and Prediction Direction Reconstruction Performance for 8x8 and 16x16 Motion Estimation Block Sizes

Sequence, ME Block Size	Frame Interval	% of Directions Correctly Predicted, Total Blocks	Reconstructed Motion Vector Mean Error (pixels), for Correctly Predicted Directions		
			Forward	Backward	Interpolative
Table Tennis, 8x8	6	72.5%, 165000	1.0	1.2	1.3
	10	71.0%, 178200	1.6	1.7	1.9
Flower Garden, 8x8	6	73.0%, 165000	0.93	0.86	0.98
	10	71.7%, 178200	0.99	1.0	1.3
Air, 8x8	6	59.9%, 158400	3.6	3.5	5.6
	10	58.9%, 178200	4.8	4.5	8.0
Table Tennis, 16x16	6	74.7%, 41250	3.5	3.4	4.3
	10	69.1%, 44550	5.2	4.8	6.4
Flower Garden, 16x16	6	71.7%, 41250	1.3	1.8	6.0
	10	66.6%, 44550	1.4	2.3	9.5
Air, 16x16	6	56.5%, 39600	5.8	6.1	18.0
	10	51.5%, 41580	8.5	7.9	26.0

Error in motion vectors and prediction directions causes degradation in the reconstructed video sequence ranging from slight "flicker" in detailed areas to blatant errant blocks, depending on the sequence content and the motion estimation block size. At a frame interval of 6, the "table tennis" sequence exhibited almost no visible errors after suffering loss and undergoing reconstruction of 1% of motion vectors and corresponding prediction directions. This sequence has relatively large moving objects against a smoothly moving or stationary background, so the smooth movement assumption is valid for both ME block sizes. The scene change is handled well, due to the bi-directional prediction in the coding and the majority vote of prediction directions in the reconstruction. The "flower garden" sequence exhibits several 1-block errors in the 150 frame sequence. Errors occur as slight flicker in detailed areas of the house, where a small error in a motion vector can create an objectionable block. Thus although the "flower garden" sequence has slightly smaller error vectors than the "table tennis" sequence, the effects of these errors were greater due to the detailed content. The "air" sequence exhibited the most errors of the three sequences. The size of moving objects is smaller than that in "table tennis" and the movement is also faster. As a result, fewer blocks are moving together. The effect of temporal reconstruction errors on the "air" sequence was equivalent to the errors caused by intraframe error propagation; in "table tennis" and "flower garden," the temporal reconstruction errors were less evident than those caused by intraframe error propagation.

At a frame interval of 10 and a ME block size of 8x8, temporal errors in the "flower garden" and "air" sequences equalled and exceeded, respectively, the errors caused by spatial intraframe reconstruction, while the "table tennis" sequence exhibited minimal errors. Again, the noticeable errors in the "flower garden" sequence were primarily in the detailed sections, while the errors in the "air" sequence were around areas of fast movement. At the larger ME block size of 16x16, the "flower garden" and "air" sequences were judged to be unacceptable due to visible temporal reconstruction errors, while the "table tennis" sequence was still acceptable.

The temporal reconstruction technique produced higher quality reconstructed video than both a standard previous frame copy [16] (the video sequences were coded using forward-only prediction when evaluating this scheme) and an "intelligent" previous next frame copy, in which a block is chosen from the frame appropriate to the prediction direction reconstruction.

3.5 Summary

The video reconstruction technique performs best at the smaller frame interval of 6, when the smooth movement assumption on which the temporal reconstruction is based is valid. Under these conditions, temporal reconstruction of slow movement sequences is excellent, with minimal errors occurring in fast movement and detailed sequences. Intraframe spatial reconstruction is also a function of sequence content, with errors being most noticeable in highly detailed areas and on diagonal edges and lines. Omission of difference information causes negligible error. When all types of data are lost simultaneously, only several blatantly errant blocks in the 150 frame sequences appear and the reconstructed sequences are quite acceptable.

Using a commercially available floating point DSP chip operating at 25MHz, the complete reconstruction algorithm can compensate for up to 60% loss of all information. For portable applications, custom design that consumes less than 30 *m*Watts is feasible for reconstruction of 10% loss of information.

4.0 Conclusions

In this paper we surveyed two techniques for designing wireless video compression/decompression modules under the influence of transmission errors. To make video on demand a universal service, we need to address the following issues. Low-power implementation is far by the most important criterion for designing portable devices, which guides the development of compression algorithms and reconstruction procedures. The use of error correcting codes on top of compressed video data is a topic that warrants further study, and the combination of channel and source coding into a single platform is necessary for the development of optimum fault-tolerant compression. Furthermore, dynamic allocation of available transmission bandwidth to each end user in a wireless network should be supported, and the quantification of distortion that measures image quality according to human vision perception needs to be formally defined. The final goal is the delivery of best quality video to every user at any time, anywhere, given any transmission rate.

References

[1] E. K. Tsern, A. C. Hung, and T. H.-Y. Meng, "Video Compression for Portable Communications Using Lattice Vector Quantization of Subband Coefficients," submitted to *1993 IEEE Workshop on VLSI Signal Processing*.

[2] J. W. Woods and S.D. O'Neil, "Subband Coding of Images," *IEEE Trans. on Acoust. Speech Signal Processing*, ASSP-34:1278-1288, October 1986.

[3] J. W. Woods, editor, *Subband Image Coding*, Kluwer Academic Publishers, Boston, 1991.
[4] T. R. Fisher, "A Pyramid Vector Quantizer," *IEEE Trans. Inform. Theory*, IT-32:568-583, July 1986.
[5] T. R. Fisher, "Geometric Source Coding and Vector Quantization," *IEEE Trans. Inform. Theory*, IT-35:137-145, January 1989.
[6] H. C. Tseng and T. R. Fisher, "Transform and Hybrid Transform/DPCM Coding of Images Using Pyramid Vector Quantization," *IEEE Trans. on Comm.*, Vol. COM-35 no 1.:79-86, January 1987.
[7] R. C. Reininger and J.D. Gibson, "Distributions of the Two-Dimensional DCT Coefficients for Images," *IEEE Trans. on Comm.*, Vol. COM-31: 835-839, June 1983.
[8] M. Barlaud, P. Sole, M. Antonini, and P. Mathieu, "A Pyramidal Scheme for Lattice Vector Quantization of Wavelet Transform Coefficients Applied to Image Coding," *Proc IEEE ICASSP 92*, pp. 401-404, March 1992.
[9] T. Senoo and B. Girod, "Vector Quantization for Entropy Coding of Image Subbands," *IEEE Trans. on Image Processing*, vol. 1, no. 4, pp 526-533, October 1992.
[10] G. C. Clark, Jr., and J. B. Cain, *Error-Correction Coding for Digital Communications*, Plenum Press, New York, 1981.
[11] M. Horowitz, private communication.
[12] S. S. Hemami and T. H.-Y. Meng, "Spatial and Temporal Video Reconstruction for Non-layered Transmission," *Fifth International Workshop on Packet Video*, March 22-23 1993.
[13] CCITT SG XV, Working Party XV/1, Expert Group on ATM Video Coding, "Preliminary Working Draft of Test Model 0," *Doc. no. AVC-212*, March 1992.
[14] S. S. Hemami and T. H.-Y. Meng, "Reconstruction of Lost Transform Coefficients Using a Smoothing Criterion," *Proc. of Picture Coding Symposium 1993.*
[15] Y. Wang and Q.-F. Zhu, "Signal Loss Recovery in DCT-based Image and Video Codecs," *SPIE Conf. on Visual Communication and Image Processing*, vol. 1605, Boston, Nov. 1991, pp. 667-678.
[16] F. C. Jeng and S. H. Lee, "Concealment of Bit Error and Cell Loss in Inter-frame Coded Video Transmission," *Int. Conf. Comm. '91*, Denver, Colorado, June 1991, pp. 496-500.

11

A Low-Power Handheld Frequency-Hopped Spread Spectrum Transceiver Hardware Architecture

Jonathan Min, Ahmadreza Rofougaran, Victor Lin, Michael Jensen, Henry Samueli, Asad Abidi, Gregory Pottie, and Yahya Rahmat-Samii

Integrated Circuits and Systems Laboratory
Electrical Engineering Department
University of California, Los Angeles

Abstract - *The overall goal of this research project is to develop low-power personal communications transceiver hardware technologies, coupled with advanced systems techniques such as antenna diversity, channel coding, and adaptive power control, for achieving robust wireless digital data transmission over multipath fading channels. A frequency-hopped spread spectrum (FH/SS) code division multiple access (CDMA) technique was chosen over other multiple access schemes because a) it provides an inherent immunity to multipath fading; and b) the signal processing is performed at the hopping rate, which is much slower than the chip rate encountered either in a direct sequence (DS) CDMA or time division multiple access (TDMA) system, thereby potentially resulting in much lower receiver power consumption. Furthermore, frequency-shift keying (FSK) modulation with noncoherent detection in a frequency-hopped system results in a much simpler transceiver architecture as compared with coherent amplitude and phase modulation methods commonly used in DS and TDMA systems. Architectural innovations as well as advanced circuit techniques will be incorporated into the design of a portable handset to achieve minimum power and size without sacrificing robustness in performance. A 3-V CMOS implementation is being developed for the entire transceiver including the 900 MHz radio frequency (RF) front-end. Multiple miniature antennas will also be integrated into the handset to achieve the maximum diversity benefit for robust data transmission. The hardware and system technologies developed for the transceiver design will be general enough to be applicable to a wide variety of commercial and military wireless communications applications such as cellular and micro-cellular radios and telephones, wireless LANs, and wireless PBX systems. In order to validate the design techniques being proposed, a prototype all-CMOS transceiver handset with dual antenna diversity is being developed to demonstrate pedestrian-to-pedestrian communication over the 902-928 MHz band. This paper overviews the proposed FH/SS transceiver hardware architecture. The system rationale, analog and digital circuits, and antenna design techniques for the transceiver are also presented in the following sections.*

I. System Architecture

1) System Rationale

The UCLA personal communications project has the objective of passing data at rates up to 160 kb/s between two low-power handsets, using the 902-928 MHz band. The principal channel impairments in multi-user radio systems are attenuation due to shadowing, multipath fading, and interference from other radios. To overcome these impairments without resorting to high transmitter power, the proposed architecture incorporates many advanced techniques. Two antennas will be used to provide polarization/space diversity. Frequency hopping combined with error control coding will be implemented to provide both frequency and interferer diversity, further reducing the effects of multipath fading. Adaptive power control will be employed so that the minimum transmitted power required for reliable communication is used. Finally, an adaptive data rate permits a longer transmission range for lower rate messages, without increasing the peak transmit power.

The most damaging type of fading is Rayleigh fading, and it is also frequently encountered. Whereas in a Gaussian channel the bit error rate (BER) for a well-designed system drops exponentially with SNR, for a Rayleigh fading channel the BER declines only linearly with SNR [1]. Thus, a huge power penalty must be paid unless diversity techniques are used to mitigate the effects of multipath propagation. Diversity is the technical term for reception of different versions of the same information, with different fading levels. With L^{th} order diversity, the probability of error declines as the Lth power of SNR. Diversity may be achieved in any combination of the space (antenna), time, or frequency domains. A combination of techniques for each domain provides robust performance and is an economical means of achieving a high aggregate diversity order. This permits receiver performance close to that achievable in the additive white Gaussian noise channel.

Frequency diversity may be achieved by hopping many times per symbol. In a synchronous system, it is possible to arrange the hopping patterns so that no interference is generated by users within the same cell, and in addition, the hopping patterns will intersect those of users in surrounding cells in only one position. Thus, there will be no dominant interferer. At higher data rates, it may only be possible to hop but once per symbol. In this case, channel coding is essential in providing frequency and interferer diversity protection [2]. Furthermore, channel coding also increases the effective diversity order for faster hopped systems (multiple hops per symbol), as well as providing the usual benefits of coding gain.

Frequency-hopped systems permit a large frequency/interferer diversity to be obtained with lower transmission delay and significantly less complexity than wideband direct sequence spread spectrum or time division multiple access systems. However, fast hopping generally precludes the use of a coherent receiver. Consequently, the preferred modulation is binary FSK. This implies a significant SNR versus BER performance penalty--roughly 6 dB as compared to coherent binary PSK. However, the combination of frequency diversity, synchronous access and reduced complexity compared to DS/SS CDMA more than makes up for this penalty.

2) Channel Modeling and System Simulation

The radio propagation channel can be conceptualized as a time-variant linear filter that transforms input signals into time-varying output signals. Various research groups have proposed radio channel models for use in simulations [3]. In our initial investigation, the tapped-delay line model methodology was chosen and a Rayleigh fading channel was implemented. A typical application for a propagation model is to estimate the coverage area of the communication system. In our preliminary link budget analyses several scenarios were considered. The results for a portable-to-portable communication link are summarized in Table 1.

Table 1. Estimated Coverage Areas for a 900 MHz Radio Link Between Two Portable Units

OPERATING CONDITION	RADIUS OF COVERAGE
Clear Line-of-Sight (LOS) Channel	1470 ft
LOS Channel with Tree Obstructions	620 ft
Non-LOS Channel without Tree Obstructions	695 ft

In the link budget calculations, it was assumed that the transceiver employed two branches of antenna diversity to mitigate multipath effects. Other assumptions included FSK modulation, an average BER of 10^{-3} to provide good-quality voice reproduction, and an average transmit power of 10 mW. A portable to base-station communication scenario was also considered as a way of improving the coverage area due to the improved base-station antenna performance. This situation is similar to that of the mobile cellu-

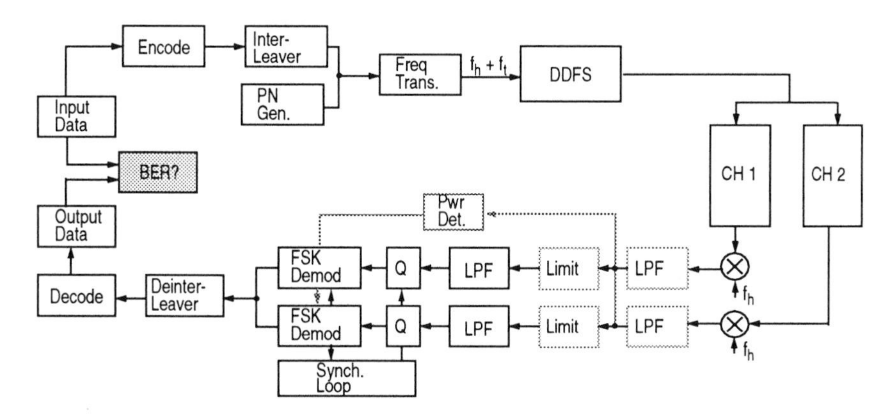

Figure 1. Simulation block diagram.

lar environment. It was found that a substantial improvement can be achieved by using a base-station to act as a switching office for relaying messages between portables. For example, by utilizing a base-station in the non-LOS channel scenario considered above, the radius of coverage increases to around 1500 feet.

The overall system simulation including the radio fading channel is in progress to evaluate the performance of the frequency-hopped spread spectrum (FH/SS) transceiver. System features such as antenna diversity and channel coding will be incorporated into the transceiver architecture and synchronization simulations. Various diversity combining techniques will be compared to trade the performance gain versus the hardware complexity. Simulations will be performed over a 26 MHz hopping bandwidth, using VANDA (Visual Analysis and Design Automation) [4], an integrated CAD environment for communication signal processing systems being developed at UCLA. The overall simulation block diagram is shown in Figure 1.

II. Hardware Architecture

1) RF/IF Architecture

A baseline RF/IF architecture employing high-performance low-power circuit innovations has been proposed for implementing the FH/SS transceiver. The combination of a direct digital frequency synthesizer (DDFS) and a digital-to-analog converter (DAC) is used to implement the fast frequency hopping. In order to simplify the designs of the DAC, the image reject filtering, the local oscillator, and the up conversion circuits, the DDFS is designed to hop only in a 0-13 MHz band. A 26 MHz hopping bandwidth is obtained by frequency doubling in the following RF stages. In addition, because of the doubling, the frequency of local oscillator (LO) used for up conversion will be 451 MHz, not 902 MHz, which results in power savings.

When the signal is up-converted, the SSB modulation technique [5] is used to reject the image of the signal. Image rejection can be achieved by taking either the real or imaginary part of the multiplication of two complex signals. This eliminates the need for filters with sharp roll-offs, and therefore lowers the overall power consumption. Image reject filtering is easily achieved in an FSK system because both the in

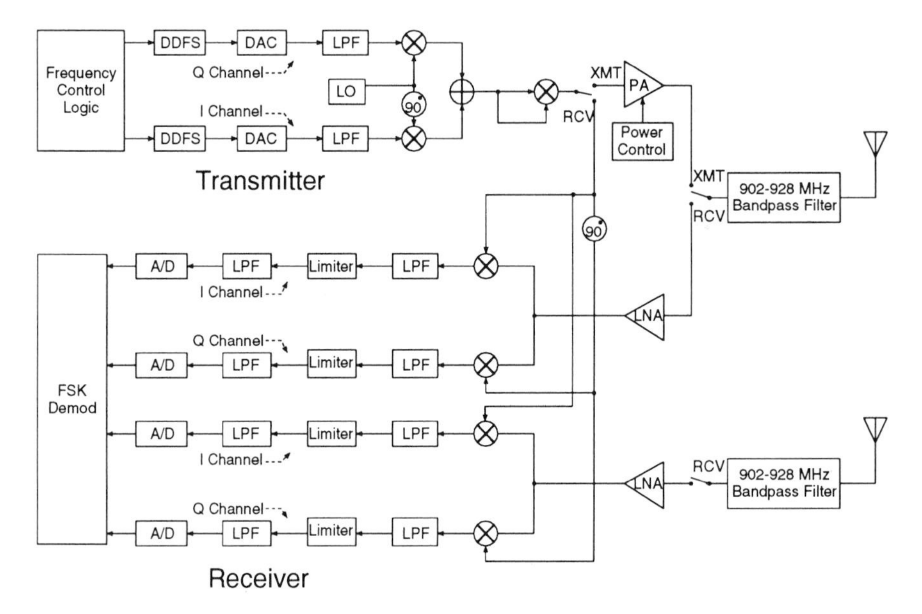

Figure 2. FH/SS RF/IF block diagram.

phase and quadrature signal components of the transmitted signal are easily and precisely generated using a quadrature DDFS, as shown in Figure 2. After rejecting the images, frequency doubling is used to convert the signal to the desired RF band (900MHz). Since we are using an FSK modulation scheme, a simple highly efficient class-C power amplifier can be used for transmission. To further minimize power consumption, the power amplifier will be integrated into a single chip along with the rest of the transceiver circuits. Finally, a low-cost off-the-shelf 902-928 MHz dielectric resonator bandpass filter will be used between the power amplifier and the antenna to reject the out-of-band harmonics and meet FCC transmission mask requirements.

Before down conversion, the received signal needs to be amplified using a low noise amplifier (LNA). The gain and noise figure of the LNA are critical factors in determining the overall transceiver performance. We have successfully developed a new technique to achieve a monolithic high-Q inductor in CMOS without altering the fabrication process [6]. This technique will be applied to design an LNA with high gain at 900 MHz. The high-Q inductor amplifier also provides extra filtering, thereby further attenuating out-of-band signals.

To minimize the number of high frequency components and thereby save power, the de-hopping and down conversion is performed in a single "direct-conversion" stage. In other words, the 900 MHz signal is de-hopped and down-converted directly to baseband. The down conversion circuit must achieve a low intermodulation distortion (IMD). One way to reduce power and lower the IMD products is to use a sub-harmonic sampling circuit which is switched at an integer (small) fraction of the RF frequency (900 MHz). This method can easily be implemented using conventional switch-capacitor techniques which are well developed for CMOS technology. These techniques not only result in low IMD products but also result in lower overall power dissipation.

2) Baseband Architecture/Algorithms

One of the key requirements for the portable communicator is low power consumption. This requires that complex signal processing, such as adaptive equalization, be avoided if possible. The hopping rate will be carefully chosen to meet this requirement [7]. Quadrature demodulation is required to efficiently use the entire 26 MHz transmission bandwidth. The automatic gain control (AGC) function is implemented using a combination of a lowpass filter (LPF), a hard limiter, and another LPF. Due to odd harmonics produced by hard-limiting the sine wave, this approach limits the modulation scheme to binary FSK rather than M-ary FSK. However, the merits from power saving and hardware simplicity obtained from the absence of an expensive (in both power and complexity) linear variable gain amplifier (VGA) and relaxation of the analog-to-digital converter (ADC) resolution (4 bits only) justifies this architecture for a portable handset design. The quadrature binary FSK modulation scheme also relieves the problem of group delay variations at different frequencies in the anti-aliasing filter. Following the ADC, the baseband signal processing is all performed digitally. This includes binary FSK demodulation, frame synchronization, PN acquisition, clock recovery, frequency tracking, and forward error correction.

Several architectural trade-off studies have been performed to determine the baseband (BB) architecture for the transceiver. An all-digital quadrature demodulation architecture using a correlation detector has been chosen over a fast fourier transform (FFT) or matched filter method. This choice is based on the fact that the correlator provides a flexible design which can easily accommodate programmable data rates, and the same hardware can be used to demodulate both data and sync hops.

A simple frame structure for the time duplexing mode (half-duplex) has been developed for master/slave configured communication transceivers. The frame structure consists of three fields: C0, C1, and C2. C0 is for frame synchronization, C1 for receiver ID, and C3 for user data. Synchronization is accomplished in three steps. The slave handset first listens for the C0 field of the frame, which is broadcasted by the master. Once frame synchronization is achieved, PN code acquisition follows simply by detecting the C1 field to match the pre-assigned receiver ID pattern. The PN code is synchronous with the frame and the pattern restarts at the same point with every new frame. The time track function is achieved by detecting the timing error and correcting the numerically controlled oscillator (NCO) accordingly. The overall frequency error is assumed to be small and therefore frequency tracking may not be necessary in a noncoherent FSK system, however, a frequency tracking architecture will be developed as a backup.

III. Advanced Analog and Digital Circuits

Various low-power analog and digital communication transceiver building blocks are under development. They range from RF analog to baseband digital circuits, all developed using CMOS technology. A 3-V, 800-MHz tuned RF amplifier in 2-µm CMOS was designed, fabricated and characterized. Several other circuits which are currently in design or fabrication are described in this section.

RF Amplifier - A low-power, low-noise RF amplifier in 2–µm N-well CMOS has been implemented and tested [6]. The amplifier has been measured to provide 14 dB of gain while dissipating only 6.9 mW (excluding output buffers) with a single 3-V supply. The amplifier has been designed to be a tuned amplifier in order to limit the noise contribution from the amplifier itself. The tuning of the amplifier was achieved by a large on-chip spiral inductor resonating with the parasitic capacitances. A novel technique has been applied to suspend spiral inductors over air-filled pits in order to attain over 100 nH of inductance on chip while providing a high self-resonant frequency. The noise figure of this amplifier was measured to be 6.1 dB (excluding input terminations), and the center frequency was 770 MHz. A new version of this amplifier is being developed in 1-µm CMOS which will be a tuned amplifier centered at 915 MHz. A lower noise figure is expected, and special attention will be given to the matching circuits between the amplifier inputs and the antenna.

Sub-Sampler Circuit - A low-distortion sub-sampler circuit using 1-μm CMOS technology is in fabrication. The sub-sampler is to be used in the receiver after the RF pre-amplifier stage to perform down conversion. To achieve the desired linearity performance, the design requires a high-speed op-amp and careful optimization of switch sizes and capacitance values. Simulation results from extracted layout show that the op-amp can achieve a 70 dB DC gain and a 500 MHz unity-gain frequency when driving a 300 fF load (with a 60° phase margin). Moreover, the power dissipation can be kept under 10 mW with a single 3-V power supply. Sub-sampler circuit simulations also show 11-bit linearity when sampling an AM signal at 40 MHz (with a carrier frequency of 400 MHz).

Frequency Synthesizer (PLL/VCO) - A low phase noise clock source is required to up-convert and down-convert the signals in the transceiver. A frequency synthesizer (FS) must be designed to generate these high clock frequencies by scaling a stable reference source. The FS must not only have low phase noise but it must also have a low power dissipation. These specifications are further complicated by the large operating temperature range and low-power supply (3-V). A phase-locked loop (PLL) was chosen to keep the phase noise to a minimum. The individual circuit blocks will be optimized for low power dissipation. The PLL consists of an external crystal used to generate the reference frequency, a phase/frequency detector, a charge pump, a voltage controlled oscillator (VCO) and a digital divider. A low phase noise design should incorporate a high gain phase/frequency detector and a low gain VCO.

Digital-to-Analog Converter - A monolithic low-power 30 Msample/s 10-bit digital-to-analog converter (DAC) with a 3-V supply is being developed. Conventionally, high-speed and high-precision DACs are implemented with current mode architectures which implies that they are usually power hungry. The major challenge for this circuit lies in finding an architecture that will provide high speed while still maintaining high linearity. The proposed DAC employs a pipelined charge redistribution algorithm implemented differentially, in which the pipelined digital code controls MOS switches that charge the capacitors to some known reference voltage. Due to the inherent nature of a switched-capacitor circuit, there is no DC standing current running through the circuit itself. Thus, it is ideally suited for the low-power application we are seeking for this project. The analog signal will traverse a series of charge redistribution stages. After 10 such stages (10 bits), the final analog output is available, and it feeds directly to a sample and hold buffer that includes a low-power op-amp. In fact, the output buffer is the only active component of the DAC; the rest consists of only passive components such as MOS switches and capacitors. Due to the high speed and low supply voltage requirements, the analog output swing is limited to only 0.5 V. This small output swing will not degrade the performance since the total thermal noise of the capacitors is less than one-half an LSB.

Direct Digital Frequency Synthesizer - A low-power CMOS quadrature direct digital frequency synthesizer (QDDFS) has been designed and is currently in fabrication. The QDDFS synthesizes 10-bit output sine and cosine waves simultaneously at 40 Msample/s. The synthesizer covers a bandwidth from DC to 20 MHz with a switching speed of 25 ns and a tuning latency of 2 clock cycles. Several techniques are employed to reduce the ROM storage [8]. A general-purpose ROM contents generation/simulation program has been developed, which can be used to optimize the ROM contents to get the best spurious response and ROM size compression. The worst-case spurious response for the proposed DDFS is -72.63 dB, which is analytically guaranteed for all output frequencies.

IV. Miniature Antenna Design

The design of miniature antennas suitable for implementation on a handheld transceiver involves investigation of radiating elements which can achieve maximum diversity performance when confined to a unit whose dimensions are less than a wavelength. This investigative process requires a detailed theoretical examination and comparison of the behavior of a wide variety of candidate antenna geometries operating in the presence of the transceiver structure. To accomplish this task, simulation tools based upon the finite-difference time-domain (FDTD) algorithm [9] and moment method have been developed which provide high accuracy while maintaining the flexibility required to model different antenna configurations.

These state-of-the-art antenna simulation tools have been used to investigate the performance of several different antenna topologies mounted on a conductive transceiver housing. The formulation allows determination of the antenna radiation pattern, gain, input impedance, and envelope correlation coefficient (diversity performance) over a wide frequency band. Figure 3 shows a representative example of the results of the FDTD analysis tool for the case of a Planar Inverted-F Antenna (PIFA)--an air-substrate modified microstrip antenna which provides efficient radiation for small physical dimensions [10]. This figure illustrates the antenna/transceiver geometry with its associated radiation pattern (normalized to the directivity) and input impedance. Numerous similar computations are being performed for other antenna geometries in an effort to identify an optimal diversity configuration.

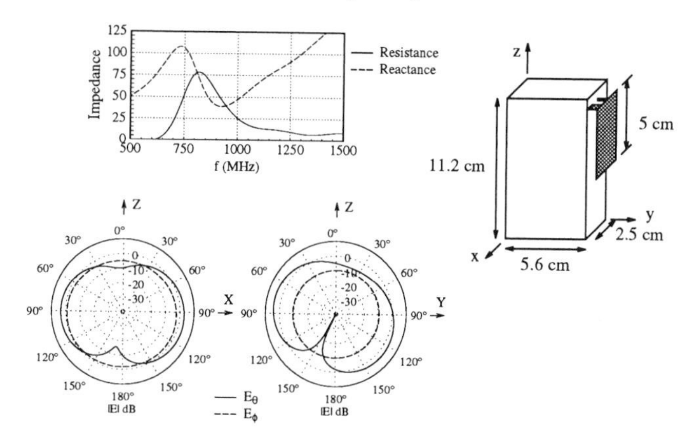

Figure 3. Planar Inverted-F Antenna on a conducting transceiver housing: geometry, input impedance versus frequency, and directivity pattern.

V. Conclusion

In this paper, the overview of the UCLA FH/SS transceiver project has been presented, from system issues to miniature antenna design. Robust operation of the handset will be achieved through the use of space diversity (dual antenna), frequency diversity (spread spectrum), and time diversity (channel coding). An efficient hardware architecture using FH/SS techniques has been proposed, utilizing direct conversion, frequency doubling, sub-harmonic sampling, and parallel interleaving. An open-loop hard-limiting automatic gain control (AGC) method is chosen to simplify hardware and reduce power consumption. Various RF/BB building blocks for the transceiver have been identified and are being developed. Highly integrated design of these low voltage custom analog and digital CMOS components reduces the use of discrete components, which enables the transceiver to achieve the minimum power consumption.

Acknowledgment

This work is supported by DARPA contract DAAB07-92-R-C977.

References

[1] J.G. Proakis, *Digital Communications*, 2nd Ed. New York: McGraw-Hill, 1989, pp. 719-728.

[2] G.J. Pottie and A.R. Calderbank, "Channel Coding Strategies for Cellular Radio," *IEEE International Symposium on Information Theory*, San Antonio, TX, Jan. 1993.

[3] T.S. Rappaport, "Statistical Channel Impulse Response Models for Factory and Open Plan Building Radio Communication System Design," *IEEE Transactions on Communications*, pp. 794-807, May 1991

[4] P. Tjahjadi, P. Yang, B. Wong, B. Chung, E. Cohen, and R. Jain, "VANDA - A CAD System for Communication Signal Processing Circuits Design," *VLSI Signal Processing IV*, pp. 43-52, IEEE Press, 1990.

[5] P. O'Sullivan, R. Benton, A. Podell, and J. Wachsman, "Highly Integrated I-Q Converters for 900MHz Applications," in *Technical Digest of IEEE 1990 GaAs IC Symposium*, New Orleans, LA, pp. 287-290.

[6] J.Y-C. Chang and A.A. Abidi, and M. Gaitan, "Large Suspended Inductors on Silicon and Their Use in a 2-μm CMOS RF Amplifier," to appear in *IEEE Electron Device Letters*, vol. 14, no. 5, May 1993.

[7] A.A.M. Saleh, A.J. Rustako, L.J. Cimini, G.J. Owens, and R.S. Roman, "An Experimental TDMA Indoor Radio Communications System Using Slow Frequency Hopping and Coding," *IEEE Transactions on Communications*, vol. 39, pp. 152-162, Jan. 1991.

[8] H.T. Nicholas and H. Samueli, "A 150-MHz Direct Digital Frequency Synthesizer in 1.25-μm CMOS with -90 dBc Spurious Performance," *IEEE Journal of Solid-State Circuits*, vol. 26, pp. 1959-1969, Dec. 1991.

[9] M.A. Jensen and Y. Rahmat-Samii, "FDTD Analysis of PIFA Diversity Antennas on a Handheld Transceiver Unit," *IEEE Antennas and Propagation Society International Symposium*, Ann Arbor, MI, Jun. 1993.

[10] T. Taga and K. Tsunekawa, "Performance Analysis of a Built-in Planar Inverted-F Antenna for 800MHz Band Portable Radio Units," *IEEE Journal on Selected Areas in Communications.*, vol. SAC-5, pp. 921-929, 1987.

12

An Overview of Broadband CDMA

G.R. Lomp & D.L. Schilling
InterDigital Communications Corp.
Great Neck, New York

L.B. Milstein
University of California - SD
LaJolla, California

1.0 Introduction

Broadband code division multiple access (BCDMA) refers to a technique which uses 10 MHz of bandwidth in the cellular frequency band to overlay the existing AMPS system with a CDMA network of users. In this paper, we provide an overview of this overlay, and in particular demonstrate that the interference between the two sets of users can be kept sufficiently small so that significant increases in capacity are achievable.

2.0 System Description

As is well known, the AMPS system consists of analog FM waveforms which occupy 30 KHz of bandwidth and which are separated from one another in a given cell which employs a three-sectored base antenna by 630 KHz; in a 10 MHz bandwidth, there are 16 such signals. In order to increase the capacity, a BCDMA system has been proposed which will overlay the existing AMPS system. This means that no AMPS users will have to be displaced in order to accommodate the initial set of BCDMA users. And while the number of BCDMA subscribers is initially constrained due to the presence of the AMPS subscribers, as the AMPS usage declines, the capacity of the BCDMA system will significantly rise.

The obvious advantages of such a broadband overlay, in addition to the favorable transition plan, are the flexibility for higher quality voice (relative to what is achievable with the narrower CDMA system) as well as data rate on demand (up to 144 Kbits/sec), and superior performance in any environment where the mulitpath delay spread is small (thus implying that the coherence bandwidth of the fading channel is large); the obvious examples of channels of this type are indoor channels and urban canyon areas in major cities.

In order for these advantages to be achievable, the interference between the two sets of users must not be excessive. Note that there are four distinct links which must be addressed, namely the two mobile to opposite base links, (e.g., the AMPS mobiles must not jam the BCDMA bases) and the two base to opposite mobile links (e.g., the BCDMA bases must not jam the AMPS mobiles). Qualitatively, this is accomplished in the following manner: In both the BCDMA base transmitter and the BCDMA base receiver, notch filters are used. By notching out a sufficient number of 30 KHz slots where the AMPS users are located, the signal from a BCDMA base will not jam the AMPS mobiles; similarly, by employing interference suppression notch filters in the BCDMA base station receivers, the AMPS mobiles will not jam the BCDMA base. With respect to the interference from an AMPS base station, by setting the ratio of received AMPS power to received BCDMA power (from the co-located BCDMA base station) at the BCDMA mobile in the worst-case location (which occurs at the intersection of three cells) to a suitable value (say, for example, 6 dB), the processing gain of 1000 enjoyed by the BCDMA receiver (assuming 8 Kbit/sec voice, the processing gain is 8MHz/8KHz = 1000) renders the AMPS signals harmless. Finally, to prevent the BCDMA mobiles from jamming the AMPS bases, the ratio of AMPS power/user to BCDMA power/user, as seen at an AMPS base station, should be about 17dB; this, in conjunction with the processing gain as seen by an AMPS signal, namely 8MHz/30KHz = 267, ensures proper reception of the AMPS signal.

3.0 Simulation Results

Because the dynamic range of the power control for the BCDMA system is much greater than that of the AMPS system, the interference from AMPS mobiles to BCDMA base stations could be quite high. For this reason, a computer simulation was run whereby users were randomly located in cells, and random shadowing effects from a lognormal density were assigned to the users; further, sectored antennas with a front-to-back ratio of 15 dB were used in deriving the results. To minimize the interference from the AMPS mobiles, the 50 largest AMPS signals, as seen by the BCDMA base, were notched out.

Table 1 shows three sets of result from that simulation, corresponding to three different values of the radius of a cell. For each radius, capacity estimate are shown as a function of the number of active AMPS users per sector (assuming a three-sectored AMPS cell); results are shown for both three-sectored and six-sectored BCDMA antennas, and all BCDMA capacity estimates assume a voice activity factor of 0.5. It is seen from Table 1 that, for example, for a cell radius of 7,500 ft., if 12 AMPS users are active per sector, and if three-sectored BCDMA antennas are employed, then 38 BCDMA users per sector can simultaneously operate, for a total of 150 users (both AMPS and BCDMA combined) per cell.

AMPS USAGE USERS/SECTOR	CDMA CAPACITY USERS/SECTOR	TOTAL 3 SECTORS	6 SECTORS
16	25	123	246
12	38	150	300
8	100	324	648
4	230	700	1400
0	488	1464	2930
7,500' RADIUS			

COMPUTER SIMULATION
B-CDMA OVERLAY 10Mchips/s, 8kb/s

TABLE 1

13

Performance of Direct-Sequence Code-Division Multiple-Access Using Trellis-Coded Orthogonal Signalling

Scott L. Miller
University of Florida
Department of Electrical Engineering, CSE 422
Gainesville FL 32611

Abstract

This paper examines the performance of trellis-coded modulation with orthogonal signal sets. In a previous work, the author obtained efficient methods for the design and analysis of such codes. This work is mainly concerned with the application of these results to a Gaussian noise channel and a Rayleigh fading channel with non-coherent reception. A technique based on the Chernoff bound for bounding the two-codeword error probability in a form that is convenient for use with transfer function bounding is developed and tested for several different channels. It is found to give very good results. The motivation for wanting to use orthogonal (or other wideband) signal constellations is a DS-CDMA system where the signal set is already wideband. The results are then also applied to a multi-user (DS-CDMA) channel.

1 - Introduction

It is fairly well known that when compared on the basis of capacity for a Gaussian noise channel, frequency division multiple acces (FDMA), time division multiple acces (TDMA), and code division multiple acces (CDMA) all turn out to be equivalent [6, p516]. The reason for much of the interest in CDMA is that it offers robustness against channel impairments other than the Gaussian noise, in particular the many types of interference that are typically present in a mobile communication channel. In this paper we would like to point out another reason for wanting to consider CDMA even for a Gaussian noise channel. Although FDMA, TDMA, and CDMA all have the same capacity, it should be asked: for which of these schemes will it be easier to achieve this capacity? To help answer this question, consider the fact that one way to approach the capacity of a Gaussian noise channel is through the use of large dimensional signalling (orthogonal or bi-orthogonal for example). Some of the drawbacks associated with such a signalling scheme are the large complexity and bandwidth required. In a CDMA application, bandwidth is not really a problem since a narrowband signalling format is going to be spread into a larger bandwidth anyway. By choosing the signal set to consist of various cyclic shifts of a specific PN sequence, the resulting signal constellation is nearly orthogonal and can be demodulated with two identical matched filters as opposed to a bank of M matched filters [7]. Thus it can be quite practical to implement an M-ary orthogonal (or bi-orthogonal) signal set in a CDMA scenario while this could not be done in TDMA or FDMA due to bandwidth limitations. As a result, it may be easier to come close to the information theoretic limits in a CDMA environment than with FDMA or TDMA.

Another way of looking at the problem is the following. Since CDMA forces each user to occupy a large bandwidth, why not use a signalling scheme which inherently takes advantage of the wide bandwidth, so long as is doesn't give up the advantages gained by using the spread spectrum. Using the M-ary orthogonal signalling scheme described above does jsut that. The rest of this paper then deals with applying trellis coded modulation (TCM) to an orthogonal signal set. It was shown in [8] that using traditional convolutional codes was superior (or at least as good as) using TCM in a CDMA application. However, in expanding to an orthogonal signal set, the signals do not get packed closer together as with a two-dimensional scheme, and so it is expected that the orthogonal based TCM will have better performance.

2 - Trellis-Coded Orthogonal Signalling in a Gaussian Noise Channel

In this section we consider the use of rate-1 trellis-coded modulation using an M-ary orthogonal signal set with $M=2^n$. This is implemented as a rate 1/n binary convolutional code followed by a mapping from binary n-tuples to the M-ary orthogonal constellation as illustrated in Figure 2.1. The important aspect of this encoder is the matrix **G** which combines K consecutive input bits to form an n bit output (for a constraint length K code). Ryan and Wilson [1] have found, through computer search, matrices **G** which produce optimum codes. In their work as well as our own, optimum is defined in terms of the information weight spectrum of the code, $N_s(w)$, which is defined as the number of information "1"'s on all paths through the code trellis with a Hamming weight of w. In [2], the author has given a procedure for finding "good" codes which in many cases resulted in the same codes found by computer search.

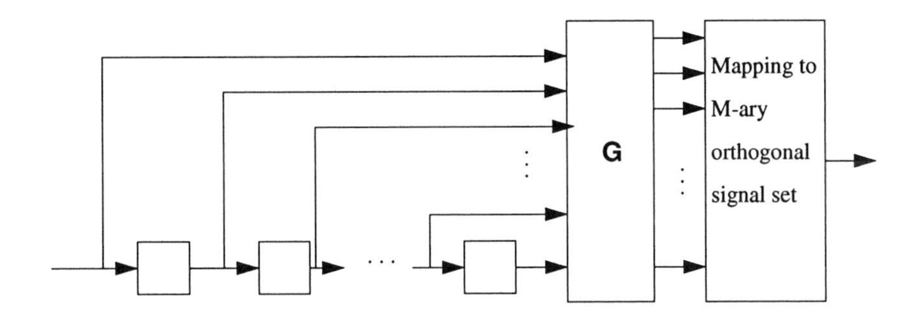

Figure 1 - Block diagram of trellis encoder for orthogonal signalling

The results of the work in [2] also led to the characterization of a "typical good code" in terms of a transfer function. In particular it is shown that a "good" code will have a transfer function which follows very nearly

$$T(W,I) = \frac{W^K I(1-W)}{1-W(1+I(1-W^{n-1}+W^{n-2}(1-W^{K-n})))} . \tag{1}$$

As usual, by taking the two-dimensional Taylor Series expansion of T(W,I), the coefficient of W^iI^j results in the number of paths with weight i and j information "1"'s. The information weight spectrum is then given by the terms in the Taylor Series expansion of $\frac{\partial}{\partial I}T(W,I)\Big|_{I=1}$ which for the specific case of (1) results in

$$\frac{\partial}{\partial I}T(W,I)\Big|_{I=1} = \frac{W^K(1-W)^2}{[1-W(2-W^{n-1}+W^{n-2}(1-W^{K-n}))]^2} . \tag{2}$$

The remainder of this section deals with the application of this result.

The probability of bit error for the trellis-coded system is given by

$$P_e \le \sum_{w=d_{free}}^{\infty} N_s(w) P_2(w) . \tag{3}$$

where d_{free} is the free distance of the code and $P_2(w)$ is the two-codeword error rate for orthogonal signalling with a Hamming distance w. Traditionally, to apply transfer function results requires that $P_2(w)$ can be written in the form of

$$P_2(w) = Az^w , \tag{4}$$

so that

$$P_e \leq \sum_{w = d_{free}}^{\infty} N_s(w)\,(Az^w) = A\frac{\partial}{\partial I}T(W,I)\bigg|_{I=1,\,W=z} . \tag{5}$$

Usually, the actual two-codeword error probability is not of the form of (4), but a sufficiently tight bound of that form can often be found.

For a Gaussian noise channel with coherent demodulation [5]

$$P_2(w) = Q\left(\sqrt{w\frac{E}{N_o}}\right) \leq Q\left(\sqrt{d_{free}\frac{E}{N_o}}\right) exp\left(\frac{d_{free}E}{2N_o}\right)\left[exp\left(-\frac{E}{2N_o}\right)\right]^w , \tag{6}$$

or

$$P_e \leq Q\left(\sqrt{d_{free}\frac{E}{N_o}}\right) exp\left(\frac{d_{free}E}{2N_o}\right)\frac{\partial}{\partial I}T(W,I)\bigg|_{I=1,\,W=exp(-\frac{E}{2N_o})} . \tag{7}$$

If non-coherent demodulation with a square law metric is used on a Gaussian noise channel, then the two-codeword error probability becomes [2]

$$P_2(w) = \frac{exp\left(-w\frac{E}{2N_o}\right)}{2^{2w-1}} \sum_{j=0}^{w-1} C(w,j)\left(w\frac{E}{2N_o}\right)^j , \tag{8a}$$

where

$$C(w,j) = \frac{1}{j!}\sum_{i=0}^{w-j-1}\binom{2w-1}{i} . \tag{8b}$$

This exact result does not easily fit into the transfer function bound and so a reasonable bound of the form of (4) is sought. This simplest bound of this form is the Chernoff bound

$$P_2(w) \leq (F(z_o))^w , \tag{9a}$$

where

$$F(z_o) = \frac{1}{1-z_o^2} exp\left(-\frac{E}{N_o}\frac{z_o}{1+z_o}\right) , \tag{9b}$$

$$z_o = \frac{-(\frac{E}{N_o}+1) + \sqrt{(\frac{E}{N_o}+1)^2 + 4\frac{E}{N_o}}}{2}$$

and (9c)

It is found that this bound is a little to loose to be useful, but that a slight modification leads to very good results. For a particular code with free distance, d_{free}, the bound used for $P_2(w)$ only needs to be valid for $w \geq d_{free}$. The Chernoff bound can be tightened by adding a constant out in front

$$P_2(w) \leq A_{d_{free}} (F(z_o))^w \quad , \qquad w \geq d_{free}, \tag{10a}$$

where
$$A_{d_{free}} = \frac{P_2(d_{free})}{(F(z_o))^{d_{free}}} \quad . \tag{10b}$$

It seems that using this adjusted Chernoff bound together with equation (5) often leads to better results than using the exact value of $P_2(w)$ in equation (3), since to evaluated equation (3) requires a truncation of the infinite series. To compare the fidelity of these various bounding techniques, Figure 2 shows a comparison of equation (5) using the Chernoff bound of (9) and the adjusted Chernoff bound of (10) along with equation (3) truncated to 1,2,4,and 8 terms using the exact value of $P_2(w)$ given in (8). For large SNR's, the truncated series agrees very closely with the adjusted Chernoff bound while the Chernoff bound is about 0.5 to 1dB off. For smaller SNR's, the truncated series and the adjusted Chernoff bound begin to disagree. By examining the terms in the truncated series, it can be seen that in those cases the series has not sufficiently converged and thus the truncated series is unrelaible. The code used for this comparison is a constraint length, K=4, binary-to-8-ary code, whose transfer function can be obtained from (1).

Some further comparisons are illustrated in Figures 3 and 4 (again for the case of non-coherent detection with a square law metric). Figure 3 shows how varying the size of the signal set effects the performance for a fixed constraint length, while Figure 4 shows the effect of varying the constraint length for a fixed constellation size. It is seen that, as is usually the case with trellis-coded-modulation, there comes a point where there is little to gain by expanding the size of the signal set further. In the example shown in Figure 3 where the constraint length is fixed at K=5, this seems to occur when the size of the constellation reaches M=8. Note that this is four times the uncoded binary constellation. This is slightly different from the general rule of thumb for two-dimensional signal sets which says to double the size of the signal constellation [3]. It is expected that this phenomenon will become even more prevelant when the input is changed from binary to multiple bits (see [2], Section 2). It should be pointed out that the approximate transfer functions as given by equations (1) and (2) were used in all cases in Figures 3 and 4 with the exception of the K=5 binary-to-4-ary code (top curve in Figure 3). It was found that for this particular code, the

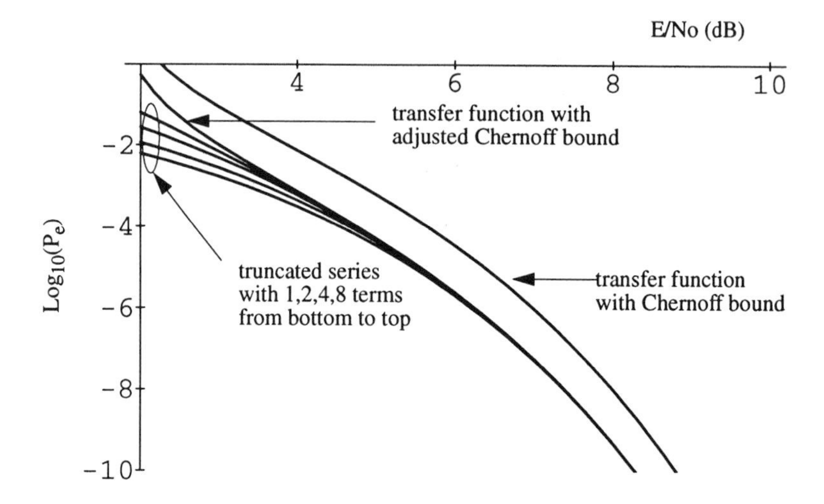

Figure 2 - comparison of performance bounds for K=4, rate-1 binary to 8-ary code.

actual information weight spectrum did not match very well that predicted by equation(2). By comparing the actaul information weight spectrum (as given partially in [1]) with what is predicted by equation (2), it was found that multiplying equation (2) by a factor of 3 resulted in an approximate information weight spectrum that matched the true information weight spectrum quite well.

3 - Trellis-Coded Orthogonal Signalling in a Rayleigh Fading Channel

The techniques applied to analyzing the performance of the orthogonal trellis-coded modulation can easily be applied to other types of channels as well. In a mobile radio environment, the channel is likely to exhibit slow Rayleigh fading. It is assumed the sufficient interleaving is used so that the fading is independent from symbol to symbol. In this case, the expression of (3) can be used with the exact two codeword error probability which is given by

$$P_2(w) = \left(2+\frac{E}{N_o}\right)^{-w}\sum_{j=0}^{w-1}\binom{w+j-1}{j}\left(1-\frac{1}{2+\frac{E}{N_o}}\right)^{j} \quad . \tag{11}$$

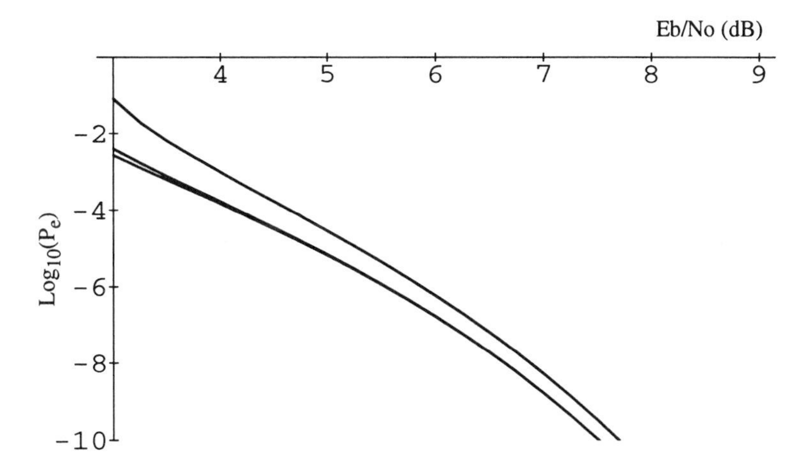

Figure 3 - Error Probability for K=5 binary to M-ary code over Gaussian noise channel with non-coherent (square law) detection; M=4, 8, 16 from top to bottom.

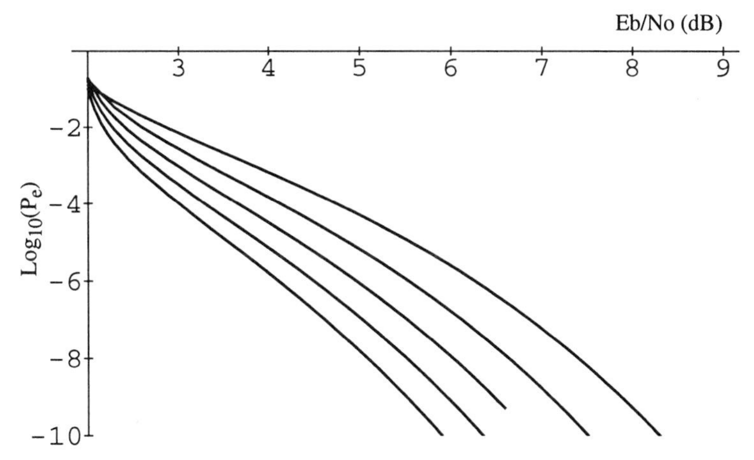

Figure 4 - Error Probability for binary to 16-ary code over Gaussian noise channel with non-coherent (square law) detection; K=4,5,6,7,8 from top to bottom.

The Chernoff bound for this two codeword error probability works out to be

$$P_2(w) \le \left(\frac{4\,(1+\frac{E}{N_o})}{(2+\frac{E}{N_o})^2} \right)^w . \tag{12}$$

As with the case of the Gaussian noise channel, this Chernoff bound tends to be rather loose, but an adjusted version seems to work quite well. In a manner entirely analogous to equation (10), the two-codeword error probability for the Rayleigh fading channel can be bounded very nicely by

$$P_2(w) \le P_2(d_{free}) \left(\frac{4\,(1+\frac{E}{N_o})}{(2+\frac{E}{N_o})^2} \right)^{w-d_{free}} , \qquad w \ge d_{free}. \tag{13}$$

Curves similar to Figures 2-4 can be generated for the Rayleigh fading channel using the above results. As an example, Figure 5 shows the performance of the same codes as used in Figure 4 for the case of independent Rayleigh faading.

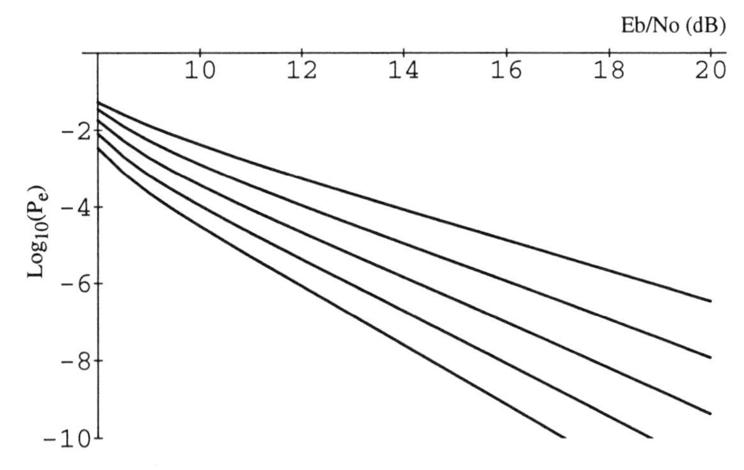

Figure 5 - Error Probability for binary to 16-ary code over Rayleigh fading channel with non-coherent (square law) detection; K=4,5,6,7,8 from top to bottom.

4 - Application to Direct-Sequence Code-Division Multiple-Access

Accurately tracking the effects of multiple-access interference in a DS-CDMA system can be quite tedious. Instead we take the approach of [4] and approximate the multiuser interference as Gaussian with a zero mean. All users signals are assumed to be received at an equal level (perfect power control for the case of the Gaussian noise channel), and the sequences used are approximated by truly random sequences for the purposes of calculating multiuser interference. It is easily found under these assumptions that the results of the previous sections can be applied if the value for E_b/N_o is replaced by an effective E_b/N_o given by

$$\left(\frac{E}{N_o}\right)_{eff} = \frac{\frac{E}{N_o}}{1 + \frac{2(K-1)}{3N}\frac{E}{N_o}} \quad . \tag{14}$$

In equation (14), K is the total number of users and N is the number of chips per symbol in the DS-SS waveform.

In order to keep the bandwidth of various systems fixed, the number of chips per symbol must be adjusted according to the size of the signal constellation and the type of code used. If N_b is the number of chips per bit for a binary uncoded system, then an uncoded system using M-ary modulation can use $N_b \log_2(M)$ bits/symbol. If binary-to-M-ary codes are used, the value of N stays at N_b since theis scheme also sends one information bit per symbol. When standard convolutional codes are used with a code rate of r, the value of N becomes rN_b for a binary signal set and $rN_b \log_2(M)$ for an M-ary convolutional code. To see the effects of the diffeing number of chips per symbol, Figure 6 shows the performance of several uncoded systems. In this figure, the binary system (M=2) used 127chips/bit, and as a result the M=4 and M=16 used 255 and 511 chips per symbol respectively. In all cases there were K=25 users (note K is being used here to represent the number of users, not the constraint length of the code). Note that 25 users is just too much for the binary system to handle (with only 127 chips per bit) and the error floor is unacceptably high. By going to larger signal constellations, the number of chips/per symbol is increased which reduces the level of the error floor.

Finally, Figure 7 shows the performance of the same codes as in Figure 4 but this time with multiple users (K=25, N=127 for all cases). Note that the performance is quite good and that the error flor has been lowered well below 10^{-10} in all cases. In the cases illustrated in Figures 6 and 7, the coded systems are quite superior to the uncode system, even the the uncoded system enjoys an advantage of a larger number of chips per symbol.

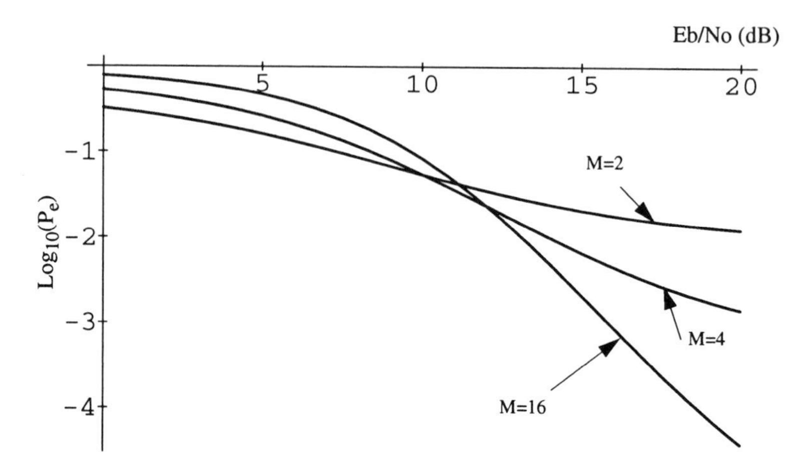

Figure 6 - Error Probability for uncode M-ary orthogonal CDMA; K=25, M=2,4,16, N=127, 255, 511, non-coherent detection, Gaussian noise channel.

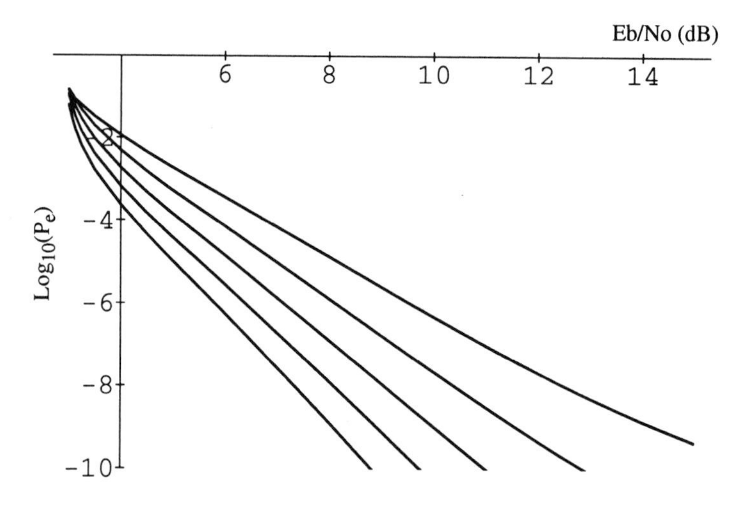

Figure 7 - Error Probability for binary to 16-ary code over Gaussian noise channel with non-coherent (square law) detection; 25 users, N=127chips/bit, K=4,5,6,7,8 from top to bottom.

References

[1] Ryan, W.E., Wilson, S.G., "Two Classes of Convolutional Codes Over GF(q) for q-ary Orthogonal Signalling, IEEE Transactions on Communications, vol. 39, no. 1, January 1991, pp30-40.

[2] Miller, S.L., "Design and Analysis of Trellis Codes for Orthogonal Signal Sets," to appear in IEEE Transactions on Communications.

[3] Ungerboeck, G., "Channel Coding with Multilevel/Phase Signals," IEEE Transactions on Information Theory, vol. IT-28, no. 1, Jan. 1982, pp55-66.

[4] Pursley, M.B., "Performance Evaluation for Phase-Coded Spread-Spectrum Multiple-Access Communication - Part I: System Analysis, "IEEE Trans. on Comm., vol. COM-25, no. 8, Aug. 1977, pp795-799.

[5] Viterbi, A.J., Omura, J.K., Principles of Digital Communication and Coding, McGraw-Hill, 1979.

[6] Blahut, R.E., Digital Transmission of Information, Addison-Wesley, 1990.

[7] Miller, S.L., "An Efficient Channel Coding Scheme for Direct Sequence CDMA Systems," in Proceedings of MILCOM '91, pp1249-1253.

[8] Bodreau, G.D., Falconer, D.D., Mahmoud, S.A., "A Comparison of Trellis Coded Versus Convolutionally Coded Spread Spectrum Mutiple-Access Systems," IEEE Journal on Selected Areas in Communications, vol. 8, No. 4, May 1990, pp628-640.

14

Noncoherent Spread Spectrum Receiver with Joint Detection and Synchronization using modified Matched FIR Filters

P.G. Schelbert, W.J Burmeister, and M.A Belkerdid

University of Central Florida
Electrical and Computer Engineering Department
Orlando, FL 32816-2450

-Abstract-

A noncoherent direct sequence spread spectrum DPSK receiver (DPSK-DSSS) is presented. The receiver is designed to operate in heavy multipath environment with a high number of multipath components and variable fade rates.

Asynchronous despreading of the received signal by two complementary modified matched FIR filters is followed by a joint data detection, and bit synchronization. Data detection is based on the nonlinear postdetection diversity combining of the two FIR filter outputs .

Bit synchronization is achieved by estimating the moment of the combined nonlinear weighted matched filter outputs over a window. The receiver is insensitive to various power delay profiles with severe multipath components. Fast synchronization within several databits is achievable which results in low delay of the datastream and fast startup. The performance of the receiver is evaluated using the discrete channel model proposed by Saleh et. al [1] using parameters extracted from measurements made in a large factory hall [2]. The variation of the BER performance is less than 2 dB over a range of different multipath channels assuming a delay spread of 1/10 the databit time.

I. Introduction

A wide range of different receiver structures that combat multipath effects are known. [3] gives an excellent overview of several methods. The well known RAKE receiver structure is nearly ideal as a multipath diversity combiner under the assumption that the (complex) multipath gains and delays are known. The estimation of the these variables, however, causes severe problems particularly in a relatively fast changing multipath environment. Also, conventional synchronization schemes for carrier synchronization, chip synchronization, and bit-clock synchronization require a significant amount of

* *This work was funded in part by the Swiss National Science Foundation and the AGEN Stiftung Berne, Switzerland.*

time to acquire lock. The approach presented here, known as postdetection multipath combining, integrates the energy that is spread over a time slot by the multipath channel. With noncoherent asynchronous operation, only bit synchronization is required. The power delay profile need not be estimated since it is tracked by the bit synchronization algorithm.

II. Multipath Channel Model

The discrete multipath channel approach, extensively studied in [3] [1], is used here. The complex baseband impulse response *g(t)* of the channel can be expressed as

$$g(t) = \sum_{k=1}^{k=K} g_k \, \delta(t - \tau_k) \, e^{j\Phi_k} \tag{1}$$

where g_k is the complex gain, τ_k the delay and Φ_k the random phase of the k-th path. The distribution of the phase is assumed to be uniform over [0, 2π].

The path gains g_k and delays τ_k are extracted from measured data [2]. The measurements [2] include an underground shopping mall, airport departure hall, a large office room, and a six story high factory hall. Similar measurements are also reported in [4]. The center frequency of the channel measurements is 900 MHz. The time resolution is 10 ns. Details of the measurement equipment and results are given in [2].

The strongest multipath effects are observed in the factory hall. Therefore, these channel characteristics will be considered here for the mulipath performance of the DPSK-DSSS spread spectrum system. The overall channel statistics for this factory environment are: Mean-Excess-Delay M = 122.3 ns (standard deviation = 51.1 ns), and RMS Excess-Delay-Delay-Spread S = 119 ns (standard deviation = 24.8 ns). The mean -Excess-Delay M is defined as

$$M = \int_0^\infty \tau \, g(\tau) \, d\tau \Big/ \int_0^\infty |g(\tau)|^2 d\tau \tag{2}$$

and the delay spread S is defined as

$$S = \sqrt{\left\{ \int_0^\infty (\tau - M)^2 |g(\tau)|^2 \, d\tau \Big/ \int_0^\infty |g(\tau)|^2 d\tau \right\}} \tag{3}$$

with *g(t)* the measured channel impulse response.

It was decided to choose representative measured impulse responses of spatially separated locations and to fit them to the above discrete channel model. Analysis of the data showed eleven discrete paths to be sufficient. The identification of the path gains g_k, and the associate delays τ_k, was performed in a manner similar to that in [1]. The eleven strongest paths were identified, and their complex gains and delays assigned to the corresponding gains and delays. The delay of the first path was

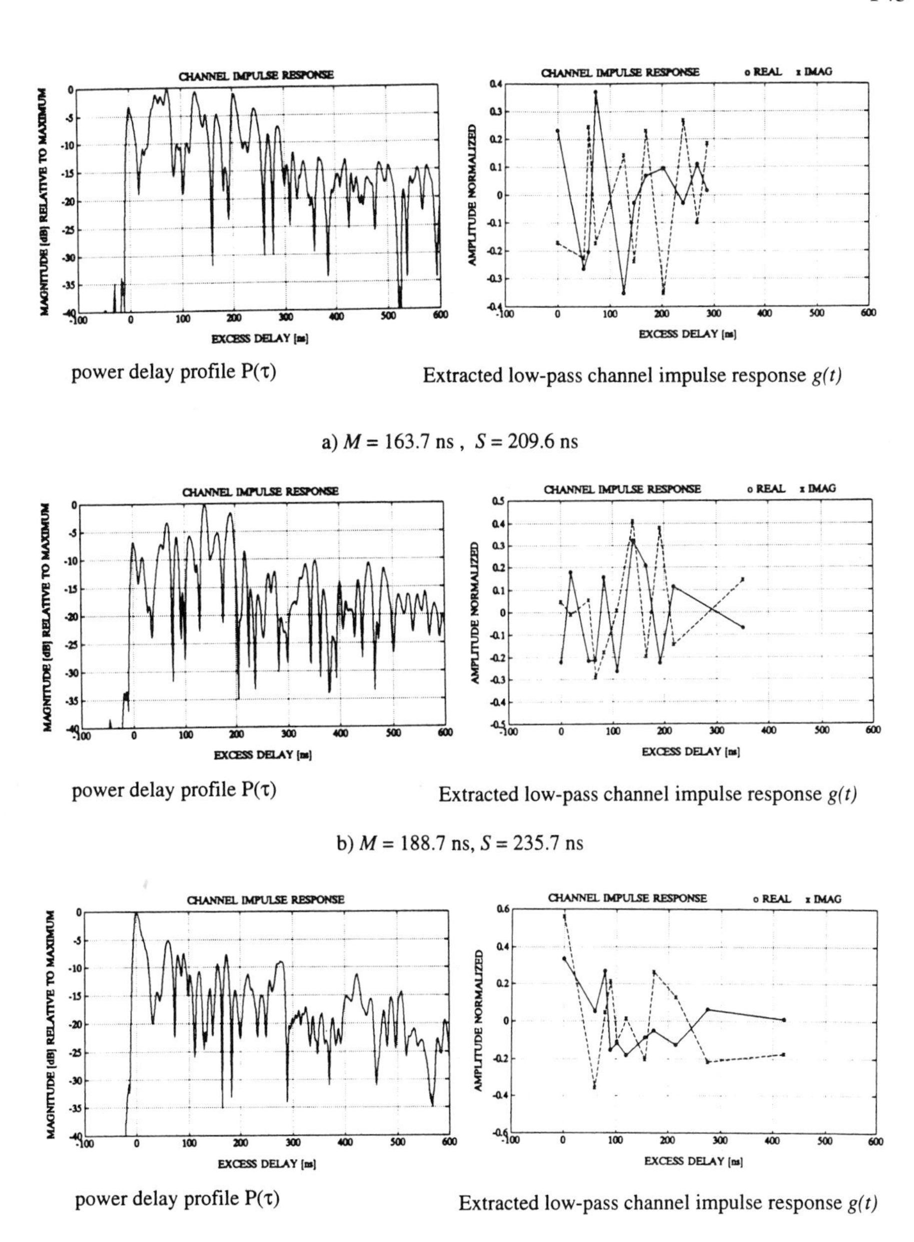

Fig.1 Measured power delay profiles of a six story high factory hall [2].

set to zero. The multipath power gain G was normalized to unity according to

$$G = \sum_k a_k^2 \equiv 1 \tag{4}$$

Fig.1a through c) shows three measured power delay profiles $P(\tau) = |g(\tau)|^2$ and the corresponding extracted complex lowpass channel impulse response $g(t)$ [2].

III. Transmitter

The transmitted signal for the DSSS-DPSK system can be expressed as

$$s(t) = Re \quad [x(t)\, e^{j2\pi f_c t}] \tag{5}$$

where f_c is the carrier frequency, and $x(t)$ is the baseband DS signal. For DPSK modulation, $x(t)$ *is* given by

$$x(t) = \sqrt{2P}\, b(t) W(t) u(t) \tag{6}$$

Where P is the power of the transmitted signal, $W(t)$ is the chip waveform of duration T_c , and $b(t)$ is the differentially encoded data $d(t)$ with duration T_b , and values { -1, 1}. With the L-bit spreading code $u(t)$, the duration of each chip is $T_b = L\ T_c$. The chip waveform is assumed to be a rectangular window of duration T_c. Fig.2 shows the block diagram of the transmitter.

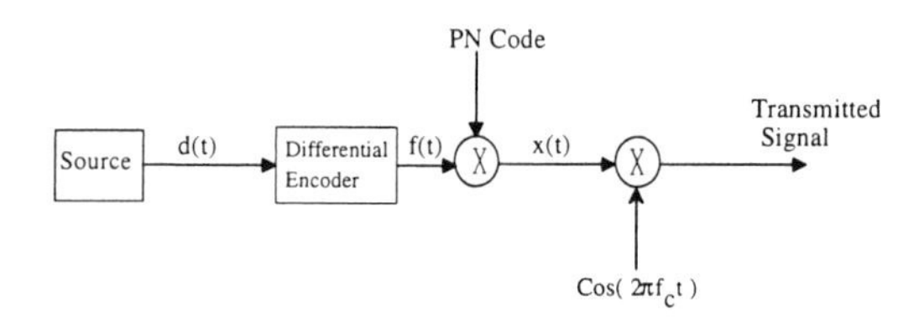

Fig.2 Transmitter block diagram

IV. Receiver Structure

The transmitted signal in (6) is assumed to be distorted by the channel given by (1) and additive white gaussian noise $n(t)$ (AWGN) of the two-sided power density spectral of $N_0 / 2$. The received signal $r(t)$ is given then by

$$r(t) = s(t) * g(t) + n(t) \tag{7}$$

where * denotes convolution.

Fig.3 shows the structure of the receiver. The received signal is divided and fed into two correlators. These correlators serve as noncoherent asynchronous despreaders as well as differential demodulators. The upper correlator A is matched to a waveform corresponding to a differentially encoded bit sequence $b(t)$ of {1 , 1} whereas the second correlator B is matched to the waveform corresponding to $b(t)$ with values {1 , -1}.

For each resolvable multipath, the outputs of the correlators deliver a correlation peak corresponding to the amplitude of the received multipath component. The multipaths are resolvable as long as $\tau_k - \tau_l > 1/B$ with B the bandwidth of the transmitted signal (i.e. as long as $\tau_k - \tau_l > T_c$) and $k \neq l$.

With impulse responses $h_A(t)$ and $h_B(t)$, and corresponding correlators A and B, the signal entering the decision circuit is:

$$z_{A,B}(t) = \int_{t-T_W}^{t} | [s(t) * g(t) + n(t)] * h_{A,B}(t) |^n \, dt \tag{8}$$

The window integrator is a moving averager which integrates the signal over the time window T_w. The integration window is set to the maximum expected excess delay of the multipath channel. The objective of the window integrator is to collect all the energy in the expected time slot T_w. The power of the signal is calculated by setting n = 2. The moving averager outputs are sampled once during each bit duration T_b. The sampling instant is estimated by the bit synchronization algorithm described below.

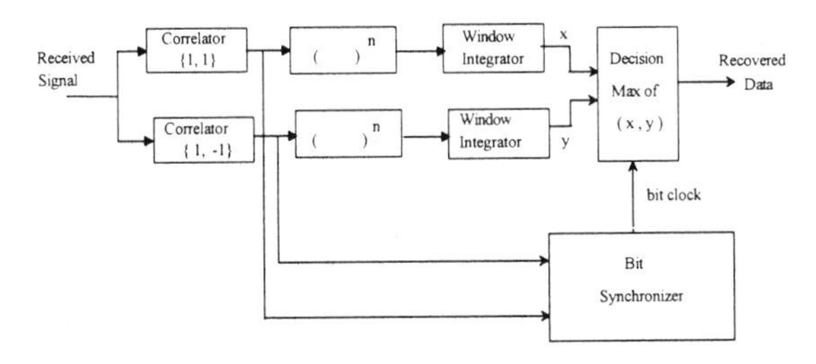

Fig.3 DSSS receiver: Block diagram.

1. Despreading and Demodulation

The despreading and differential demodulation is achieved by the two correlators. The two correlators are FIR-filters with the impulse responses $h_A(t)$ respective $h_B(t)$.

Two types of FIR-filter impulse responses have been studied. One approach uses matched filters, and the other uses inverse filters. Matched filter are known to maximize signal to noise ratio under white gaussian noise conditions. The autocorrelation of binary pseudorandom sequences (PN) have nonzero time sidelobes. The matched filter therefore does not minimize these sidelobes. An inverse filter minimizes the time sidelobes and their energy but sacrifices signal to noise ratio. The impulse response of an inverse filter corresponding to a PN-code, with values { 1, -1}, has nonzero coefficients over all time but with weights which are decreasing exponentially [5]. However, a truncation to a finite length results in a realizable FIR-filter. The performance figure for a PN-code u, and the matched filter is the processing gain $G_p = L$. The processing gain of a filter with impulse response h is $G_{u,h}$

$$G_{u,h} = [C_{u,h}(0)]^2 \;\; / \;\; \{u_{\max}^2 \sum_{i=-\infty}^{\infty} h^2(i)\} \tag{9}$$

with $C_{u,h}$ the aperiodic crosscorrelation of u and h. The processing gain $G_{u,h}$ for any other than the matched filter will be lower than the processing gain G_p. The noise degradation of the filter h is expressed by the noise enhancement factor $\kappa = G_p / G_{u,h} = L / G_{u,h}$ [6]. The relative sidelobe level is given by the peak correlation to max partial correlation ratio Δ

$$\Delta = \frac{C_{u,h}(0)}{\max[|C_{u,h}(k)|]} \qquad -\infty \leq k \leq \infty \qquad k \neq 0 \tag{10}$$

The sidelobe energy is expressed by the Golay [7] Merit factor F_g

$$F_g = [C_{u,h}(0)]^2 \;\; / \;\; \{\sum_{k=-\infty}^{\infty} C_{u,h}(k)^2\} \tag{11}$$

The matched filter for the sequence u(i) i = 1...L, is given by $h_m(i) = u(L+1-i)$. The inverse filter is given by $h_{iv}(i) = DFT^{-1}\{1/U(\omega)\}$ and $U(\omega) = DFT\{U(i)\}$ where *DFT* denotes the discrete Fourier Transform. The continous time versions of the above filters are denoted $h_m(t)$ and $h_{iv}(t)$ respectively.

The differential demodulators are now matched to a two bit time interval of the transmitted signal $x(t)$ according to the matched and inverse matched filter impulse responses $h_A(t)$,and $h_{Ai}(t)$ respectively given by

$$h_A(t) = h_m(t) + h_m(t - T_b) \qquad h_B(t) = h_m(t) - h_m(t - T_b) \tag{12}$$

$$h_{Ai}(t) = h_{iv}(t) + h_{iv}(t - T_b) \qquad h_{Bi}(t) = h_{iv}(t) - h_{iv}(t - T_b) \tag{13}$$

The correlator $h_A(t)$ despreads and demodulates the two differentially encoded databits with pattern {1 , 1}, and whereas the correlator $h_B(t)$ recognizes the pattern { 1, -1}.

Fig.4 shows the inverse filter for the 31-bit m-sequence from Pursley [8] (Fig A.1, K = 45). Fig.5 shows the truncation effect on the sidelobe level Δ, the sidelobe energy F_g, and the noise enhancement factor κ for the 31-bit m-sequence. The sidelobe level can be lowered to any desired level. The noise enhancement factor reaches an asymptotic value of about 2.1 dB. The sidelobe energy decreases monotonically with increased inverse filter length. A compromise leads to a filter length in this specific case of 63 bits. The impulse response of the correlator $h_A(t)$, based on the 31-bit inverse filter (length 63 bits), is shown in Fig.5.

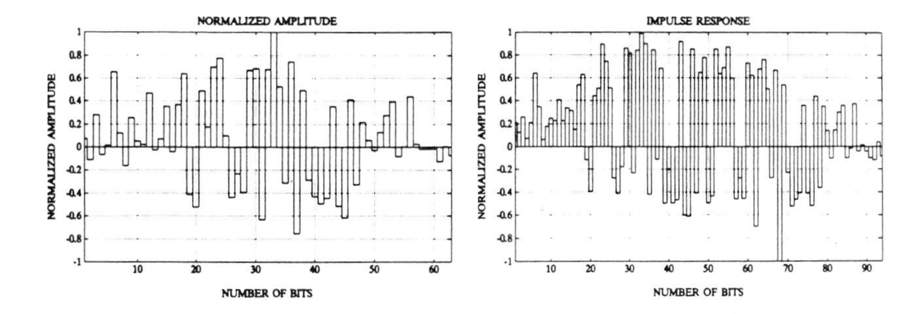

Fig.4. Impulse response $h_{iv}(t)$ of the inverse filter, truncated to 63-bits, code 31-bit m-sequence.

Fig.5. Differential demodulator $h_A(t)$ (FIR filter): Impulse response. Code 31-bit m-sequence.

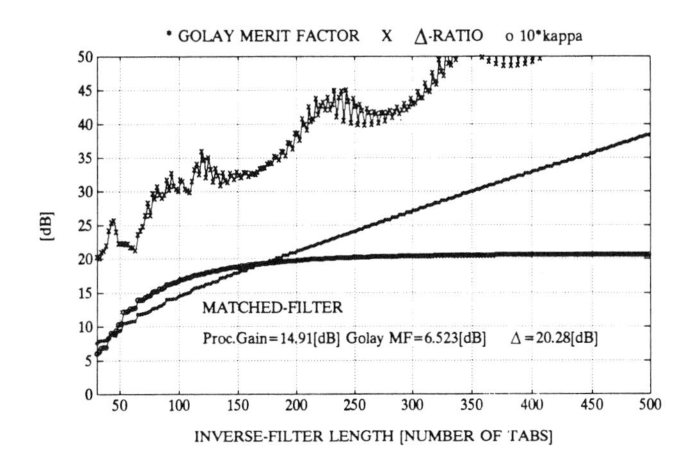

Fig.6. Golay Merit factor F_g, Processing gain G_p and noise enhancement factor κ for the inverse filter based on the 31-bit sequence. Matched filter: $G_p = 14.91$ dB, $F_g = 6.52$ dB, $\Delta = 20.28$ dB, Inverse Filter, length 63 bit: κ=1.27 dB, $F_g = 10.25$ dB, $\Delta = 22$ dB.

Fig.7 shows the responses of the matched filter based differential demodulators $h_A(t)$ and $h_B(t)$ to the arbitrary chosen data sequence $d(t)$. Fig.8 shows the same configuration but with differential demodulators derived from inverse filters. Fig.8 shows, on the average, a lower and a more uniform sidelobe level. As can be seen from Fig.7 as well as from Fig.8, the two parallel demodulators are complementary filters.

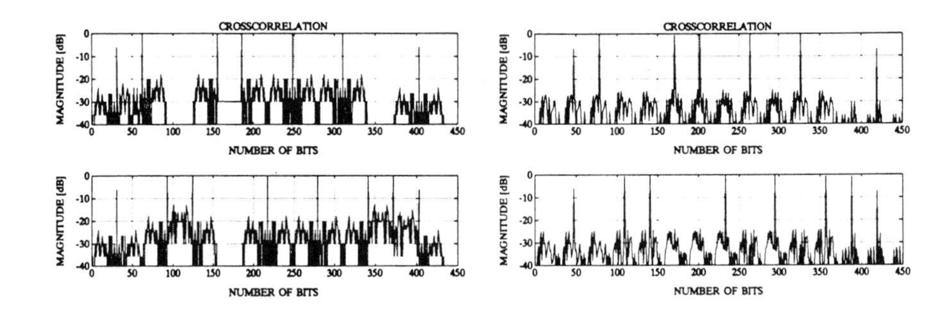

Fig.7 Matched filter based differential demodulators. Response to a data sequence $d(t)$ = 10011010100
PN-code: 31-bit m-sequence.
Upper: Output of demodulator $h_A(t)$
Lower: Output of demodulator $h_B(t)$

Fig.8 Inverse filter based differential demodulators. Response to a data sequence $d(t)$ = 10011010100
PN-code: 31-bit m-sequence.
Upper: Output of demodulator $h_A(t)$
Lower: Output of demodulator $h_B(t)$

2. Bit Synchronization and Decision

Due to the noncoherent and asynchronous operation, carrier recovery, and chip synchronization is not required. However, bit synchronization is still necessary. The following assumptions are made for the bit synchronization algorithm. The channel is not changing during m databits. The data bit time T_b is known. The operation begins with the summation of the demodulator outputs $y_A(t)$ and $y_B(t)$, and the averaging of the samples which are T_b apart over a time mT_b. A windowed moving first order moment computation localizes the center of the energy on the time axis. Each time a center of energy is detected, a bit clock pulse is generated. Proper time alignment has to be found while taking into account the demodulator delays, the window integrator time, and the bit synchronizer time offset.

The averaging is performed by successive multiplicative weighting of samples spaced T_b apart over m stages and integrating over a window corresponding to the maximal expected delay τ_k.

$$y_{av}(t) = \int_{t-T_w}^{t} \prod_{k=1}^{m} [y_A(t-kT_b) + y_B(t-kT_b)]^n dt \qquad (14)$$

where $y_A(t) = r(t) * h_A(t)$, and $y_B(t) = r(t) * h_B(t)$. The power weighting factor n is set to 2.

$$t_c = \int_{t-T_{w1}}^{t} t|y_{av}(t)|dt \;\; / \;\; \{ \int_{t-T_{w1}}^{t} |y_{av}(t)|dt \;\; \} - (2t - T_{w1})/2 \qquad (15)$$

For $t_c=0$, a local center of energy is assumed to be detected and a clock impulse is generated. The clock triggers the decision device, which follows the rule for the detected databit $\hat{d}_k$

$$\text{if} \quad z_A(t) > z_B(t) \;\; \text{then}\; \hat{d}_k = 1 \qquad \text{and} \;\; \text{if} \quad z_B(t) > z_A(t) \quad \text{then} \quad \hat{d}_k = 0 \qquad (16)$$

V. Performance in Multipath Environment

The performance of the receiver is evaluated using the three power delay profiles given in Fig.1. The selected profiles show strong multipath effects. The data rate is assumed fast in comparison to the receiver movement (i.e. the phase changes Φ_k are minor). This permits the angle Φ_k of (1) to remain constant during one simulation cycle. The bit error probability (BER) is calculated using Monte-Carlo simulation. The model is implemented using the commercial software package SPW [9]. The data rate has been chosen to be 427.3 kbits/s . The chip time is T_c= 78.4ns using the 31-bit m-sequence as spreading code. The relative delay spread $\mu = S_{mean} / T_b$ then becomes $\mu = 0.1$. S_{mean} denotes the average delay spread of the three power delay profiles.

The integration window T_w was set equal to $\tau_{k\max} = 420\, ns$ ($0.18\, T_b$) and $T_{w1} = T_b = 2.34\, \mu s$. The length m of the synchronizer was chosen to be m = 6. The performance of the synchronizer for both the externally supplied bitclock, and the estimated bitclock are plotted in Fig.9 - Fig.12. Fig.9 shows the BER for AWGN and the ideal channel $g(t)$ = 1. In Fig.10 through Fig.12, depict the BER performance for the three different multipath profiles shown in Fig.1a-c. The curves marked FP are the BER curves for a theoretical DPSK receiver using only the energy of the first path (FP). The DPSK curve, shown as reference, is the BER curve for an ideal DPSK receiver using the full transmitted energy. The BER achieved using the estimated bitclock is slightly degraded, especially in the high noise region, compared to perfect clock case. The BER performance is dependent on the power delay profile. The degradation is less than 2 dB relative to the performance for a perfect channel. A considerable improvement over the performance of a first path receiver can be seen. The synchronizer needs the received signal for a time equal to mT_b in order to achieve the best possible performance during startup. The databit delay in the full system is approximately the delay of the synchronizer plus the delay of the differential demodulators. The receiver structure has a rather fast synchronization time when compared to receivers with conventional synchronization structures.

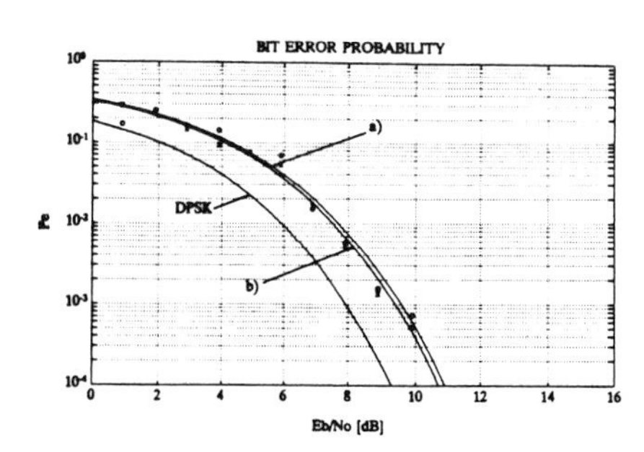

Fig.9. BER for a perfect channel. Performance under AWGN only.
FP: first path DPSK receiver.
a) Perfect synch. Fit to the simulated points"o" .
b) Estimated bitclock. Fit to simulated points "x".

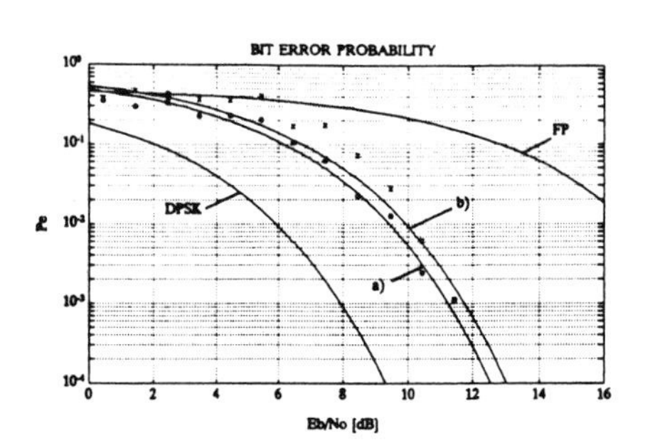

Fig.10. BER for the channel in Fig.1a) with M = 163.7 ns and S = 209.6ns.
FP: first path DPSK-receiver.
a) Perfect synch. Fit to simulated points "o"
b) Estimat. bitclock. Fit to simulat. points "x".

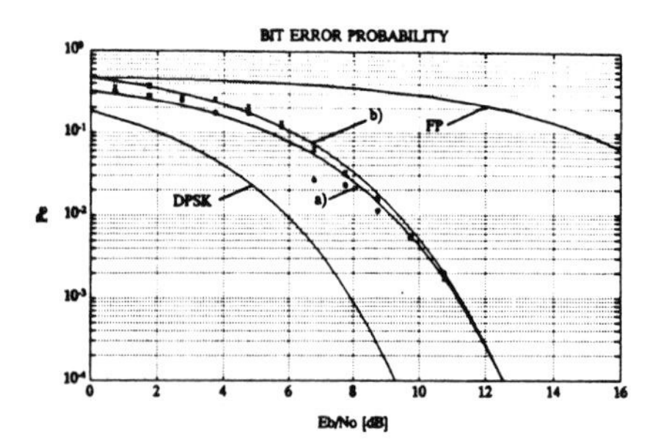

Fig.11. BER for the channel in Fig.b) with M = 188.7 ns and S = 235.7 ns
FP: first path DPSK-receiver.
a) Perfect synch. Fit to simulated points "o"
b) Estimat. bitclock. Fit to simulat. points "x".

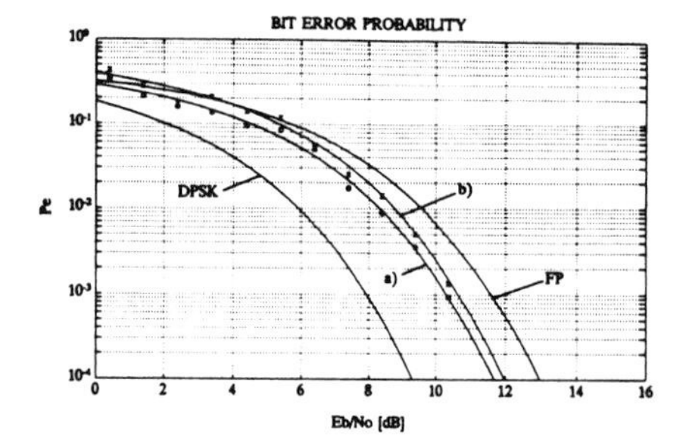

Fig.12. BER for the channel in Fig.1c) with M = 142.4 ns and S = 221.2 ns
FP: first path DPSK-receiver.
a) Perfect synch. Fit to simulated points "o"
b) Estimat. bitclock. Fit to simulat. points "x".

VI. Conclusions

The presented receiver allows exploitation of multipath gain. The noncoherent asynchronous principle used allows a very fast synchronization of the receiver, and a very low total delay of the databits. Although the performance does not reach the performance of a perfect RAKE receiver, the simplicity, and the insensitivity to different severe multipath power delay profiles show a vast improvement over a receiver without diversity gain (first path receiver). The receiver is especially useful for packet radio applications where fast synchronization and good diversity gain is required. The receiver is self adaptive, and will follow changing multipath profiles as long as the phase changes during the m databits, over which the synchronizer averages, are insignificant.

Acknowledgments

The authors are grateful to A. Radovic, E. Zollinger and P. Leuthold for providing the multipath channel data.

References

[1] A.A.M. Saleh, R.A. Valenzuela, "A Statistical Model for Indoor Multipath Propagation", IEEE JSAC, Vol. SAC-5, No.2, pp.128-137, Feb. 1987.

[2] Andrej Radovic, Ernst Zollinger, "Measured Time Variant Characteristics of Radio Channels in the Indoor Environment", Mobile Radio Conference MRC 1991, pp.267-274, 13.-15. Nov. 1991.

[3] G.L.Turin, "Introduction to Spread-Spectrum Antimultipath Techniques and their Application to urban Digital Radio", Proc. IEEE, Vol.68, pp.328-353, March 1980.

[4] Th. S. Rappaport, "Characterization of UHF Multipath Radio Channels in Factory Buildings", IEEE Transactions on Antennas and Propagation, Vol.37, NO. 8, pp. 1058-1069, Aug. 1989.

[5] J.P. de Weck, J. Ruprecht, "Real-Time Estimation of Very Frequency Selective Multipath Channels", IEEE Globecom 90, pp. 226-231, Dec. 1990.

[6] J. Ruprecht, F.D. Neeser, and M. Hufschmied, " Code Time Multiple Access: An Indoor Cellular System", IEEE Vehicular Technology Conference VTC"92, pp. 736-739, May 1992.

[7] M.J. E. Golay, " Sieves for Low Autocorrelation Binary Sequences", IEEE Transactions on Information Theory, Vol. IT-23, No. 1, pp. 43-51, Jan. 1977.

[8] M.B. Pursley, H.F.A. Roefs, "Numerical Evaluation of Correlation Parameters for Optimal Phases of Binary Shift-Register Sequences", IEEE Transactions on Comm., Vol. COM-27, No.10, pp.1597-1604, Oct. 1979.

[9] Comdisco Inc., "SPW Signal Processing Work System Manuals, Version 2.8", Comdisco Inc, Foster City CA, April 1991.

15

Optimum Acquisition Method Using Parallel Partial Correlator for Spread Spectrum Communication

Daeho KIM, Hoyoung KIM
ETRI, KOREA

Abstract

In spread spectrum communication systems, synchronization acquisition and its tracking affects system performance heavily. Then fast acquisition time, hardware simplicity, and low false alarm probability are of important concerns. Obviously, so many different synch schemes have been developed that we can choose suitable one depending upon purpose.

Especially concerning synch acquisition, correlator size should be carefully considered because it is directly related to acquisition time, hardware complexity, and false alarm probability. Therefore, in this paper, we present parallel partial correlator for synch acquisition and show the performance of this through computer simulation also. The Performance of partial correlator is evaluated in terms of acquisition time, index of discrimination, in AWGN as well as Rayleigh fading enviroment. It is shown that partial correlator reduces acquisition time and hardware complexity while maintaining low false alarm probability.

1. INTRODUCTION

In spread spectrum communication systems, transmission bandwidth is much wider than information signal's bandwidth, and the receiver despread the receiving spreaded siganl, through those process SS system has processing gain which enable SS system robust against interference and noise. The transmitter spreading TX bandwidth by multiply spreading code and receiver despreading RX bandwidth by multiply the same code used in transmitter. In these processing, synchronization affects system performance heavily especially acquisition method effect on acquisition time, hardware complexity, and tracking so carefull interest must be done. Those system like the half duplex SS transceiver and TDD SS systems, fast acquisition time is most important concern so transmitted reference or preamble methods are suggested and used depending on purpose.

This paper focus on fast acquisition using parallel partial correlator by use of PN code characteristics which in syncronization acquisition mode correlation window is smaller than the PN code period. For fast acquisition and H/W simplicity we chose PN code period L=127 and generated using stage=7 LFSR(linear feedback shift register) and to avoid initial shift register setting special generation ploynomial was selected. The PN code grneration polynomial is

$$G(x) = (x^7 + x^1 + 1) \quad (1)$$

The PN code generator is shown in Fig. 1. Through the simulation we analyze partial correlator output as Index of Discrimination vs correlator input SNR and correlation window size then find optimum correlator window size suitable to communication enviroment.

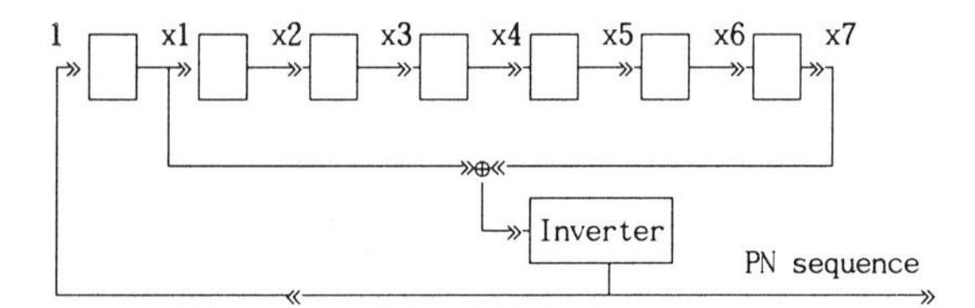

Fig. 1. A 7-stage shift register sequence generator for generation of a 127-bit PN sequence.

2. ACQUISITION USING CORRELATOR

In spread spectrum communication sync acquisition is the most important factor which determine system performance. When receiver despreading the receiving spreaded signal by multiply the same code used in spreading, if not synchronized less than 1 chip than receiver don't know input signal being. So to despread the spreaded received signal PN code generator in receiver must be synchronized to PN code generated in transmitter less than 1 chip and this is sync acquition. Most important factor which determine sync acquisition performance is accuracy and time. Accuracy in acquisition is determined by the difference of correlation value between synchronized and not synchronized i.e. correlation value difference has some value even though bit error occur in transmission for reduce false alarm probability. The autocorrelation value is showh in Fig.2 and autocorrelation function is

$$
\begin{aligned}
R_C(d) &= \sum_{n=0}^{n=L-1} \{c(n) \cdot c(n+d)\} \\
&= 2^{n-1} \qquad \text{for } d=0,\ L,\ 2L,\ \ldots \qquad (2) \\
&= -1 \qquad \text{Otherwise}
\end{aligned}
$$

The difference between the correlation value of a code at zero shift and the maximum correlation value at any nonzero shift(up to one period) is called the ID(*Index of Discrimination*). The larger the index of discrimination the better the code is suited for synchronization. For a maximal-length code, the index of discrimination is ID = $(2^n-1) - (-1) = 2^n$. Acquisition time in sync acquisition is

$$
T_A = \frac{T_c L}{1-P_{fd}} + (L-1)T_c P_{fa} - \frac{T_c L}{2} \qquad (3)
$$

where T_c is the time of M chip test per one time so if transmission rate=R_c, then $T_c=M/R_c$, P_{fd} is false dismissal probability, and P_{fa} is false alarm probability. From (2),(3) the longer the PN code period the better for acquisition but the longer time take. So tradeoff between fast acquisition time and acquisition accuracy is necessary in the half duplex transceiver or TDD system which require fast acquition time. So some method to reduce acquisition time and hardware complexity while maintaining acquisition accuracy.

3. ACQUISITION USING PARALLEL PARTIAL CORRELATOR

To achieve fast acquisition time and hardware simplicity while acquisition accuracy we use parallel partial correlator. So if there exist enough partial ID(ID_p) for communication enviroment which correlation value of a code at zero shift and maximum correlation value at any nonzero shift for correlator size K($n<K\leq L$, n=# of shift register, L=PN code period) then we can design partial correlator which reduce hardware complexity and acquisition time while maintaining accuracy. Partial correlation function is

$$
\begin{aligned}
R_C(d) &= \sum_{n=0}^{n=K-1} \{c(n) \cdot c(n+d)\} \\
&= K \qquad \text{for } d=0,\ L,\ 2L,\ \ldots \qquad (4) \\
&= S \qquad \text{Otherwise}
\end{aligned}
$$

and

$$ID_p = K - S_{max} \quad (5)$$

where S_{max} is maximum correlation value at any non zero shift. ID_p of partial correlator vary with correlation window K and communication enviroment. When bit error occurs then the nonzero shift partial correlation value K decrease 2 times of number of bit errors and S_{max} increases in general, so S_{max} value can greater than the value K when communication enviroment is become poor and correlation window size is too small. The partial correlation window size K must be variable to maintain low false alarm probability for various enviroments. In communication enviroment relative index of discrimination(IDr) is

$$ID_r = \frac{A_{max} - S_{max}}{A_{max}} \cdot 100(\%) \quad (6)$$

where A_{max} is correlation value of a code at zero shift when bit error occurs. The greater ID_r the better the correlator is suited for the acquisition and the more complex the hardware and the more take acquisition time. Fig. 3 shows paralle partial correlator using DMF(digital matched filter). A matched filter computes the correlation between the signal and the reference code once per chip. Consequently , code tracking is automatic and code acquisition is instantaneous, but can only be used with short sequence or parts of sequence because of high power and complexity. So optimum correlation size in various enviroments are of important concern.

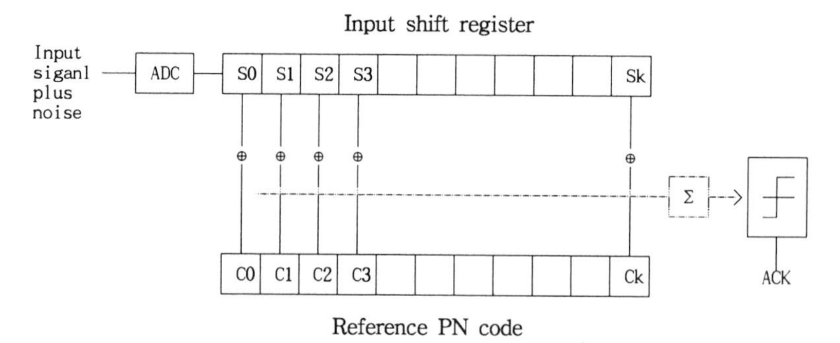

Figure 3 Parallel partial correlator using digital matched filter.

4. SIMULATION RESULTS

Through the computer simulation to find optimum correlation window size we analyze partial correlator output as relative index of discrimination for input SNR in AWGN enviroments. The simulations are repeated 1000 times for each window size and averaged. The results are presented in Figure 4 through 6 plus Tables 1 and 2 and individual simulation results deviation is ±3% from averaged values. The following observations can be made:

o Parallel partial correlator can be used for acquisition in various enviroment.

o At constant ID_r, required correlation window size increase nonlinearly with the input SNR. For example 70% ID_r, window size L=40 for 9 dB, L=60 for 5 dB, L=85 for 1 dB.

o Half of PN code period window size conform 60 % ID_r for noisy enviroments.

o Above than window size L=50, ID_r increase is saturated, so unconditionally increasing window size L is not advisable.

o At constant window size L=64, ID_r = 84 % for 9 dB, ID_r=78 % for 5 dB, ID_r=72 % for 3 dB can be obtained.

Table 1. ID_r for input SNR and correlation window size K. [%]

Input SNR[dB]	Average # of bit error	Partial correlation wnidow size(K)					
		24	32	64	88	96	127
1	16.9	8.18	36.9	64.2	71.5	72.8	79.1
3	10.1	31.7	49.1	72.4	78.6	80.3	85.3
5	4.7	39.3	55.3	77.7	84.8	86.5	93.6
7	1.5	40.5	59.9	81.9	88.6	90.1	98.3
9	0.27	41.1	62.1	84.1	90.6	91.5	99.7

Table 2. Partial correlation window size[K] for obtaining ID_r vs input SNR.

Input SNR[dB] / ID_r	1	3	5	7	9
25[%]	24	19	13	13	13
50[%]	43	35	26	26	26
75[%]	105	79	53	51	50
90[%]	-	-	120	100	86

In the Rayleigh fading enviroments if fc=900 MHz, mobile unit speed is 30 km/h, threshold level -15 dB, then average duration of fade t(-15 dB) = 2.88 ms and level crossing rate n(-15 dB) = 10.76 s^{-1}. If transmission rate is 1 Mcps then approximately 2880 bit will be lost during each period in which 22 PN period. Under these conditions, it will be necessary to repeat the 127-bit PN sequence at least 24 times to ensure synchronization of the received signal. This requirement is based on the following equation;

$$\frac{127 + 2880 \text{ bit}}{127} \approx 24$$

So the minimum PN sequence repeation time is 24*127 us = 3.05 ms and then reliable sychronization can be made.

5. CONCLUSION

This paper presents the results of computer simulation of the parallel partial correlator in DSSS communication system in which K, the correlation window size is small than L, the period of PN code for the sync acquisition which enable fast acquisition and reduce hardware complexity. Simulation results show that parallel partial correlator has good performance in sync acquisition, for example K=35 for SNR=3 dB, K=26 for SNR=5 dB is enough to attain ID_r = 50 % in WGN enviroments. These parallel partial correlation method may be useful for fast synchronization in DSSS communication systems.

6. REFERENCES

[1] The American Radio Realy League, Inc, Spread spectrum sourcebook, 1991.

[2] Robert C. Dixon, Spread spectrum systems 2nd ed., Jhon Wiely & Sons, 1984.

[3] Ferrel G. Stremler, Introduction to communication Systems, Addison Wesley, 1982.

[4] W.C.Y. Lee, Mobile Communications Engineering, McGraw Hill, 1982.

[5] R.L. Pickholtz, D.L. Schilling. "Theory of spread spectrum communications- A Tutorial," IEEE Trans. Comm. Vol COM-30, pp. 855-884, May 1982.

[6] Laurence B. Milsein, John Gevargiz, and Pankaj K. Das, "Rapid acquisition for direct sequence spread spectrum communications Using parallel SAW convolvbers," IEEE Trans. Comm. VOL COM-33, NO. 7, July 1985.

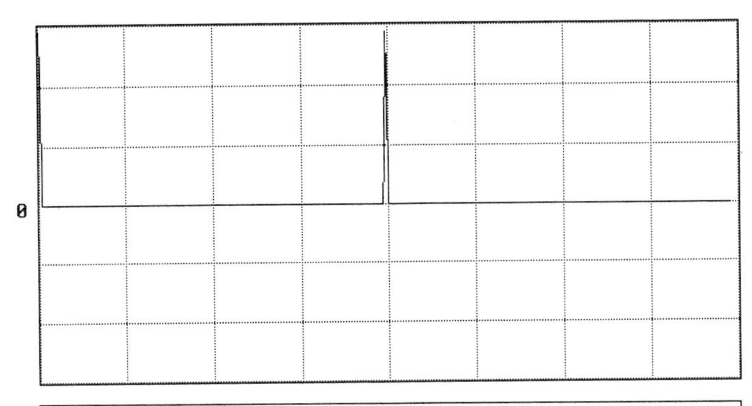


```
Max = 127.000000   Min = -1.000000

Number of one  ==> 63
Number of zero ==> 64

Auto correlation coefficient
Polar binary   ==>  Max = 127   Min = -1
```

Figure 2. The correlation of a 127-bit PN sequence

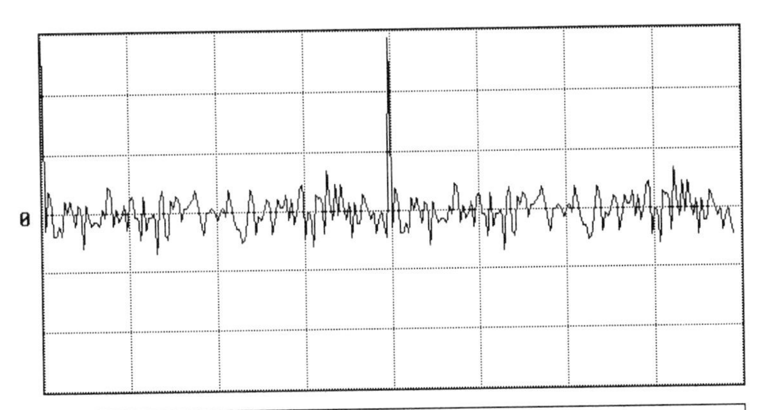


```
Max = 76.000000   Min = -18.000000

Input PN code SNR ==> 2 dB
Correlator length ==> 96

Zero shift value  ==> 76.000000
Alpha max value   ==> 18.000000
Index of Disc.    ==> 76.315789[%]
```

Figure 4. The partial correlation of a 127-bit PN sequence(K=96, SNR=2dB)

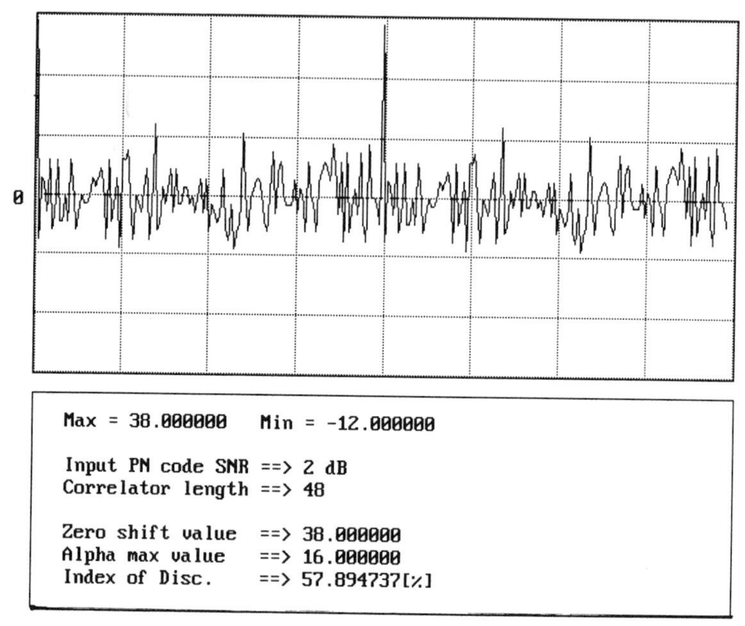

Figure 5. The partial correlation of a 127-bit PN sequence(K=48, SNR=2dB)

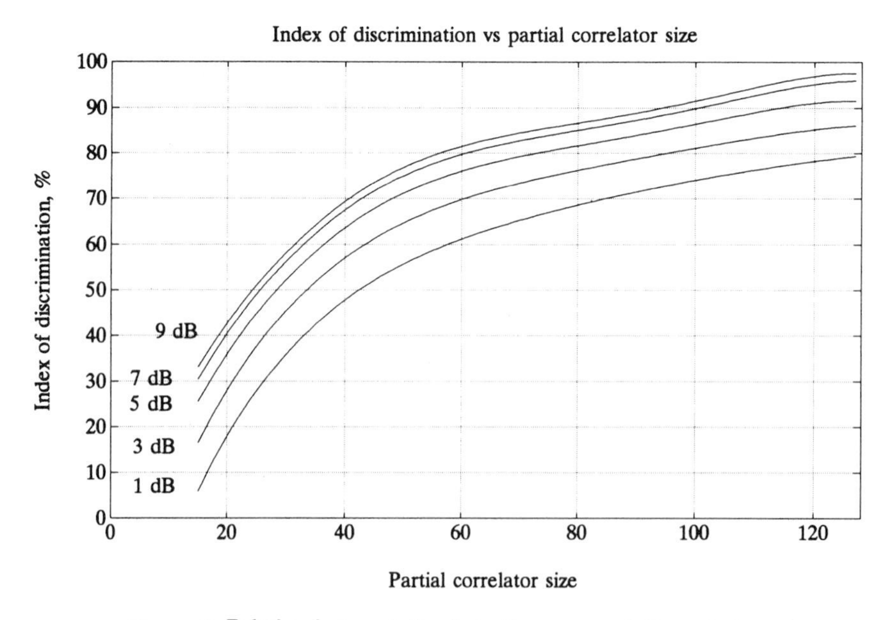

Figure 6. Relative index of discrimination vs correlation window size and input SNR

16

Importance Sampling Methodologies for Simulation of Wireless Communication Links

Wael A. Al-Qaq, Michael Devetsikiotis and J. Keith Townsend
Center for Communications and Signal Processing
Department of Electrical & Computer Engineering
North Carolina State University, Raleigh, North Carolina 27695-7914

Abstract

Characteristics that make bit error rate (BER) analysis of wireless communication links difficult include: multipath Rayleigh fading channels, diversity combining, adaptive equalization, intersymbol interference (memory), error control coding, and trellis coded modulation. Although Monte Carlo simulation is a useful and general approach for BER analysis of many voice-band systems with these impairments, the technique is not feasible for systems with low BER's due to long runtimes.

Importance sampling (IS) techniques offer the potential for large speedup factors for Monte Carlo simulation of low BER systems. The basic idea behind importance sampling is to artificially modify, or "bias" the probability distributions to yield more events of interest (i.e., bit errors) than occur naturally. Appropriate "weights" are then used to yield an unbiased estimator. It is crucial to employ a proper IS biasing scheme and to choose "good" biasing parameter values, otherwise the required simulation run length for a given accuracy could be larger than for conventional Monte Carlo.

This paper summarizes IS methodologies we have developed [1, 2] for Monte Carlo simulation of communications systems with the following characteristics: links with linear adaptive equalizers and slowly-varying channels, and links with both coherent and non-coherent envelop detection with diversity combining in the presence of nonselective Rayleigh fading. Simulation results are also given for these systems.

1 Introduction

Performance analysis of time-varying digital communication links is a difficult problem. This difficulty is mainly attributed to time variations in the channel impulse response and the complexity of the algorithms used in mitigating the effects of these time variations. In these communication links, adaptive equalization, diversity schemes, error control coding, and trellis coded modulation can be used to provide the low BER's required for future wireless data applications. Formula-based analysis of the bit error rate (BER) of these links is not always tractable. Monte Carlo simulation must be used to obtain bit error rate (BER) estimates of the system. Although Monte Carlo simulation is a useful and general approach for BER analysis of many voice-band systems with these impairments, the technique is not feasible for systems with low BER's due to long runtimes.

[1]This research was sponsored in part by an IBM Fellowship. Details of the methodologies presented here can be found in [1, 2].

Importance sampling (IS) techniques offer the potential for large speed-up factors for BER estimation using MC simulation [3, 4, 5, 6]. Before [1], IS techniques had not been used to speed-up MC simulation of communication links with time-varying channels and the techniques mentioned above. The basic idea behind importance sampling is to artificially modify, or "bias" the probability distributions to yield more events of interest (i.e., bit errors) than occur naturally. Appropriate "weights" are then used to yield an unbiased estimator. It is crucial to employ a proper IS biasing scheme and to choose "good" biasing parameter values, otherwise the required simulation run length for a given accuracy could be larger than for conventional Monte Carlo.

In contrast to earlier methods for locating near-optimal IS parameter values for the simulation of *static* communication channels [7, 8], this paper summarizes IS methodologies we have developed [1, 2] for Monte Carlo simulation of communication systems with the characteristics found in wireless communication links. These characteristics include: links with linear adaptive equalizers and slowly-varying channels, and links with both coherent and non-coherent envelop detection with diversity combining in the presence of nonselective Rayleigh fading. Simulation results are also given for these systems.

The first IS methodology discussed is denoted here as the "twin system" (TS) methodology [1] which we use for MC simulation of communication links characterized by time-varying channels and continuously adapting linear equalizers. This methodology was developed to resolve a fundamental complication in the inherent nature of continuously adapting equalizers which possess a cumulative memory that extends back to time zero. The deleterious effects of large memory on IS are well known. Also, the adaptive equalizer taps, and thus the optimal IS parameter values, are a function of time.

A key feature of the TS method is that biased noise samples are input to the adaptive equalizer, but the equalizer is only allowed to adapt to these samples for a time interval equal to the memory of an equivalent system model. The memory for this equivalent system model remains constant. At the end of this interval, a decision is collected. The adaptive equalizer taps are then reset to values which would have resulted had there been no IS. A new optimal IS biasing value for the next vector of samples is calculated, and the process (a "biasing cycle") is repeated. Equalizer tap values are transferred from the original system to the twin system at the beginning of each biasing cycle.

The second IS methodology used for MC simulation of a communication system with diversity combining and slow nonselective Rayleigh fading channel is denoted here as the stochastic gradient descent (SGD) algorithm [2]. In this algorithm, minimization of the variance of the IS BER estimator is achieved by using a descent in the cost function (variance of the IS estimator) in a direction suggested by the gradient of the cost function with respect to the IS parameters. Since for most practical cases, the gradient of the variance of the IS estimator is not available in a closed form expression, an unbiased estimate of the gradient of the variance is used.

The SGD algorithm can be started at an arbitrary point where there is a sufficient "raw" (i.e., unweighted) error count. This feature yields a better starting estimate of the gradient of the variance to be minimized. Also contributing to a better gradient estimate is the fact that the SGD methodology is used in conjunction with an IS methodology derived in [9]. These two

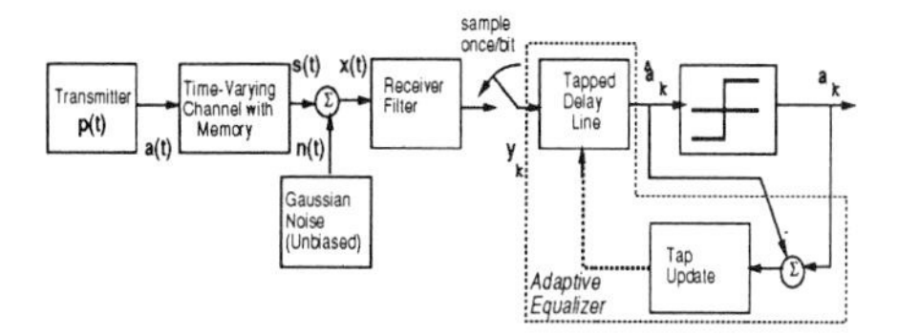

Figure 1: Block diagram of a digital communication system which includes a time-varying channel with memory and an adaptive equalizer [1].

methodologies were combined using a conditional importance sampling scheme. In this scheme, the thermal additive white Gaussian noise (AWGN) is conditionally biased using a mean translation (MT) biasing scheme [6, 9], while the multiplicative Rayleigh attenuation is biased using a variance scaling (VS) biasing scheme [4] (since MT can not be used to bias a Rayleigh distributed random variable). In most practical communication systems with Rayleigh fading channels, the Rayleigh fading process is more dominant than thermal AWGN.

Experimental results are presented that show runtime speed-up factors of 2 to 7 orders of magnitude for a static linear channel with memory, and of 2 to 4 orders of magnitude for a slowly varying, random linear channel with memory for the TS methods. Also, speed-up factors of 2 to 7 orders of magnitude for diversity systems in nonselective Rayleigh fading are achieved using the SGD algorithm.

2 Adaptive Systems

2.1 Equivalent Adaptive System Model

Fig. 1 shows the block diagram of a digital communication system in which a linear LMS adaptive equalizer is used to compensate for the distortion caused by the transmission medium (channel). In order to develop a model of our adaptive system suitable for use with IS, we need to express the final output $\hat{a}_k$ as a function of the random inputs $x(l)$, $l = 0, \ldots, k$, to the receiver filter. This function must have components in both the analog and the digital parts of the receiver, posing certain modeling difficulties. We overcome these difficulties as follows: Let M_2 be the receiver filter memory in bits. Denote the receiver filter taps by $\mathbf{q} = [q_1, q_2, \ldots, q_{M_2L}]$ where L is the number of samples per bit. Let $\mathbf{S}$, $\mathbf{N}$, and $\mathbf{X}$ denote the $1 \times \hat{M}$ *vectors of bit-length vectors* $\mathbf{s}(i)$, $\mathbf{n}(i)$, and $\mathbf{x}(i)$, respectively, with $\mathbf{X}(k) = \mathbf{S}(k) + \mathbf{N}(k) = [\mathbf{x}(k) \,|\, \mathbf{x}(k-L) \,|\, \ldots \,|\, \mathbf{x}(k - (\hat{M}-1)L)]$ and $\mathbf{x}(i) = \mathbf{s}(i) + \mathbf{n}(i) = [x(i), \ldots, x(i-L+1)]$, $i = k, k-L \ldots, k-(\hat{M}-1)L$, where $\hat{M} = M + M_2 - 1$ bits (corresponding to $\hat{M} \times L$ samples). Let $\mathbf{A}$ be the $\hat{M}L \times M_2$ matrix

$$\begin{bmatrix} \mathbf{q}^T_{M_2L\times 1} & \mathbf{0}_{L\times 1} & \cdots & \mathbf{0}_{L\times 1} \\ \mathbf{0}_{L\times 1} & \mathbf{q}^T_{M_2L\times 1} & \cdots & \mathbf{0}_{L\times 1} \\ \cdot & \mathbf{0}_{L\times 1} & \cdots & \cdot \\ \cdot & \cdot & \cdots & \cdot \\ \cdot & \cdot & \cdots & \mathbf{0}_{L\times 1} \\ \mathbf{0}_{L\times 1} & \mathbf{0}_{L\times 1} & \cdots & \mathbf{q}^T_{M_2L\times 1} \end{bmatrix}$$

then the equivalent adaptive system response at instant k, $\mathbf{h}_{eq}(k)$, which combines the receiver, the sampler, and the adaptive equalizer is given by the following recursive equation

$$\mathbf{h}_{eq}(k) = \mathbf{h}_{eq}(k-1) + \mu e(k-1)\mathbf{X}(k-L)\mathbf{A}\mathbf{A}^T \tag{1}$$

with $\hat{a}_k = \mathbf{X}(k)\mathbf{h}_{eq}^T(k)$. Thus, $\mathbf{h}_{eq}(k)$ is a function of $\mathbf{X}(k-L)$ and $\mathbf{h}_{eq}(k-1)$. But $\mathbf{h}_{eq}(k-1)$ is a function of $\mathbf{X}(k-2L)$ and $\mathbf{h}_{eq}(k-2)$ and so on, therefore we can conclude that

$$\mathbf{h}_{eq}(k) = \mathbf{g}(\mathbf{x}(0), \ldots, \mathbf{x}(k-L)) = \mathbf{g}'(\mathbf{X}_C(k-L), \mathbf{X}(k-L)) \tag{2}$$

where $\mathbf{g}$ and $\mathbf{g}'$ are functionals dependent on $\mathbf{A}$ and μ and $\mathbf{X}_C(k) = [\mathbf{x}(k-\hat{M}L)|\ldots|\mathbf{x}(0)]$ is the vector "complementary" to $\mathbf{X}(k)$.

2.2 The Importance Sampling Estimator

For the time-varying communication system in question, we assume that the discrete random process $I(\mathbf{X}(k)\mathbf{h}_{eq}^T(k))$ is ergodic for $k_0 < k \leq K$, where k_0 is chosen to be greater than the training period and K is some large number. $I(.)$ is an indicator function equal to 1 when the detected bit is in error, and 0 otherwise. Using this ergodicity assumption, we show in [1] that an IS-based, MC estimator of the probability of error P_e is

$$\hat{P}_e = \frac{1}{K-k_0}\sum_{k=k_0+1}^{K} I(\mathbf{X}^*(k)\mathbf{h}_{eq}^{*T}(k)) \prod_{j=0}^{j=k} w_{x(j)}(x^*(j)) \tag{3}$$

where $\mathbf{X}^*(k)$ denotes the biased vector $[\mathbf{x}^*(k)\,|\,\mathbf{x}^*(k-L)\,|\,\ldots\,|\,\mathbf{x}^*(k-(\hat{M}-1)L)]$ and the asterisk of $\mathbf{h}_{eq}^*(k)$ implies that the equalizer taps *have been adapting* to the *biased* random numbers $x^*(0), \ldots, x^*(k)$. $\prod_{j=0}^{j=k} w_{x(j)}(x^*(j)) = \prod_{j=0}^{j=k} f_{x(j)}(x^*(j))/f_{x(j)}^*(x^*(j))$ is the IS cumulative weight function when the sequence of input data $x^*(j), j = 0, 1, ..., k$, are independent.

3 Twin System IS Methodology

3.1 Twin System IS

To overcome the difficulties introduced by the increasing cumulative memory involved in (3) we proposed a block biasing approach in [1], namely the *twin system* (TS) method: When only $\mathbf{X}(k)$ is biased, denoted by $\mathbf{X}^*(k)$, $\mathbf{h}_{eq}^*(k)$ is the value assumed by the *resulting biased random vector of taps*: $\mathbf{h}_{eq}^*(k) = \mathbf{g}(\mathbf{x}(0), \ldots, \mathbf{x}(k-\hat{M}L), \mathbf{x}^*(k-(\hat{M}-1)L), \ldots, \mathbf{x}^*(k-L))$. Thus, at each instant k the random vector $\mathbf{X}(k)$ is biased while the random vector $\mathbf{X}_C(k)$ is left unmodified. Therefore, under the ergodicity assumption, we have the following IS estimator of the probability of error P_e

$$\begin{aligned}\hat{P}_{e,TS} &= \frac{1}{K-k_0}\sum_{k=k_0+1}^{K} I(\mathbf{X}^*(k\hat{M}L)\mathbf{h}_{eq}^{*T}(k\hat{M})) \times \\ &\quad w_{\mathbf{X}_C(k\hat{M}L),\mathbf{X}(k\hat{M}L)}(\mathbf{X}_C(k\hat{M}L), \mathbf{X}^*(k\hat{M}L))\end{aligned} \tag{4}$$

Furthermore, if the random vectors $\mathbf{X}_C(k)$ and $\mathbf{X}^*(k)$ are statistically independent, the weight function becomes $w_{\mathbf{X}_C(k),\mathbf{X}(k)}(\mathbf{X}_C(k), \mathbf{X}^*(k)) = w_{\mathbf{X}(k)}(\mathbf{X}^*(k)) = f_{\mathbf{X}(k)}(\mathbf{X}^*(k))/f_{\mathbf{X}(k)}^*(\mathbf{X}^*(k))$ since $\mathbf{X}_C(k)$ is left unbiased. The terms in the summation of (4) are *not independent* since the sequence

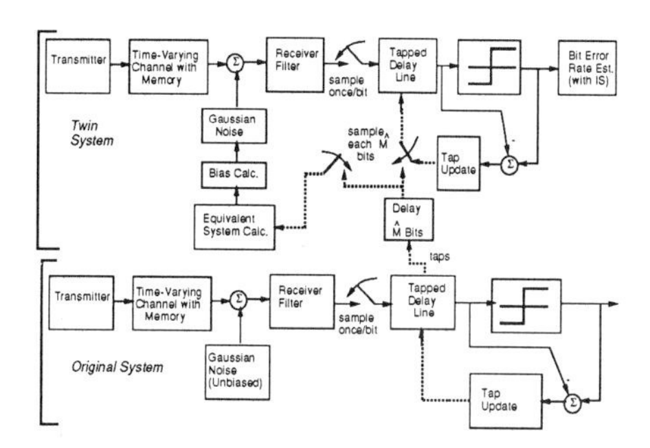

Figure 2: Block diagram of the twin system simulation approach. Equalizer taps from the original, unbiased system are used by the twin system to reset the equalizer at the beginning of each biasing cycle [1].

of random vectors $\{\mathbf{h}_{eq}^{*T}(k\hat{M})\}$, $k_0 < k \leq K$ is not independent. This will contribute to an increase in the MC estimator variance. To improve the estimator variance, we let decisions be spaced $\hat{M}$ bits (corresponding to $\hat{M}L$ samples) apart to guarantee that the sequence of random vectors $\{\mathbf{X}(k\hat{M}L)\}$ is i.i.d. [5]. We call the above estimator of (4) the *twin system* (TS) estimator, since it can be implemented as follows: Maintain two "parallel" versions of the system model. The *original* system is allowed to evolve without IS biasing. In the so-called "twin system", biased noise samples are input to the adaptive equalizer, but the equalizer is only allowed to adapt to these samples for a time interval equal to the memory $\hat{M}$ of the equivalent system $\mathbf{h}_{eq}$. Fig. 2 shows the block diagram of our simulation approach. It is important to note that, for the TS estimator in (4) to be statistically unbiased, the modified signal $\mathbf{X}^*(k)$ must be allowed to go through the process of adaptation, as suggested by the form of $\mathbf{h}_{eq}^*$.

3.2 Optimal IS Biasing for the TS Estimator

The choice of $f^*_{\mathbf{X}(k)}$ determines the magnitude of the estimator variance and the resulting speed-up factor over conventional MC. We restrict our attention to the *translation* biasing scheme, which has demonstrated the best improvement in the past for similar (although not time-varying or adaptive) models [6, 10, 7, 8]. In the translation scheme, we define the biased random vector as $\mathbf{X}^*(k) = \mathbf{X}(k) + \mathbf{c}(k)$. We call $\mathbf{c}(k)$ the translation vector at the kth instant. The optimal value C_{opt} under these conditions, was also found in [6].

In our setting, the optimal translation vector needs to be re-evaluated for every block of data, because the channel impulse response, and the mean vector are potentially changing with time.

4 Experimental Results: Twin System

In order to verify the effectiveness of applying the proposed IS scheme to adaptive systems, several simulation experiments were performed. The technique was applied to a system with a 6-tap LMS adaptive equalizer. The step size of the algorithm was found for each experiment by estimating the power of the signal at the input of the equalizer.

System	n_d	$\hat{P}^{*}_{e,TS}$	Speed-up
Static $\alpha = 0.94$	833	1.6×10^{-5}	3.8×10^{2}
Static $\alpha = 0.928$	1300	2.7×10^{-8}	1.5×10^{5}
Static $\alpha = 0.92$	1500	9.7×10^{-11}	3.4×10^{7}
Time-Varying $\alpha = 0.925$	833	1.0×10^{-5}	10^{2}
Time-Varying $\alpha = 0.9$	1500	2.7×10^{-9}	1.82×10^{4}

Table 1: Simulation data and speed-up factors for static and time-varying systems with an adaptive equalizer. Importance sampling based on the twin system approach was used in the simulations [1].

Two channels were considered, a static linear channel, and and a randomly time-varying channel. The static channel had a normalized impulse response $v(l) = \alpha^l/[\sum_{l=0}^{M_1 L-1} \alpha^{2l}]^{1/2}$, $l = 0, \ldots, M_1 L-1$ with $\alpha < 1$ a user-selected parameter.

The time-varying channel consisted of a uniformly distributed multiplicative factor followed by a static linear channel similar to the above: $s(l) = v(l) * a(l)r(l)$ where $r(l)$ was uniformly distributed, centered at 1.0, with a total spread of 0.6, $a(l)$ is the discrete-time transmitted signal, and $v(\cdot)$ is the same as above. Our artificial random "fading" was made *slow* by holding the same value of $r(l)$ for the entire duration of a bit.

In both types of channels, a memory of two bits ($M_1 = 2$) was considered. Baseband binary signaling was used in the examples, with levels of $-A$ and $+A$ corresponding to 0 and 1 data values respectively, with a signal level $A = 5$. The Gaussian noise was zero mean with variance $\sigma^2 = 0.5$. The receiver filter was a one bit ($M_2 = 1$) matched filter. As stated earlier, the equalizer memory length was $M = 6$. Throughout our experiments we used $L = 9$ samples per bit. The signal-to-noise ratio (SNR) for the example systems was $SNR \approx 27$dB. In each experiment, the adaptive equalizer was initialized to zero, then trained for a short period of time (100 $\sim$ 200 bits).

In the twin system experiments, $\hat{P}_{e,TS}$ were obtained according to (4) for each channel configuration. Decisions were spaced $\hat{M}$ bits (where $\hat{M}$ is the memory of the equivalent system $\mathbf{h}_{eq}$). The noise vector $\mathbf{N}(k)$ was biased along the direction of the equivalent impulse response $\mathbf{h}_{eq}(k)$.

Fig. 1 summarizes our experimental observations. In calculating confidence coefficients and speed-up factors we replaced the "true" BER (which was unknown) by the TS estimate. Between 833 and 1500 decisions per run (n_d), with 50 runs were made for each system in order to obtain the sample variance, $\hat{\sigma}^2$.

The measured overhead of the twin system technique (twin system implementation, optimal translation calculations etc.) was modest, reducing the net speed-up by roughly a factor of 2 ($\approx 40/72$) for a 6-tap equalizer, and a factor of approximately 3.5 ($\approx 41/145$) for a 10-tap equalizer.

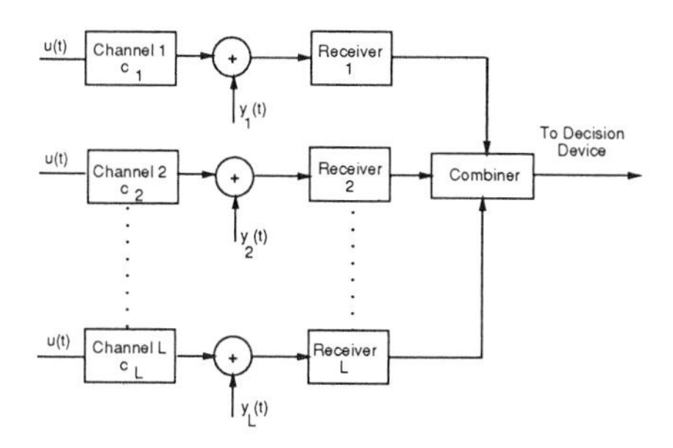

Figure 3: Model of a digital communication system with diversity reception [2].

5 Diversity Systems with Nonselective Rayleigh Fading

5.1 System Description

Fig. 3 shows the block diagram of a digital communication system in which Lth order diversity reception is used to compensate for the distortion caused by a non-selective fading channel and additive noise. A frequency non-selective channel results in a multiplicative distortion of the transmitted signal. Assuming that the channel fades slowly, the multiplicative process may be regarded as a constant for the duration of at least one signaling interval. Thus, if the transmitted signal is $u(t)$, then the received equivalent complex lowpass signal on the ith diversity over one signaling interval $(0 \leq t \leq T)$ is given by $r_i(t) = c_i u(t) + y_i(t)$, where $i = 1, \ldots, L$. For all i, $1 \leq i \leq L$, $c_i = \mathbf{x}_i \exp(-j\phi_i)$ is the fading gain of the ith fading channel which is a complex Gaussian random variable (CGRV) with $E\{c_i\} = 0$ and $E\{|c_i|^2\} = E\{\mathbf{x}_i^2\} = 2\sigma^2$. The random variable $\mathbf{x}_i$ is Rayleigh distributed and the random variable ϕ_i is uniformly distributed on the interval $[-\pi, \pi]$. For $i \neq j$, $\{c_i\}$ and $\{c_j\}$ are independent and identically distributed (i.i.d.) CGRV sequences. The complex additive noise on the ith antenna is given by $y_i(t)$, where each $y_i(t)$ is a complex AWGN with $E\{y_i(t)\} = 0$ and $E\{y_i(t)y_i^*(t+\tau)\} = 2Y_0\delta(\tau)$ (* denotes the complex conjugate), with $y_i(t)$ and $y_j(t)$ being independent processes for $i \neq j$.

For coherent detection, we consider binary PSK signaling. Let $\mathbf{X}_c = [\mathbf{x}_1, \mathbf{x}_2, \ldots, \mathbf{x}_L]$ be the vector of all the i.i.d. Rayleigh random variables (RRV's), and $\mathbf{Y}_c = [\mathbf{y}_1, \mathbf{y}_2, \ldots, \mathbf{y}_L]$ be the vector of all the i.i.d. zero-mean GRV's with variance $2Y_0 \int_0^T |u(t)|^2 dt = 2Y_0\varepsilon$ (the subscript "c" implies coherent detection). Then the output decision variable for a maximal ratio combiner U_c is given by

$$U_c = G_c(\mathbf{X}_c, \mathbf{Y}_c) = \sum_{i=1}^{L} g_c(\mathbf{x}_i, \mathbf{y}_i) \tag{5}$$

where $g_c(\mathbf{x}_i, \mathbf{y}_i) = \varepsilon \mathbf{x}_i + \mathbf{y}_i$.

For noncoherent detection, we consider envelope detection [11] with binary orthogonal FSK signaling. Let the sequences $\{\mathbf{x}_{1i}\}$ and $\{\mathbf{x}_{2i}\}$ be i.i.d. RRV's, with $E\{\mathbf{x}_{1i}^2\} = 2\sigma^2$ and $E\{\mathbf{x}_{2i}^2\} = 2Y_0\varepsilon$, and let the sequences $\{\mathbf{y}_{1i}\}$ and $\{\mathbf{y}_{2i}\}$ be i.i.d. zero-mean GRV's, with $E\{\mathbf{y}_{1i}^2\} = E\{\mathbf{y}_{2i}^2\} = Y_0\varepsilon$. Then if we let $\mathbf{X}_{nc} = [\mathbf{x}_{11}, \mathbf{x}_{21}, \mathbf{x}_{12}, \mathbf{x}_{22}, \ldots, \mathbf{x}_{1L}, \mathbf{x}_{2L}]$ and $\mathbf{Y}_{nc} = [\mathbf{y}_{11}, \mathbf{y}_{21}, \mathbf{y}_{12}, \mathbf{y}_{22}, \ldots, \mathbf{y}_{1L}, \mathbf{y}_{2L}]$ the decision variable for the noncoherent case U_{nc} (the subscript "nc" implies noncoherent envelope detection) will be given by

$$U_{nc} = G_{nc}(\mathbf{X}_{nc}, \mathbf{Y}_{nc}) = \sum_{i=1}^{L} g_{nc}(\mathbf{x}_{1i}, \mathbf{x}_{2i}, \mathbf{y}_{1i}, \mathbf{y}_{2i}) \tag{6}$$

where for $i = 1, \ldots, L$

$$g_{nc}(\mathbf{x}_{1i}, \mathbf{x}_{2i}, \mathbf{y}_{1i}, \mathbf{y}_{2i}) = \sqrt{(\varepsilon \mathbf{x}_{1i} + \mathbf{y}_{1i})^2 + \mathbf{y}^2{}_{2i}} - \mathbf{x}_{2i} \tag{7}$$

5.2 The Importance Sampling Estimator

To simplify the notation we drop the subscripts "c" and "nc" unless the discussion in this Section is related to a specific detection case (coherent or noncoherent). For the communication system in question, we show in [2] that an unbiased estimator $\hat{P}$ of the probability of a detection error P (assuming 1 was transmitted) is given by

$$\hat{P} = \frac{1}{N_X} \sum_{j=1}^{N_X} \hat{P}(X(j)) w_{\mathbf{X}}(X(j)) \tag{8}$$

where $w_{\mathbf{X}}(X) = f_{\mathbf{X}}(X)/f^*_{\mathbf{X}}(X)$, and $\hat{P}(X(j))$ is the conditional IS estimator given $X(j)$,

$$\hat{P}(X(j)) = \frac{1}{N_Y} \sum_{i=1}^{N_Y} I(G(X(j), Y(i,j))) w_{\mathbf{Y}|\mathbf{X}}(Y(i,j)|X(j)) \tag{9}$$

where $w_{\mathbf{Y}|\mathbf{X}}(Y|X) = f_{\mathbf{Y}|\mathbf{X}}(Y|X)/f^*_{\mathbf{Y}|\mathbf{X}}(Y|X)$, and $I(G(X(j), Y(i,j)))$ is an indicator function equal to 1 when $G(X(j), Y(i,j)) < 0$, and 0 otherwise. The estimator in (8) suggests the following conditional IS scheme: Independent samples $X(j), j = 1, \ldots, N_X$ are drawn from a marginal biased density $f^*_{\mathbf{X}}(X)$. For each fixed j, independent samples $Y(i,j)$, $i = 1 \ldots, N_Y$, are drawn from a biased conditional density $f^*_{\mathbf{Y}|\mathbf{X}}(Y|X)$.

For a fixed $\alpha_o > 0$, the empirical precision of the estimator in (8) may be found by using the sample variance estimator

$$\hat{V^*}\{\hat{P}\} = \frac{1}{N_X^2} \sum_{j=1}^{N_X} \hat{P}^2(X(j)) w^2_{\mathbf{X}}(X(j)) - \frac{\hat{P}^2}{N_X} \tag{10}$$

The simulation is terminated when the condition $\sqrt{\hat{V^*}\{\hat{P}\}}/\hat{P} < \alpha_o$ is satisfied.

Our goal is to select the simulation distributions $f^*_{\mathbf{Y}|\mathbf{X}}(Y|X)$ and $f^*_{\mathbf{X}}(X)$ to minimize the variance of the estimator in (8), namely $V^*\{\hat{P}\}$. In [2] we show that the asymptotically optimal $f^*_{\mathbf{Y}|\mathbf{X}}(Y|X)$ is given by $f^*_{\mathbf{Y}|\mathbf{X}}(Y|X) = f_{\mathbf{Y}}(Y - \mu(X))$.

For the coherent detection case $\mu_c(X) = [\mu_1, \ldots, \mu_L]$ where $\mu_i = -\frac{1}{L}\sum_{k=1}^{L} \varepsilon x_k$, and for the noncoherent detection case, $\mu_{nc}(X) = [\mu_{11}, \mu_{21}, \ldots, \mu_{1L}, \mu_{1L}]$ where $\mu_{1i} = -\frac{1}{L}\sum_{k=1}^{L} \varepsilon x_{1k} - x_{2k}$, and $\mu_{2i} = 0$.

6 The Stochastic Gradient Descent (SGD) Algorithm

After choosing $f^*_{\mathbf{Y}|\mathbf{X}}(Y|X)$, our next goal is to determine the marginal simulation distribution $f^*_{\mathbf{X}}(X)$ in order to minimize $V^*\{\hat{P}\}$, which for most practical cases, does not have a closed form expression. We choose to use a robust stochastic algorithm of the Robbins-Monro (RM) type. In the context of discussing this algorithm, variance scaling (VS) is considered as a biasing scheme (however, this algorithm applies to other parametric biasing schemes as well). Thus, our task reduces to specifying the IS variance parameters of the random vector $\mathbf{X}$, namely, the parameter vector σ^* that will minimize the cost function $V^*\{\hat{P}\}$.

For simplicity, note that the notations $V^*\{\hat{P}\}$ and $w_{\mathbf{X}}(X)$ do not explicitly show the dependence of the variance and weight function, respectively, on the parameter vector σ^*. Let the gradient of $V^*\{\hat{P}\}$ with respect to σ^* be denoted as $\nabla_{\sigma^*}V^*\{\hat{P}\}$. In seeking a minimum, the optimization algorithm we use involves a descent of the cost function $V^*\{\hat{P}\}$ in a direction given by $\nabla_{\sigma^*}V^*\{\hat{P}\}$. Since in most practical applications, a closed form expression of $\nabla_{\sigma^*}V^*\{\hat{P}\}$ is not available, we use an unbiased estimate $\widehat{\nabla}_{\sigma^*}V^*\{\hat{P}\}$ of the gradient. Replacing the deterministic gradient with its unbiased stochastic estimate, results in the following stochastic gradient descent (SGD) algorithm

$$\sigma^*(k+1) = \sigma^*(k) - \beta(k)\widehat{\nabla}_{\sigma^*}V^*\{\hat{P}\}|_{\sigma^*=\sigma^*(k)} \tag{11}$$

where $\beta(k)$ is the step size taken at the kth iteration. The algorithm in (11) is of the RM type [12, 13]. Our approach is to use (11) to specify the optimal IS parameters, namely σ^*_{opt}. The almost sure convergence (i.e., $\text{Prob}\{\lim_{k\to\infty}\sigma^*(k) = \sigma^*_{opt}\} = 1$) of an RM type of algorithm like the one in (11) is proved in [12] provided that a proper step size $\beta(k)$ is chosen. In [14], the RM algorithm was used to perform a stochastic steady-state optimization of regenerative systems with a step size $\beta(k) = 1/k$.

In [2] we show that $\nabla_\sigma V^*\{\hat{P}\}$ can be estimated using the following *unbiased* estimator

$$\widehat{\nabla}_{\sigma^*}V^*\{\hat{P}\} = \frac{1}{N_X N_G}\sum_{l=1}^{N_G}\hat{P}^2(X(l))w_{\mathbf{X}}(X(l))\nabla_{\sigma^*}w_{\mathbf{X}}(X(l)) \tag{12}$$

For an Lth order diversity system, observe that the decision variable in (5) is the sum of the i.i.d. r.v.'s $\{g_c(\mathrm{x}_i, \mathrm{y}_i)\}$. Therefore, since the random variables $\{\mathrm{y}_i\}$ are equally biased, the random variables $\{\mathrm{x}_i\}$, $i = 1, \ldots, L$ are equally biased as well. This reduces an L-dimensional search in the space of σ^* to a one-dimensional search. Similarly, by applying the above argument to the decision variable in (6), a $2L$-dimensional search is collapsed to a 2-dimensional search.

7 The SGD Algorithm and Experimental Results

Let the signal to noise ratio (SNR) be defined as $\gamma_c = \frac{1}{2}E\{\mathrm{x}_i^2\}/2Y_0\varepsilon$, and $\gamma_{nc} = \frac{1}{2}E\{\mathrm{x}_{1i}^2\}/2Y_0\varepsilon$, for the coherent and noncoherent detection cases, respectively. For a given SNR per diversity and a fixed $\alpha_o > 0$, the algorithm starts with a first order diversity and progresses to higher order diversities. Starting with a first order diversity and with an initial simulation distribution $f^*_{\mathbf{X},1}(X)$, which corresponds to $k = 1$ and $\sigma^*(1)$, we proceed as follows:

- For $k = 1, 2, \ldots$
  - For $l = 1, 2, \ldots, N_G$
    * Sample $X(l)$ from $f^*_{\mathbf{X},k}(X)$
    * Compute $w_{\mathbf{X}}(X(l))$
    * Compute $\mu(X(l))$
    * For $i = 1, \ldots, N_Y$
      · Using $\mu(X(l))$, sample $Y(i)$ from $f_{\mathbf{Y}}(Y - \mu(X(l)))$
      · If an error is detected, compute the weight $w_{\mathbf{Y}|\mathbf{X}}(Y(i)|\mathbf{X} = X(l))$
      · Next i
    * Compute $\hat{P}(X(l))$ using (9)
    * Next l
  - Compute $\hat{P}$ using (8), and $\hat{V}^*\{\hat{P}\}$ using (10)
  - if $\alpha(k) = \sqrt{\hat{V}^*\{\hat{P}\}}/\hat{P} \leq \alpha_o$ stop, else
  - Compute $\widehat{\nabla}_{\sigma^*} V^*\{\hat{P}\}|_{\sigma^*=\sigma^*(k)}$ using (12), and and $\sigma^*(k+1)$ using (11)
  - Next k

In general, the above algorithm may be used to simulate an Lth order diversity system with the starting point being the optimal simulation density of a diversity system of order $L-1$ or $L-2$. This approach will effectively place the starting point in a neighborhood close to the optimal, thereby reducing the number of iterations required to locate the near-optimal IS parameters. Fixed step sizes of $\beta_c \leq 10^{-2}\sigma^*(1)/||\widehat{\nabla}_{\sigma^*} V^*\{\hat{P}\}|_{\sigma^*=\sigma^*(1)}||$, and $\beta_{nc} \leq 10^{-2}\min\{\sigma_1^*(1), \sigma_2^*(1)\}/||\widehat{\nabla}_{\sigma^*} V^*\{\hat{P}\}|_{\sigma^*=\sigma^*(1)}||$, for coherent and noncoherent detections, respectively, were experimentally observed to guarantee the convergence of the SGD algorithm. Choosing β is a design issue based on trade-off between accuracy and speed.

Since using a suboptimal estimate of $\nabla_\sigma V^*\{\hat{P}\}$ as an initial value for the algorithm in (11) is sufficient to provide near-optimal IS parameters, only $N_G = 100$ decisions were used in (12) to estimate P and $\nabla_{\sigma^*} V^*\{\hat{P}\}$ at each iteration. It was empirically observed that the variance of $\widehat{\nabla}_{\sigma^*} V^*\{\hat{P}\}|_{\sigma^*=\sigma^*(k)}$ improves as k increases (i.e., as the minimum is approached).

The results of applying the above algorithm to second and fourth order diversity systems for both detection cases, along with the optimal IS parameters are shown in Table 2 for a per-diversity SNR of 20 dB. For each case, $N_X \times N_Y = 1000 \times 1$ was used in the simulations. Since the asymptotically optimal translation $\mu(X)$ is determined in closed form, rather than numerically as in [9], the cost of sampling $\mathbf{Y}$ is insignificant. Therefore, we chose $N_Y = 1$ for the simulations. This choice of N_Y is justified by recalling that, for most practical cases, $\mathbf{Y}$ is not the main contributor of randomness in the communication system being considered. In each case, the transmit filter was normalized ($\varepsilon = 1$). $\hat{V}^*\{\hat{P}\}$ was calculated by averaging an ensemble of $N_E = 20$ estimates of P.

The speed-up factor (Sp) was calculated according to

$$Sp = \frac{N_{MC}}{N_X N_Y N_E + N_{oh}}$$

Diversity Order	σ^*_{opt}	$\hat{P}$	Speed-up(Sp)
Coherent			
2	.0763	2.483×10^{-5}	5.418×10^{3}
4	.083	3.1804×10^{-9}	1.86×10^{7}
Noncoherent			
2	[.0802, .0705]	7.9825×10^{-5}	3.92×10^{2}
4	[.0844, .0686]	2.5928×10^{-8}	6×10^{5}

Table 2: Simulation data and speed-up factors for coherent and noncoherent reception with diversity. A SNR of 20 dB per diversity was used in the simulations [2].

N_{MC} is the conventional MC number of decisions required to attain the same accuracy as our IS scheme. N_{MC} was computed based on a 95% confidence interval [5]. N_{oh} is the overhead of the SGD algorithm in number of decisions and is given by $N_{oh} = N_L \times N_G$, where N_L = total number of iterations needed to locate σ^*_{opt} of an Lth order diversity system. The number of iterations required to determine the starting point is also included in N_L. The overhead factor N_{oh} includes the computational effort required to locate σ^*_{opt} for a first order diversity, which is used as $\sigma^*(1)$ for a second order diversity. This is overly conservative if one is interested in performance of the system for *all* orders of diversity up to the L-th order. From the tables, observe the large improvement factors over conventional MC. Empirically, the overhead of the SGD algorithm reduced the speed-up results by a factor ranging from 36 to 81.

8 Conclusion

In this paper we have summarized IS-based methodologies for speeding up Monte Carlo simulations of communication links with adaptive systems and diversity combining. The objective of this work is to provide a means for performance analysis of wireless communication systems with low BER, such as those envisioned for future wireless data applications.

Currently, we are extending the formulation to include multipath Rayleigh fading with adaptive equalization.

References

[1] W. A. Al-Qaq, M. Devetsikiotis, and J. K. Townsend. Importance Sampling Methodologies for Simulation of Communication Systems with Time-Varying Channels and Adaptive Equalizers. *IEEE J. Select. Areas in Commun.*, 11(3):317–327, Apr. 1993.

[2] W. Al-Qaq, M. Devetsikiotis, and J. K. Townsend. Simulation of Communication Systems with Slow Rayleigh Flat Fading and Diversity Using a Stochastically Optimized Importance Sampling Technique. To be presented at the IEEE Global Telecom. Conf., GLOBECOM '93, Houston, Texas, Dec. 1993.

[3] P. Balaban. Statistical Evaluation of the Error Rate of the Fiberguide Repeater Using Importance Sampling. *Bell Sys. Tech. J.*, 55(6):745–766, Jul.-Aug. 1976.

[4] K. S. Shanmugan and P. Balaban. A Modified Monte-Carlo Simulation Technique for the Evaluation of Error Rate in Digital Communication Systems. *IEEE Trans. Commun.*, COM-28(11):1916–1924, Nov. 1980.

[5] M. C. Jeruchim. Techniques for Estimating the Bit Error Rate in the Simulation of Digital Communication Systems. *IEEE J. Select. Areas Commun.*, SAC-2(1):153–170, Jan. 1984.

[6] D. Lu and K. Yao. Improved Importance Sampling Technique for Efficient Simulation of Digital Communication Systems. *IEEE J. Select. Areas Commun.*, 6(1), Jan. 1988.

[7] M. Devetsikiotis and J. K. Townsend. A Useful and General Technique for Improving the Efficiency of Monte Carlo Simulation of Digital Communication Systems. In *Proc. of IEEE GLOBECOM '90*, San Diego, CA, 1990.

[8] M. Devetsikiotis and J. K. Townsend. An Algorithmic Approach to the Optimization of Importance Sampling Parameters in Digital Communication System Simulation. To appear in the *IEEE Trans. Commun.*

[9] J-C. Chen, D. Lu, J. S. Sadowsky, and K. Yao. On Importance Sampling in Digital Communications — Part I: Fundamentals. *IEEE J. Select. Areas in Commun.*, 11(3):289–299, Apr. 1993.

[10] R. J. Wolfe, M. C. Jeruchim, and P. M. Hahn. On Optimum and Suboptimum Biasing Procedures for Importance Sampling in Communication Simulation. *IEEE Trans. Commun.*, COM-38(5):639–647, May 1990.

[11] John G. Proakis. *Digital Communications.* New York: McGraw-Hill, 1989.

[12] M. Metivier and P. Priouret. Applications of a Kushner and Clark Lemma to General Classes of Stochastic Algorithms. *IEEE Trans. Inform. Theory*, IT-30(2):140–151, March 1984.

[13] P. W. Glynn. Stochastic Approximation for Monte Carlo Optimization. In *Proc. of the Winter Simulation Conference*, Wilson, J., Henriksen, J. and Roberts, S. (eds), IEEE Press, 1986.

[14] P. W. Glynn. Likelihood Ratio Gradient Estimation: An Overview. In *Proc. of the Winter Simulation Conference*, Thesen, A., Grant, H., Kelton, W. D., (eds), IEEE Press, 1987.

17

Markov Models for Burst Errors in Radio Communications Channels

S.Srinivas and K.S.Shanmugan
Electrical and Computer Engineering Department
University of Kansas, Lawrence, Kansas, USA

ABSTRACT *Markov Modeling of burst errors in communications channels is discussed. Fritchman Modeling and Hidden Markov Modeling techniques are considered. It is found that the Fritchman parameters are not sufficient to characterize statistically dependent error burst patterns. The Hidden Markov Modeling (HMM) algorithm is computationally inefficient because it requires an order of* N^2T *computations for an* N *state model of an observation sequence that is* T *symbols long. In this paper we propose a modified HMM algorithm that is computationally more feasible for channel modeling. The modified algorithm (CHMM) is also more applicable due to the uniqueness of its parameters.*

I Introduction

Modeling of burst errors in communications channels has been a topic of considerable interest in the past three decades. In the sixties, mobile radio communications was an active research area. Gilbert[1] proposed a two state Markov model for characterizing error burst patterns typically arising from deep signal fading. Fritchman[2] generalized Gilbert's model and could use more than two states in his model. In both these modeling techniques, the observation sequence that is composed of 0 (error free transmission) and 1 (error in transmission) bits, is assumed to be a function of an underlying stationary Markov chain.

Both these techniques are used to model slowly varying channels. Such variations might result from ionospheric conditions etc. Recently indoor wireless communication has gained a lot of popularity. Fritchman and Gilbert models are no longer sufficient to describe the complicated error burst patterns that can occur in a typical indoor multipath channel due to the channel variability caused by shorter distances and higher rates of transmission.

Another technique that can be used successfully is Hidden Markov Modeling (HMM) which has been popular in speech recognition. We propose a constrained HMM algorithm (CHMM) where computation is only needed at the beginning and ending of error burst periods. CHMM also reduces the number of local maxima of the observation sequence probability in the parameter space. Thus the model obtained is more robust and unique than the model that standard HMM algorithms yield.

II A Review of Gilbert and Fritchman Modeling

Bursty binary channels (for example channels which exhibit deep frequency fading characteristics) can be modeled using Markov models. Fritchman[2] has shown that one could model the channels by using a finite number of exponential terms in the error free length distribution. The same is also true for the burst length distribution.

In Fritchman's framework, the state space $\mathcal{S} = \{1, 2, \ldots, N\}$ is divided into two groups A and B where A is the set of good states and B is composed of bad states as in figure 1. The state transition matrix can be written as

$$P = \begin{bmatrix} P_A & P_{AB} \\ P_{BA} & P_B \end{bmatrix}.$$

Let an error free run be defined as the event $\omega = (\mathcal{O}_t = 0, t \geq 1 | \mathcal{O}_0 = 1)$ where $\mathcal{O}_t$ is the

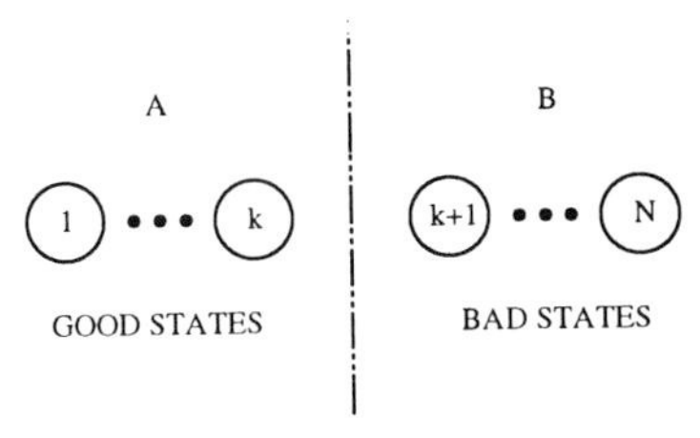

Figure 1: State space partitioning

symbol observed at time t. Using standard notation, let $(0^m|1)$ denote the event of observing m error free transmissions following an error. Similarly $(1^m|0)$ denotes m errors following a good period. Fritchman proved that the error free distribution $P(0^m|1) = P(\mathcal{O}_t = 0, 1 \leq t \leq m | \mathcal{O}_0 = 1)$ is specified by k weighted exponentials where k is the number of good states (cardinality of A). Similarly the cluster distribution, defined by $P(1^m|0) = P(\mathcal{O}_t = 1, 1 \leq t \leq m | \mathcal{O}_0 = 0)$ is characterized by $N - k$ weighted exponentials. Specifically

$$\begin{aligned} P(0^m|1) &= \sum_{i=1}^{k} f_i \lambda_i^{m-1} \\ P(1^m|0) &= \sum_{i=k+1}^{N} f_i \lambda_i^{m-1} \end{aligned} \tag{1}$$

where $(\lambda_1, \ldots, \lambda_k)$, $(\lambda_{k+1}, \ldots, \lambda_N)$ are the eigenvalues of P_A and P_B respectively. f_i are given in terms of the stationary (initial) probabilities p_i and P_{ij}.
From (1) one can express the probability of observing exactly m zeros (or ones) as:

$$\begin{aligned} P(0^{m-1}|1) - P(0^m|1) &= \sum_{i=1}^{k} f_i \lambda_i^{m-1}(1 - \lambda_i) \\ P(1^{m-1}|0) - P(1^m|0) &= \sum_{i=k+1}^{N} f_i \lambda_i^{m-1}(1 - \lambda_i) \end{aligned}$$

Thus the Fritchman parameters can be interpreted as state transition probabilities where the states represent the eigenstates of the P_A and P_B matrix, i.e.

$$\tilde{P} = \left[\begin{array}{ccc|ccc} \lambda_1 & & & & & \\ & \ddots & & & \tilde{P}_{AB} & \\ & & \lambda_k & & & \\ \hline & & & \lambda_{k+1} & & \\ & \tilde{P}_{BA} & & & \ddots & \\ & & & & & \lambda_N \end{array}\right] \tag{2}$$

Since there are $k(N-k)$ transition probabilities from good to bad states in P, the $k \times (N-k)$ independent parameters in $\tilde{P}_{AB}$ are equivalent to the corresponding probabilities P_{ij}. Similarly the $(N-k) \times k$ probabilities in $\tilde{P}_{BA}$ correctly represent the transitions from bad to good states. Thus the equivalent[1] model $\tilde{P}$ can used to characterize the channel.
The Fritchman model is not unique in situations that require more than one error state. This is so because the error free run distribution and the error length distribution do not specify the statistical dependence of the error free runs and the error periods. There are channel models that need more than one error state to correctly simulate the error patterns. One such instance involving synchronization losses is discussed in [4]. In general any channel whose error burst lengths are distributed with more than one mode (i.e. more than one characteristic length) needs to be modeled with more than one error state.

III Constrained Hidden Markov Modeling

Hidden Markov Modeling has been used extensively in speech recognition algorithms[6],[7]. A modified HMM algorithm can also be used efficiently to model bursty channels. The modifications are necessary since typical bit error rates of 10^{-6} require the processing of millions of samples to observe just a few bit errors.

The Constrained HMM (CHMM) algorithm discussed in this paper can be used to model channels with more than one error state. Another advantage of using HMM techniques is that the state duration distribution can be specified as non exponential.

Let $\mathbf{P}$ be the $N \times N$ transition probability matrix, $\mathbf{B}$ be an $N \times M$ matrix such that B_{ik} is the probability of observing the kth symbol of the set of observation symbols while in state i. M is the cardinality of the observation symbols set. To keep the notation simple, let $M = 2$, so that the two symbols are 0 and 1. Also let π be an $N \times 1$ vector of initial state probabilities. It can easily be shown that by increasing the number of states in the model (to at most $2N$), the elements in $\mathbf{B}$ can set to 0 or 1 (i.e. $B_{i0} = 1$, $i = 1, \ldots, k$ and $B_{i0} = 0$, $i = k+1, \ldots, N$). In essence this is achieved by splitting every state i whose symbol transmission is not deterministic into two states as shown in figure 2. The model can thus be characterized by the doublet $\lambda = (\mathbf{P}, \pi)$. Furthermore the state transition matrix can be assumed to be of the form of $\tilde{P}$ in (2) because of Fritchman's results. Gilbert[3] showed that there are atmost L independent observable parameters in a partitioned model such as the one in figure 3. L is given by:

$$L = N^2 - \sum_{s=0}^{s=M} n_s^2$$

n_s is the number of states in subset s. The set of states $\mathcal{S}$ is partitioned into disjoint subsets $\mathcal{S}_i, i = 1, \ldots, M$ where M is the number of elements in the observation symbol set. Furthermore the output at any time is a deterministic function of the state at that time and is given by:

$$\mathcal{O}(s) = i \quad \text{for } s \in \mathcal{S}_i$$

where s is a function of time. Essentially the number of independent parameters is the total number of parameters (N^2) minus the (unobservable) transition probabilities within each subset of states $\mathcal{S}_i$. Thus there are $2k(N-k)$ independent parameters for a $(k, N-k)$ partition, which is the number of independent parameters in $\tilde{P}$.
The HMM algorithm improves the estimate of λ by maximizing the probability of observing

[1] Two models are equivalent if and only if the probability $P(\mathcal{O})$ remains the same for every observation sequence $\mathcal{O}$ for both models.

the sequence from the model $P(\mathcal{O}|\lambda)$ at each iteration step. $P(\mathcal{O}|\lambda)$ is evaluated using a modified forward backward recursion algorithm. The reestimation algorithm that is commonly used in speech recognition programs is the Baum Welch Algorithm[11].

M is assumed to be 2 from now on so that there are only two symbols 0 and 1 allowed in the observation sequence which can hence be written as:

$$\mathcal{O} = 0^{n_0[0]}1^{n_1[0]}0^{n_0[1]}1^{n_1[1]} \ldots 0^{n_0[C]}1^{n_1[C]}.$$

where C is the total number of transitions that occur from the good to bad states (and vice versa). It is assumed that $n_0[i] \geq 1$ and $n_1[i] \geq 1$ for all i.

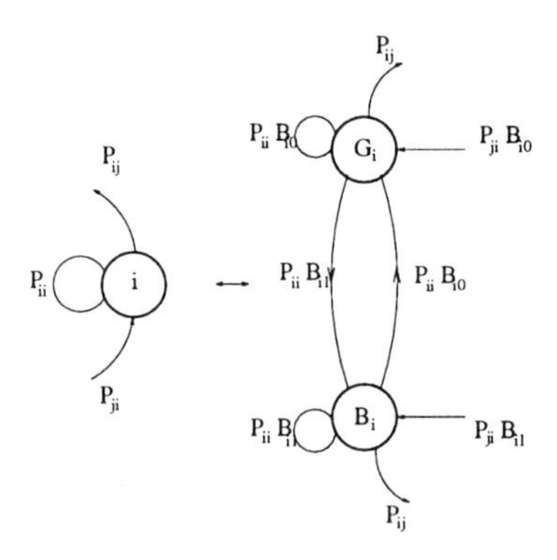

Figure 2: State splitting resulting in $B_{i0} = 0$ *or* 1

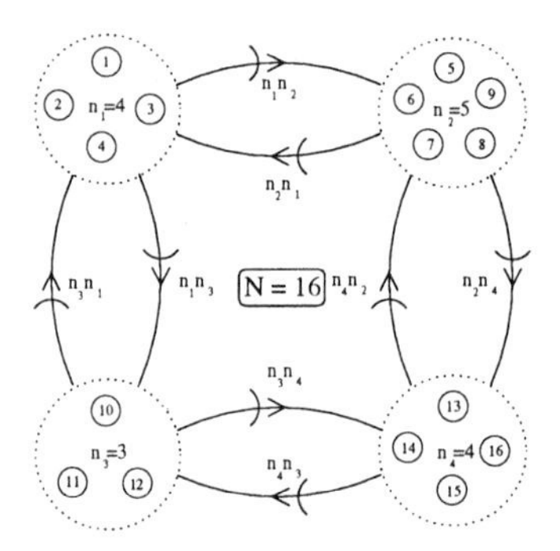

Figure 3: A Markov model partitioned into $M = 4$. $n_i n_j$ next to each path represents the number of such paths.

CHMM takes advantage of the special structure of $\tilde{P}$ which causes the underlying Markov chain to be deterministic at all times except when there is a change in the observation. Thus the model remains in the same state for the entire duration $n_0[i]$ or $n_1[i]$ as the case may be. All the variables are computed only at time steps involving a change of the observation symbol. The modified recursion and reestimation algorithms are described below:

<u>Forward Recursion</u>:

$$\alpha_t(i) \triangleq P(O_1 \cdots O_t\, ,\, s_t = q_i|\lambda)$$

1. Initialization:

$$\alpha_1(i) = \pi_i \tilde{P}_{ii}^{n_0[0]-1}\ \forall i = 1, \ldots, k$$
$$\alpha_1(i) = 0\ \forall i = k+1, \ldots, N$$

2. Induction:

$t = 2.$
$\forall\, 0 \leq c \leq C,$
$\alpha_t(i) = 0\ \forall i = 1, \ldots, k.$
$$\alpha_t(i) = \sum_{j=1}^{N} \alpha_{t-1}(j) \tilde{P}_{ji} \tilde{P}_{ii}^{n_1[c]-1}\ \forall i = k+1, \ldots, N$$
(Skip the next 4 lines for c = C).
$t = t + 1.$
$$\alpha_t(i) = \sum_{j=1}^{N} \alpha_{t-1}(j) \tilde{P}_{ji} \tilde{P}_{ii}^{n_0[c+1]-1}\ \forall i = 1, \ldots, k.$$
$\alpha_t(i) = 0\ \forall i = k+1, \ldots, N.$
$t = t + 1.$

3. Termination:

$$\Theta = t.$$
$$P(\mathcal{O}|\lambda) = \sum_{i=1}^{N} \alpha_\Theta(i)$$

<u>Backward Recursion</u>:

$$\beta_t(i) \triangleq P(O_{t+1} \cdots O_T | s_t = q_i\, ,\, \lambda)$$

1. Initialization:

$$\beta_\Theta(i) = 1,\ \forall i = 1, \ldots, N.$$

2. Induction:

$\forall\, C \leq c \leq 0,$
$$\beta_t(i) = \sum_{j=k+1}^{N} \beta_{t+1}(j) \tilde{P}_{ij} \tilde{P}_{jj}^{n_1[c]-1}\ \forall i = 1, \ldots, N.$$
(Skip the next 3 lines for c = 0).
$t = t - 1.$
$$\beta_t(i) = \sum_{j=1}^{k} \beta_{t+1}(j) \tilde{P}_{ij} \tilde{P}_{jj}^{n_0[c]-1}\ \forall i = 1, \ldots, N.$$
$t = t - 1.$

In order to reestimate the model λ using the Baum Welch method, one defines the following two quantities:

$$\begin{aligned} \gamma_t(i) &= P(s_t = i|\mathcal{O}, \lambda) \quad \forall\, i \in (1, N) \\ \zeta_t(i,j) &\triangleq P(s_t = q_i, s_{t+1} = j|\mathcal{O}, \lambda) \end{aligned}$$

It can be easily verified that $\gamma_t(i)$ and $\zeta_t(i,j)$ can be written in terms of the forward and backward variables $\alpha_t(i)$ and $\beta_t(j)$ as:

$$\begin{aligned} \gamma_t(i) &= \frac{\alpha_t(i)\beta_t(i)}{P(\mathcal{O}|\lambda)} \\ \zeta_t(i,j) &= \frac{\alpha_t(i)\tilde{P}_{ij}\tilde{P}_{jj}^{n_1[c]-1}\beta_{t+1}(j)}{P(\mathcal{O}|\lambda)} \quad t = 2c+1 \\ \zeta_t(i,j) &= \frac{\alpha_t(i)\tilde{P}_{ij}\tilde{P}_{jj}^{n_1[c]-1}\beta_{t+1}(j)}{P(\mathcal{O}|\lambda)} \quad t = 2c. \end{aligned}$$

The Baum Welch reestimation formulae are modified to:

$$\begin{aligned} \sum_{t=1}^{\Theta-1} \gamma_t(i) &= \text{expected no. transitions from } q_i \\ \sum_{t=1}^{\Theta} \gamma_t(i) &= \text{expected no. times } q_i \text{ is visited} \\ \sum_{t=1}^{\Theta-1} \zeta_t(i,j) &= \text{expected no. transitions } q_i \text{ to } q_j \end{aligned}$$

$$\pi_i' = \gamma_1(i)$$

$$\tilde{P}_{ij}' = \frac{\sum_{t=1}^{\Theta-1}\zeta_t(i,j) + \sum_{c=0}^{C}\gamma_{2c+1}(i)(n_0[c]-1)\delta_{ij}}{\sum_{t=1}^{\Theta-1}\gamma_t(i) + \sum_{c=0}^{C}\gamma_{2c+1}(i)(n_0[c]-1)}, i \le k$$

$$\tilde{P}_{ij}' = \frac{\sum_{t=1}^{\Theta-1}\zeta_t(i,j) + \sum_{c=0}^{C}\gamma_{2c}(i)(n_1[c]-1)\delta_{ij}}{\sum_{t=1}^{\Theta-1}\gamma_t(i) + \sum_{c=0}^{C}\gamma_{2c}(i)(n_1[c]-1)}, i > k$$

where δ_{ij} is the Krönecker delta function.
Scaling the forward and backward variables is essential at each transition time step. It can easily be verified that the scaling technique described in [7] can also be used for the modified algorithm. However numerical underflow problems may still occur if the exponents $(n_0[i], n_1[i])$ are large. Thus it might be necessary to split the power into convenient numbers for scaling purposes.

IV Modeling Examples

We used CHMM to model channels with 3 and 4 states. k, the number of good states was 2 in both models. In order to establish the equivalence of the CHMM model with the actual model, we need to have the first run length $n_0[0] = 1$. This is because π converges to a deterministic vector, thus the markov chain starts deterministically from one of the good states. Hence the initial run is modeled by a single exponential instead of k terms.

We used a 3 state model P to generate a sequence of observations and used CHMM to obtain an equivalent model. The initial model used in CHMM was:

$$\tilde{P}_i = \left[\begin{array}{cc|c} 0.990 & 0.000 & 0.010 \\ 0.000 & 0.600 & 0.400 \\ \hline 0.100 & 0.400 & 0.500 \end{array}\right].$$

$$P = \left[\begin{array}{cc|c} 0.9157 & 0.0025 & 0.0819 \\ 0.3409 & 0.4744 & 0.1847 \\ \hline 0.1900 & 0.2954 & 0.5146 \end{array}\right]$$

$$\tilde{P} = \left[\begin{array}{cc|c} 0.9176 & 0.000 & 0.0824 \\ 0.000 & 0.4739 & 0.5261 \\ \hline 0.4176 & 0.0678 & 0.5146 \end{array}\right].$$

The eigenvalues of the P_A matrix are 0.9176 and 0.4725 and are thus comparable to the diagonal terms in the CHMM model. The length of the observation sequence used was 12765 symbols which included 1000 transitions from good to bad states and vice versa. We compared the values of $P(\mathcal{O}|\lambda)$ for the two models for several input sequences. The results are summarized in table 1.

Length of Sequence	Number of Transitions	$ln[P(\mathcal{O}\|\lambda)]$ True Model	CHMM
13212	1000	-4734.41	-4734.40
13286	1000	-4750.63	-4750.61
26121	2000	-9431.01	-9430.97
18912	1500	-7021.21	-7021.20
63720	5000	-23416.88	-23416.87

Table 1. Comparison of $P(\mathcal{O}|\lambda)$.

We also obtained a CHMM model for a sequence generated by a 4 state model. Specifically

$$P = \left[\begin{array}{cc|cc} 0.990 & 0.003 & 0.002 & 0.005 \\ 0.120 & 0.600 & 0.180 & 0.100 \\ \hline 0.150 & 0.200 & 0.500 & 0.150 \\ 0.040 & 0.030 & 0.130 & 0.800 \end{array}\right]$$

$$\tilde{P} = \left[\begin{array}{cc|cc} 0.991 & 0.000 & 0.001 & 0.008 \\ 0.000 & 0.592 & 0.058 & 0.350 \\ \hline 0.370 & 0.193 & 0.437 & 0.000 \\ 0.092 & 0.053 & 0.000 & 0.854 \end{array}\right].$$

The eigenvalues of the P_A matrix are 0.991 and 0.599 while the eigenvalues of the P_B matrix are 0.445 and 0.855. Thus there is a slight difference in the smaller eigenvalues and the diagonal terms in $\tilde{P}$ matrix. The input sequence was nearly 5 million long and had 50000 transitions from good to bad states (and vice versa). Table 2 shows the equivalence of CHMM and the true model.

Length of Sequence	$ln[P(\mathcal{O}\|\lambda)]$ True Model	CHMM
6460	-709.32	-709.58
7675	-771.34	-771.66
7844	-754.28	-754.97
7776	-752.21	-752.59

Table 2. Comparison of $P(\mathcal{O}|\lambda)$.

Most of the computational burden lies with the recursion algorithms. It can be shown[6] that the order of complexity is N^2T where T is the total number of time steps. Thus in the 4 state example, CHMM needs only $N^2 \times 5 \times 10^4$ order of operations whereas the standard HMM needs $N^2 \times 5 \times 10^6$ operations.

The minimum number of states needed in the model can be determined from the given observation sequence. It has been observed that as the number of states N is increased, the probability $P(\mathcal{O}|\lambda)$ increases until the required number of parameters are used in the model. Increasing N beyond this does not increase $P(\mathcal{O}|\lambda)$ significantly. As an example we modeled a 740858 symbol observation sequence that had 20000 symbol transitions. The sequence was generated by the following 4 state Markov model:

$$P = \left[\begin{array}{cc|cc} 0.990 & 0.001 & 0.003 & 0.006 \\ 0.120 & 0.600 & 0.180 & 0.100 \\ \hline 0.150 & 0.200 & 0.500 & 0.150 \\ 0.040 & 0.030 & 0.130 & 0.800 \end{array}\right].$$

The eigenvalues of P_A are 0.9903 and 0.5997 and the eigenvalues of P_B are 0.8549 and 0.4451. The stationary probability vector is given by $[\ 0.8939\ 0.0205\ 0.0278\ 0.0579\ ]^T$. The values of $P(\mathcal{O}|\lambda)$ for the different models are listed below:

(N,k)	(2,1)	(3,2)	(4,2)	(5,3)	(6,3)	(7,4)
$log(P(\mathcal{O}\|\lambda))$	-79419	-74509	-74426	-74426	-74426	-74425

The models with $N > 4$ are equivalent to the (4,2) model and can be reduced to the (4,2) model by combining appropriate states. For instance combining states 2 and 3 of the (5,3) model results in the (4,2) model. Basic probability rules apply while combining states. For instance two states such that transitions between the states are disallowed can be merged as shown in figure 4. The various models and their reduced (4,2) forms are listed in table 3. The reduced (4,2) equivalent model for the (5,3) model is obtained by merging states 2 and 3. Similarly the reduced (4,2) form for the (6,3) model is got by combining state 2 with state 3 and state 4 with state 5. Finally the (4,2) form for the (7,4) model is obtained by combining states 2,3 with state 4 and state 5 with state 6. Furthermore the state indexing in the equivalent models could be a permutation of the state numbering in the original (4,2) model. For example state 3 of the reduced form of the (6,3) model is the same as state 4 of the original (4,2) model and state 4 is the same as state 3 of the original (4,2) model.

The stationary probability vectors p are also listed in the table. It can be seen that for all the models, the stationary probability of an error free transmission ($\sum_{i=1}^{k} p_i$) is the same.

V Conclusions

We successfully implemented a modified HMM algorithm that can be used to model bursty communication channels effectively. Our proposed modification of the algorithm allows us to obtain channel models with much less computational burden. In addition to the savings in CPU time, the modifications also add some desirable properties like uniqueness. A challenging problem that needs to be addressed is how to obtain a model with the least number of states[8]-[10]. Recently an algorithm has been suggested to obtain the minimum number of parameters for a model[10]. Thus an equivalent model with the least number of states can be obtained thereby allowing us to reduce the computations drastically and increase the uniqueness of the model.

<table>
<tr><th>(N,k)</th><th>λ</th><th>p</th><th>$\lambda_{reduced}$</th></tr>
<tr><td>(2,1)</td><td>$\begin{bmatrix} 0.985 & 0.015 \\ 0.162 & 0.838 \end{bmatrix}$</td><td>$\begin{bmatrix} 0.917 \\ 0.083 \end{bmatrix}$</td><td>-</td></tr>
<tr><td>(3,2)</td><td>$\left[\begin{array}{cc|c} 0.991 & 0.000 & 0.009 \\ 0.000 & 0.595 & 0.405 \\ \hline 0.103 & 0.059 & 0.838 \end{array}\right]$</td><td>$\begin{bmatrix} 0.905 \\ 0.012 \\ 0.083 \end{bmatrix}$</td><td>-</td></tr>
<tr><td>(4,2)</td><td>$\left[\begin{array}{cc|cc} 0.991 & 0.000 & 0.001 & 0.008 \\ 0.000 & 0.595 & 0.063 & 0.342 \\ \hline 0.326 & 0.204 & 0.471 & 0.000 \\ 0.093 & 0.053 & 0.000 & 0.854 \end{array}\right]$</td><td>$\begin{bmatrix} 0.905 \\ 0.012 \\ 0.004 \\ 0.080 \end{bmatrix}$</td><td>-</td></tr>
<tr><td>(5,3)</td><td>$\left[\begin{array}{ccc|cc} 0.991 & 0.000 & 0.000 & 0.001 & 0.008 \\ 0.000 & 0.607 & 0.000 & 0.027 & 0.366 \\ 0.000 & 0.000 & 0.571 & 0.147 & 0.282 \\ \hline 0.320 & 0.105 & 0.095 & 0.480 & 0.000 \\ 0.093 & 0.036 & 0.016 & 0.000 & 0.855 \end{array}\right]$</td><td>$\begin{bmatrix} 0.905 \\ 0.008 \\ 0.004 \\ 0.004 \\ 0.079 \end{bmatrix}$</td><td>$\left[\begin{array}{cc|cc} 0.991 & 0.000 & 0.001 & 0.008 \\ 0.000 & 0.595 & 0.063 & 0.342 \\ \hline 0.320 & 0.201 & 0.480 & 0.000 \\ 0.093 & 0.052 & 0.000 & 0.855 \end{array}\right]$</td></tr>
<tr><td>(6,3)</td><td>$\left[\begin{array}{ccc|ccc} 0.991 & 0.000 & 0.000 & 0.007 & 0.001 & 0.001 \\ 0.000 & 0.591 & 0.000 & 0.265 & 0.050 & 0.094 \\ 0.000 & 0.000 & 0.600 & 0.351 & 0.037 & 0.012 \\ \hline 0.091 & 0.035 & 0.019 & 0.855 & 0.000 & 0.000 \\ 0.114 & 0.012 & 0.032 & 0.000 & 0.842 & 0.000 \\ 0.331 & 0.100 & 0.109 & 0.000 & 0.000 & 0.459 \end{array}\right]$</td><td>$\begin{bmatrix} 0.905 \\ 0.007 \\ 0.005 \\ 0.069 \\ 0.010 \\ 0.003 \end{bmatrix}$</td><td>$\left[\begin{array}{cc|cc} 0.991 & 0.000 & 0.008 & 0.001 \\ 0.000 & 0.595 & 0.345 & 0.060 \\ \hline 0.094 & 0.053 & 0.854 & 0.000 \\ 0.331 & 0.210 & 0.000 & 0.459 \end{array}\right]$</td></tr>
<tr><td>(7,4)</td><td>$\left[\begin{array}{cccc|ccc} 0.991 & 0.000 & 0.000 & 0.000 & 0.006 & 0.002 & 0.001 \\ 0.000 & 0.626 & 0.000 & 0.000 & 0.250 & 0.068 & 0.056 \\ 0.000 & 0.000 & 0.558 & 0.000 & 0.368 & 0.048 & 0.026 \\ 0.000 & 0.000 & 0.000 & 0.580 & 0.155 & 0.134 & 0.130 \\ \hline 0.089 & 0.028 & 0.016 & 0.013 & 0.854 & 0.000 & 0.000 \\ 0.106 & 0.011 & 0.025 & 0.004 & 0.000 & 0.854 & 0.000 \\ 0.327 & 0.103 & 0.035 & 0.069 & 0.000 & 0.000 & 0.467 \end{array}\right]$</td><td>$\begin{bmatrix} 0.905 \\ 0.006 \\ 0.004 \\ 0.003 \\ 0.058 \\ 0.021 \\ 0.004 \end{bmatrix}$</td><td>$\left[\begin{array}{cc|cc} 0.991 & 0.000 & 0.008 & 0.001 \\ 0.000 & 0.596 & 0.341 & 0.063 \\ \hline 0.093 & 0.053 & 0.854 & 0.000 \\ 0.327 & 0.206 & 0.000 & 0.467 \end{array}\right]$</td></tr>
</table>

Table 3. CHMM models for N ranging from 2 to 7.

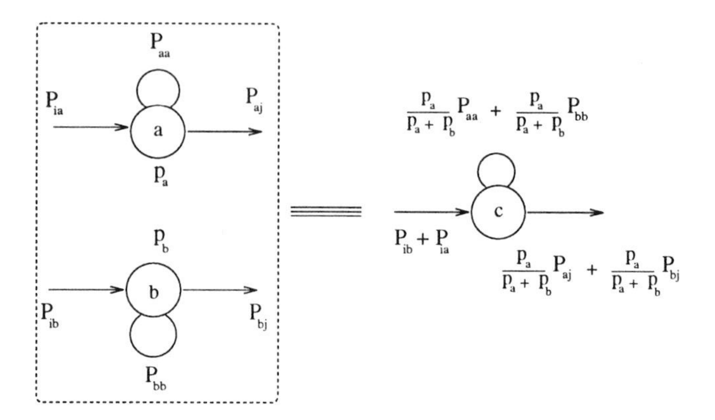

Figure 4: Combining 2 states with no transitions among them.

REFERENCES

[1] E.N. Gilbert, "Capacity of a burst-noise channel," *B.S.T.J.*, vol. 39, pp 1253-1265, September 1960.

[2] Bruce D. Fritchman, "A binary channel characterization using partitioned Markov chains," *IEEE transactions on Information Theory*, vol. IT-13, no. 2, pp 221-227, April 1967.

[3] E.J.Gilbert, "On the identifiability problem for functions of finite Markov chains," *Ann. Math. Stat.*, vol. 30, pp. 688-697, September 1959.

[4] William Turin and M.Mohan Sondhi, "Modeling error sources in digital channels," To be published.

[5] William Turin, *Performance Analysis of Digital Transmission Systems*, New York: Computer Science Press, 1990.

[6] L.R.Rabiner and B.H.Juang, "An Introduction to Hidden Markov Models," *IEEE ASSP Magazine*, January 1986, pp. 4-16.

[7] S.E.Levinson, L.R.Rabiner and M.M.Sondhi, "An Introduction to the Application of the Theory of Probabilistic Functions of a Markov Process to Automatic Speech Recognition," *B.S.T.J.*, vol. 62, No.4, April 1983, pp. 1035-1074.

[8] C.J.Burke and M.Rosenblatt, "A Markovian function of a Markov chain," *Ann. Math. Stat.*, vol. 29, pp. 1112-1122, 1958.

[9] D.Blackwell and L.Koopmans, "On the identifiability problem for functions of finite Markov chains," *Ann. Math. Stat.*, vol. 28, pp. 1011-1015, 1957.

[10] Hisashi Ito, Shun-Ichi Amari and Kingo Kobayashi, "Identifiability of Hidden Markov Information Sources and Their Minimum Degrees of Freedom," *IEEE transactions on Information Theory*, vol. 38, no. 2, pp. 324-333, March 1992.

[11] L.E.Baum, "An Inequality and Associated Maximization Technique in Statistical Estimation for Probabilistic Functions of a Markov Process," *Inequalities*, 3, pp. 1-8, 1972.

18

Application of Variable-Rate Convolutional Code for Mobile Communications

Young-Ok Park, ETRI

Abstract

An optimal application method of variable-rate convolutional coding is presented to effectively correspond to the channel environment while optimizing data transmission efficiency. The proposed method can maintain error performance at a required level regardless of the fast fluctuation of received signal envelope. And at the same time, transmission efficiency can be constantly kept by alternatively using R=1/N convolutional code, R=k/n punctured convolutional code, and R=1/2N convolutional code that made by one repetition of R=1/N code symbol according to the channel status report. To verify advanteges of the proposed method, three kinds of codes(R=1/4, 1/2, and 2/3) are applied in adaptive manners. Consequently, about 40% data transmission efficiency is achieved while maintaining the performance of the system at SNR is 10^{-3}

I. Introduction

Mobile receivers worked while moving, especially in an urban area, has to be influenced severely by Rayleigh fading in addition to AWGN(Additive White Gaussian Noise). This fading phenomenon depends upon the variation of receiver's moving speed and data transmission rate, and it causes degradation of system reliability in its received data correctness. Therefore, error correction coding is needed to protect vulnerable data from it. In mobile communication environments, convolutional coding with proper interleavers is normally used as a channel coding method, and well-known Viterbi algorithm for decoding in general. But in this paper, an optimal application method of variable-rate convolutional coding is presented to effectively correspond to the channel environment while optimizing data transmission efficiency.

II. Variable-Rate Convolutional Coding

Variable-rate convolutional code is a channel coding method that can effectively transmit the information on the widely varying mobile communication channel by using the variable code rate. It consists of R=1/N convolutional code, R=k/n punctured convolutional code based on R=1/N code, and R=1/2N code generated by repeating the code symbol of R=1/N code. So it is imortant that these three kinds of codes can be generated from only one encoder. By adaptively using one of these codes according to the communication channel environments, it is possible to achieve better transmission efficiency while maintaining the required performance.

1. Basic Operation

Variable-rate convolutional codes are generated from a R=1/N fundamental encoder. First, if the channel is affected by weak AWGN and fading, information message is encoded by a R=k/n punctured convolutional code which is from the R=1/N mother code. And as the channel condition is getting worse, code rate is changed to R=1/N. If the channel condition is the worst, encoder repeats the R=1/N code symbol once, so it generates R=1/2N code symbol in order to confront to the worst channel condition. When the selected code rate is applied to two frames and there is no bit error, the code rate is increased by one step. By doing this, transmission efficiency can be increased. Therefore, if the channel condition is good, one can get a high transmission efficiency, and even when the channel condition is not good, error rate is not seriously increased by lowering the transmission efficiency. From the receiver the measured error probability over received data is fed back to the encoder, and based on these data, encoder can determine the appropriate code rate.

2. Simulation Model

2-1 Encoder

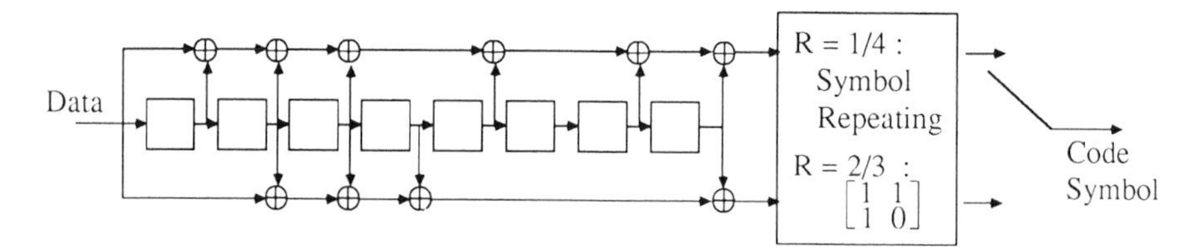

Fig.1 Variable-rate convolutional encoder

A R=1/2 convolutional code is used as a mother code, and a R=2/3 punctured convolutional code and R=1/4 code are also used in this simulation. Constraint length is 9 and the tap coefficients g_0, g_1 are 753, 561(octal) respectively. Puncturing period of the R=2/3 convolutional code is 2 and puncturing method is followed by the perforating matrix below which is known as an adequate matrix to the 16-level decision decoding. Blockdiagram of this encoder is shown as Fig. 1[7].

$$[P]=\begin{bmatrix}1 & 1\\ 1 & 0\end{bmatrix} \tag{1}$$

2-2 Encoding

Encoding procedure is shown as a sequence diagram in Fig.2. In a mobile communication environment, the required error probability of the voice is generally 10^{-3}. So code rate decision threshold level is selected at the point where error probability of 10^{-3} can be maintained even when the received power is seriously infected with the fading effects. One frame is composed of 5,000 bits and if more than 4 bits of errors are detected, then the code rate is increased in order to correct the errors powerfully. And the code rate is decreased by one step when there is no error for two frames.

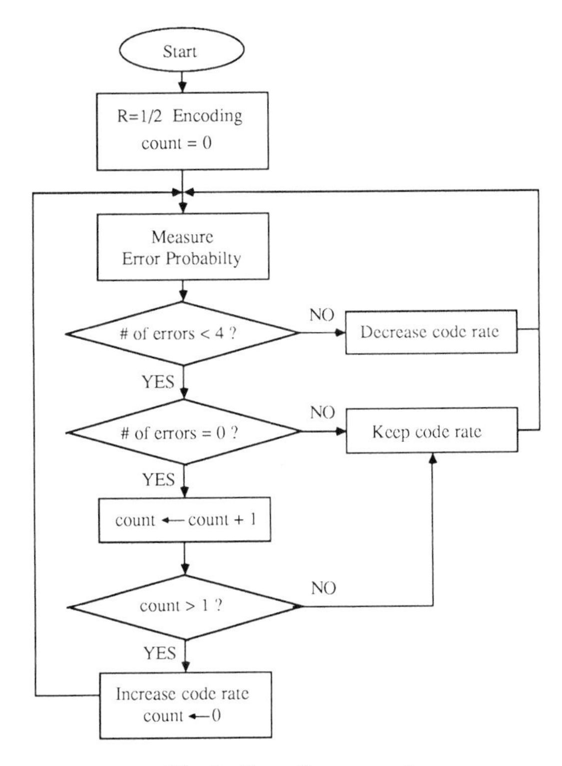

Fig.2 Encoding procedure

2-3 Decoding

Decoder directly receives the variable-rate information from the encoder. When the received code rate is R=1/4, one symbol of 4 bits is averaged with 2 bits, and resulting 2 bits are fed to the Viterbi decoder input. When the code rate is R=2/3, metric calculation within the Viterbi decoder is not performed on the punctured bits.

2-4 Rayleigh Fading Model

The coding scheme described above is simulated in a AWGN and Rayleigh fading environment. Over all simulation procedure is shown in Fig.3. F is a random variable, and its probability distribution function is Rayleigh distribution as shown in equation (2), The mean value of this pdf is $\sigma\sqrt{\pi/2}$.

$$p(F) = \frac{F}{\sigma^2} \exp\left(-\frac{F^2}{2\sigma^2}\right) \qquad (2)$$

A Rayleigh random process is generated in order to model a communication environment : assume 84km/s of receiver moving speed and 1GHz of carrier frequency. F is a sampled value from this process at a rate of 10kbps, and N is a random variable whose pdf is Gaussian distribution. Practical channel coding schemes generally use interleavers to reduce the fading effects. But in this simulation, for simple anlysis and comparision of performances of the conventional R=1/N type convolutional codes and the proposed method under the same fading and AWGN environments, interleaving and modulation is ignored.

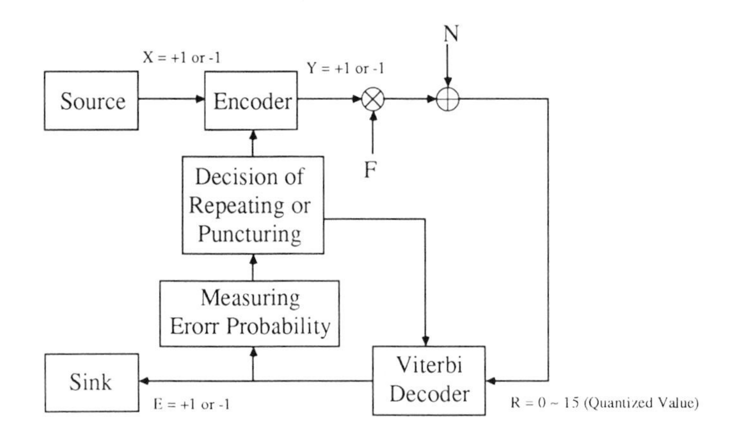

Fig.3 Simulation model

2-5 Others

The number of bipolar samples for simulation is 10^6 per specific SNR value of channel. Input value of Viterbi decoder which has affected by noise and fading effect is quantized for 16-level soft decision.

Ⅲ. Simulation Results and Analysis

The performance of a variable-rate convolutional coding under a AWGN channel is depicted in Fig. 4, and the performances under AWGN and Rayleigh fading channel is depicted in Fig. 5. In fig.4, the fluctuation of error probability when SNR is low can be improved if the rate decision threshold level is optimally selected and the difference of performances among the used codes is minimized. In Fig. 5, error probability of 10^{-3} or more is maintained when SNR is more than 10dB. This is a significantly different result comparing to the other codes. This result shows that the performance of the proposed coding is superior to the R=1/2 code when SNR is below 11dB, but comparatively inferior when SNR is more than 11dB. This is a very useful result in a mobile comunication environment that uses spread spectrum technology, because mobile voice communication requires less than 10dB of SNR and 10^{-3} of error probability. Especially in a mobile communication environment where received power level variation is large, the proposed method has an advantage of performance to the conventional coding scheme when SNR is more than 5dB. Overall performance curve shows that R=1/4 code is adopted when SNR is 0~5dB, R=2/3, 1/2, 1/4 codes are adequately used when SNR is 10~13 dB, and R=2/3 code is used when SNR is more than 14dB. Fig.6 and Fig.7 show transmission efficiencies under a AWGN and fading environment respectively. Fig. 7 shows that the transmission efficiency is about 39.8% when SNR is 10~13dB. This result is somewhat inferior to the R=1/2 code but shows an improvement to the R=1/3 code.

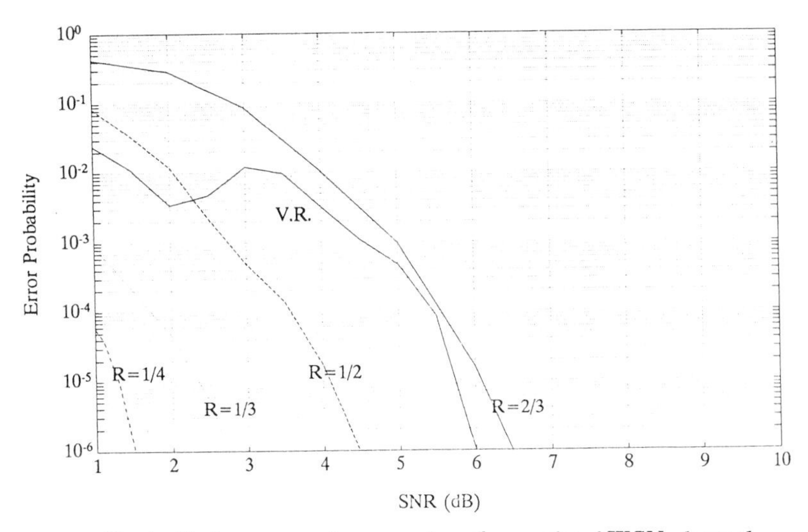

Fig.4 Performance of proposed coding under AWGN channel

< References >

[1] M. Y. Rhee, *Error Correcting Coding Theory,* McGraw-Hill, 1989.

[2] S. Lin and D. J. Costello, Jr., *Error Control Coding*, Englewood Cliffs, NJ:Prentice-Hall, 1983.

[3] J. Hagenauer, "Rate-compatible punctured convolutional codes (RCPC codes) and their applications," *IEEE Trans. Comm.*, vol.C-36, no.4,pp.389-400, April 1988.

[4] D. Haccoun and G. Begin, "High-rate punctured convolutional codes for Viterbi and sequential decoding," *IEEE Trans. comm.*, vol.C-37, no.11, pp.1113-1125, November 1989.

[5] J. Hagenauer, "The performance of rate-compatible punctured convolutional codes for future digital mobile radio," 1988 Vehcular Technology Conference, VTC-88, pp.22-29.

[6] I. M. Onyszchuz, "Truncation length for Viterbi decoding," *IEEE Trans. comm.*, vol.C-39, no.7, pp.1023-1026, July 1991.
[7] Y. Yasuda, K. Kashki, and Y. Hirata, "High-Rate Punctured Convolutional Codes for soft decision Viterbi decoding," *IEEE Trans. Comm.*, vol.C-32, no.3, pp.315-319, March 1984.

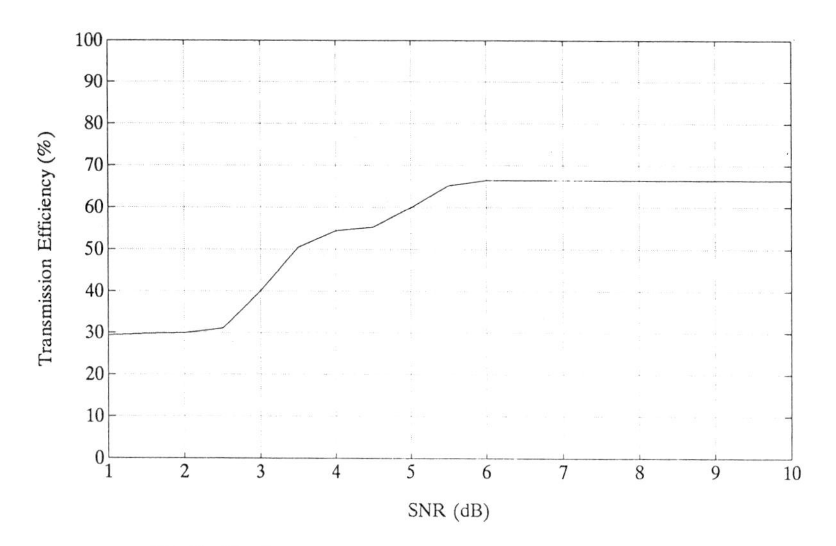

Fig.5 Performance of proposed coding under AWGN and Rayleigh fading channel

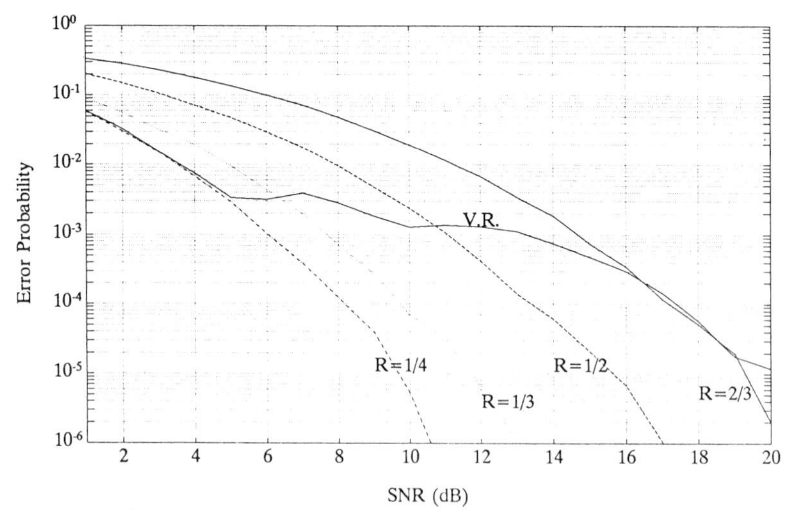

Fig.6 Transmission efficiency of proposed coding under AWGN channel

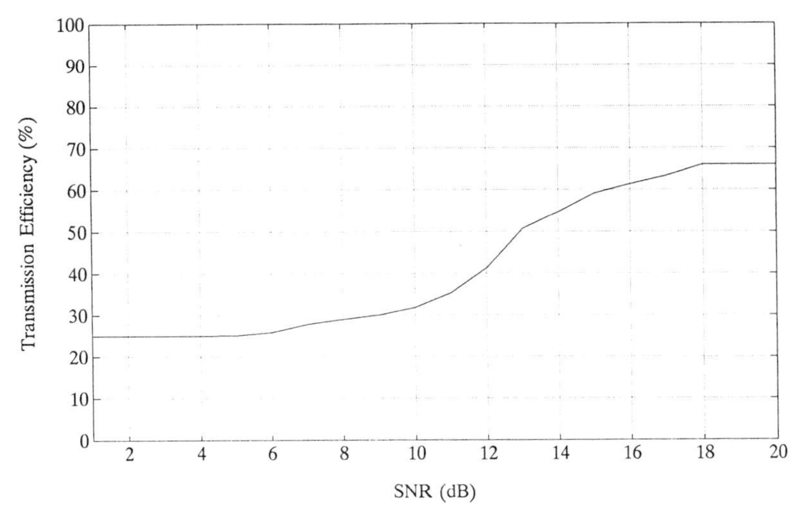

Fig.7 Transmission efficiency of proposed coding under AWGN and Rayleigh fading

19

Reservation versus demand-assignment multiplex strategies for packet voice communication systems

J. L. Sobrinho, J. M. Brázio

CAPS, Complexo I, I.S.T.
Av. Rovisco Pais, 1096 Lisboa Codex, Portugal

Abstract

Reservation and demand-assignment methods provide two different strategies for multiplexing voice sources in packet voice communication systems using speech activity detection. In this paper we describe a representative of each of these methods for use in the common framework of a TDM transmission structure. A comparative performance study indicates that about 20 % more voice sources can be accommodated in the demand-assignment method as compared to the reservation one.

1 Introduction

Speech activity detection techniques have been considered for use in packet voice communication systems as a means of expanding the number of voice sources able to time-share costly and/or scarce transmission resources without noticeable degradation in speech quality. These systems and techniques have in the past been studied in the context of wired LAN's [1], [2] and are currently receiving wide attention in the framework of short-range wireless communications [3], [4].

Speech activity detectors exploit the activity structure of voice signals. The output of the detector follows an alternate sequence of talkspurts and of silence periods spending about 40 % of the time in the active state (in talkspurts). During talkspurts a voice source generates data at a given nominal rate. Statistical multiplexing of voice sources can potentially more than double the capacity of the available transmission resources. However, statistical fluctuations in the talkspurts generation process create overload periods, during which the total nominal data rate of the active voice sources exceeds the available transmission capacity. Voice sources require prompt delivery of information, but are otherwise relatively tolerant to speech clipping. Accordingly, in most voice communication systems using speech activity detection, voice sources respond to congestion by discarding segments of their talkspurts so as to reduce the total data rate presented to the transmission resources.

In general, two different types of multiplexing strategies for packet voice transmission have been implemented or discussed in the literature from which two different types of speech discarding properties result.

In reservation methods [3], [4] active voice sources are assigned fixed fractions, if available, of the transmission resources, normally time slots in a TDM structure, on a first-come first-served basis, for the duration of their talkspurts. A newly arriving talkspurt finding the transmission resources fully utilized will be blocked until a sufficient fraction of resources is released. If the waiting time until access to the system exceeds a pre-specified value, the talkspurt will start discarding its data bits, thus incurring front-end clipping. Reservation methods have also been employed in non-packet voice communication systems using speech activity detection, namely in the TASI systems [5], [6].

In demand-assignment methods [1], [2] the total available transmission resources are at all times shared by the active voice sources in such a way that during congestion periods the data rate available to each source for transmission is smaller than its nominal data rate. As a result, voice sources will have to, more or less periodically, discard some of their data in order to satisfy the delay constraints. In this way, mid-speech clipping, evenly affecting all the active voice sources, occurs.

In this work we compare the performance of the reservation and demand-assignment methods in packet voice communication systems with speech activity detection. For the sake of uniformity, the framework of a slotted TDM transmission system is used.

In Section 2 we present a short characterization of voice sources, together with a summary of the relevant facts on the effect of clipping on perceived speech quality. In Section 3 we describe the reservation and demand-assignment methods considered. In Section 4 the analysis of the clipping performance of both methods is given. Finally, in Section 5 numerical results are presented and discussed.

2 Voice Sources

In conversational speech, a voice signal consists of an alternating sequence of talkspurts and silence periods. A speech activity detector can be used to recognize the silence periods, therefore preventing the unnecessary transmission of the corresponding data bits.

It is common to distinguish between fast and slow speech activity detection. The fast speech activity detectors can recognize the short silence periods between words and syllables in addition to the larger silence periods, occuring among phrases and sentences and during listening periods, that are detected by the slow speech activity detectors. The advantages and disadvantages of these two types of speech activity detection are discussed in [7].

In this work we consider slow speech activity detection. The activity of the voice source as classified by the speech activity detector is described by a two-state Markov process [8]. The durations of a talkspurt, T_t, and of a silence period, T_s, are exponentially distributed with means $1/\mu$ and $1/\lambda$ seconds, respectively. The speech activity factor γ measures the fraction of time that a voice source is in the active state, and is given by $\gamma = \lambda/(\mu + \lambda)$. Typical values for $1/\mu$ range between 1 s and 2 s, whereas typical values for γ range from 0.35 to 0.45 [8], [7], [9].

Front-end clipping (FEC) and mid-speech clipping (MSC) affect the perceived speech quality in different ways, and set different performance requirements to the underlying transmission mechanism.

Front-end clipping affects the leading edge of a talkspurt. The subjective impairment on perceived speech quality depends on both the fraction of talkspurts clipped and on the duration of the clipping periods. For example, the Campanella criterion, valid when slow speech activity detection is used, states that 2% of talkspurts can suffer front-end clipping exceeding 50 ms without noticeable degradation in speech quality [10]. On the other hand, with mid-speech clipping there can be more than one clip per talkspurt, and the speech clips occur at uncertain times during a talkspurt. The impairment caused by mid-speech clipping on speech quality depends on both the total fraction of speech clipped and the duration of the clipping periods. For instance, 2 % of speech can be clipped without objectionable degradation to speech quality as long as the clipping durations do not exceed 5 ms [11].

3 Statistical multiplexing of voice sources

In this Section we consider the statistical multiplexing of a number of independent voice sources. We begin by presenting the common framework underlying both the reservation and demand-assignment methods. We, subsequently, proceed with their descriptions.

3.1 TDM transmission structure

A TDM transmission structure is assumed. The time axis is divided into slots of fixed length. Each slot can accommodate a maximum of t_{pac} seconds of speech encoded at the nominal data rate of the voice sources, where t_{pac} is an integer multiple c of the slot duration. We will call a voice channel any sequence of time slots separated by t_{pac} seconds. Therefore, the number of voice channels in the system is given by c and corresponds to the maximum number of voice sources that can share the transmission resources without speech activity detection and without clipping.

In both multiplexing methods described in the sequel it is considered that whenever a voice source is allowed to send δ seconds of speech ($\delta \leq t_{pac}$) in slot i, it will correspondingly transmit in that slot the data bits generated between the instants of time $t_i - \delta$ and t_i, where t_i marks the beginning of slot i. In this way no source bit will be delayed for transmission by more than t_{pac} seconds.

3.2 Reservation method

Figure 1 illustrates the operation of the reservation multiplex strategy, taking as an example a transmission channel comprising 3 voice channels shared by 4 voice sources.

An arriving talkspurt that finds less than c active voice sources in the system will have access to the transmission channel in one of the next c time slots, allowing its initial data bits to be transmitted with a delay not exceeding t_{pac} seconds. Succeeding access points are then automatically reserved periodically every t_{pac} seconds for the duration of the talkspurt.

In Figure 1 talkspurt B started at time t_s^B. It has its first access to the transmission channel at t_6, where it transmits the initial $t_6 - t_s^B$ seconds of speech. Slots $6 + 3 \cdot k$, $k = 1, \ldots,$ will then be reserved to talkspurt B until its end.

On the other hand an arriving talkspurt that finds n active voice sources, with $n \geq c$, in the system will have to discard its initial data bits until any $n-(c-1)$ of those n voice sources become inactive. Subsequent information bits, if any, will have reserved access to the transmission channel.

In Figure 1 talkspurt D starts at t_s^D and finds 3 active voice sources. This talkspurt will be blocked until time t_e^C, when talkspurt C ends. Talkspurt C has reserved access to the transmission channel in slot 20. However, it only needs that slot to transmit its last $t_{pac} - (t_{20} - t_e^C)$ seconds of speech. Therefore, the remaining fraction of slot 20 can already be used to transmit the most recent $t_{20} - t_e^C$ seconds of speech information belonging to talkspurt D. Talkspurt D will then have reserved access to the channel every c slots.

We note that the reservation method leads to front-end clipping of talkspurts arriving during overload periods.

3.3 Demand-assignment multiplex method

Figure 2 illustrates the operation of the demand-assignment method, again taking as an example the same transmission channel comprising 3 voice channels shared by 4 voice sources.

While the system is in an underload period, i.e. with c or less active voice sources, its behaviour is the same as in the reservation method. However, when the system becomes overloaded, with $n > c$ active voice sources, they will have access to the transmission channel in a round robin order, in such a way that none of the active voice source present incurs clipping durations greater than $t_{pac}(n-c)/c$ seconds.

In Figure 2 talkspurt D is generated at t_s^D and although it already finds 3 active voice sources, it will nevertheless force its entrance at time t_{13} to transmit t_{pac} seconds of speech in slot 13. Talkspurt A will now have its instant of access to the transmission channel postponed to t_{14}. But since it can only transmit the more recent t_{pac} seconds of speech in slot 14, its oldest $t_{pac}/3$ seconds of speech have to be discarded.

In much the same way as with the reservation method, the final transmission associated with a talkspurt may leave an unused portion of a time slot that can potentially be used to transmit speech information pertaining to another talkspurt. We see in Figure 2 that talkspurt C transmits $t_{pac} - (t_{19} - t_e^C)$ seconds of speech in slot 19, where $t_{pac}/3 \leq t_{pac} - (t_{19} - t_e^C) < 2t_{pac}/3$. There are still $t_{19} - t_e^C$ seconds of speech that can be accommodated in that slot and they will be used by the talkspurt that incurs a clipping duration less than or equal to $t_{pac}/3$ seconds, which in this case is talkspurt D. Contrary to the reservation strategy, in the demand-assignment method all active voice sources are impaired by midspeech clipping during overload periods.

4 Performance Analysis

We consider l independent voice sources, each alternating between the active and the inactive states, according to the model in Section 2. These voice sources time-share a transmission channel comprising c voice channels.

The process $N(t)$, describing the evolution of the number of talkspurts in the system over time, is regenerative with possible regeneration cycles defined by successive instants of time when the

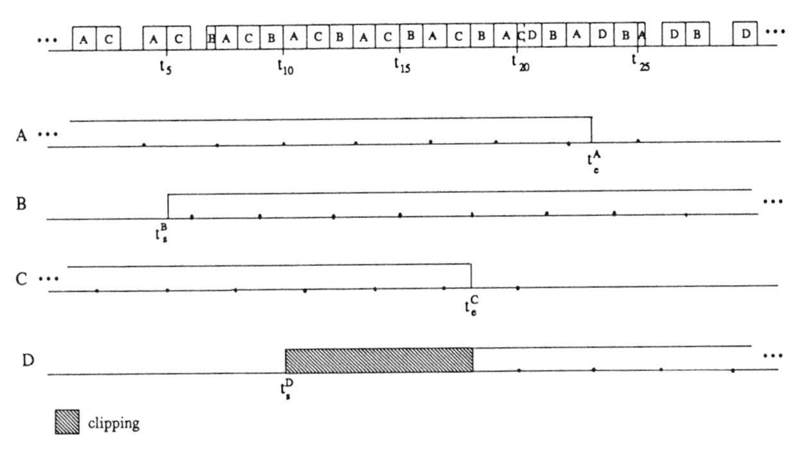

Fig. 1 - Illustration of the operation of the reservation method, where 4 voice sources share 3 voice channels.

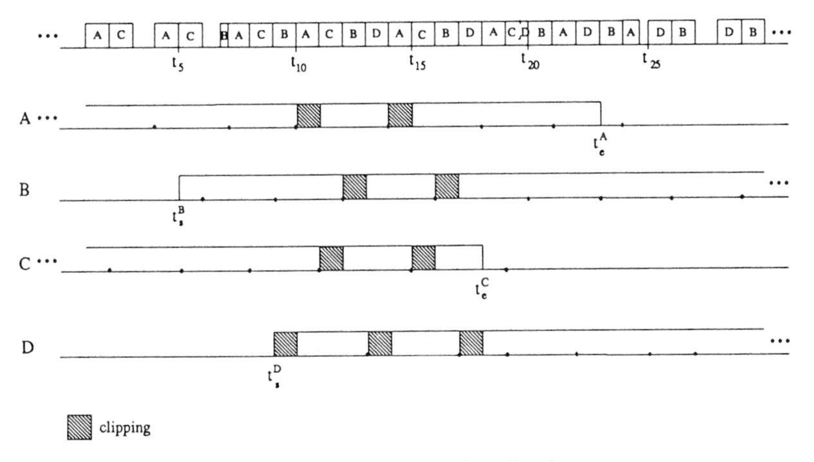

Fig. 2 - Illustration of the operation of the demand-assignment method, also with 4 voice sources sharing 3 voice channels.

number of talkspurts increases from $(n-1)$ to n $(n \leq l)$. This allow us to equate time averages and averages over the number of talkspurts to the corresponding stationary probabilistic quantities.

4.1 Reservation method

We want to determine the fraction $\mathcal{F}_{FEC}$ of talkspurts that experience front-end clipping in excess of a specified value of b seconds. This fraction coincides with the probability that, in a stationary situation, the amount B of FEC, measured in seconds, incurred by an arbitrary talkspurt exceeds b seconds.

Let N_a be the number of active voice sources found by an arriving talkspurt. We have that

$$P\{N_a = n\} = \nu_n = \binom{l-1}{n} \gamma^n (1-\gamma)^{l-1-n}, n = 0, \cdots, l-1. \tag{1}$$

Let Q be equal to zero for $N_a < c$ and for $N_a \geq c$, be equal to the time that elapses between the generation of the talkspurt and the instant when only $c-1$ of the active voice sources found by the talkspurt remain active. This instant corresponds to the first data bit of the talkspurt that could be transmitted with a delay not greater than t_{pac} seconds. The probability that $Q > q$ given that the arriving talkspurt finds n active voice sources in the system is the probability that less than $n-(c-1)$ of those voice sources become inactive during q. That is,

$$P\{Q > q | N_a = n\} = \sum_{i=0}^{n-c} \binom{n}{i} (1-e^{-\mu q})^i (e^{-\mu q})^{n-i}, n = c, \cdots, l-1. \tag{2}$$

When $N_a \geq c$ it may happen that a talkspurt is totally discarded because the value of Q exceeds the talkspurt duration T_t. Therefore, the amount of front-end clipping incurred by the talkspurt, given that $N_a = n$, is the minimum between Q and T_t, and thus

$$P\{B > b | N_a = n\} = \sum_{i=0}^{n-c} \binom{n}{i} (1-e^{-\mu b})^i (e^{-\mu b})^{n-i+1}, n = c, \cdots, l-1. \tag{3}$$

Unconditioning we obtain the value we are looking for,

$$\mathcal{F}_{FEC} = P\{B > b\} = \sum_{n=c}^{l-1} \left(\sum_{i=0}^{n-c} \binom{n}{i} (1-e^{-\mu b})^i (e^{-\mu b})^{n-i+1} \right) \nu_n. \tag{4}$$

4.2 Demand-assignment method

In this case, we want to determine the total fraction $\mathcal{F}_{MSC}$ of speech bits that are discarded. We implicitly assume that most clipping durations incurred by the talkspurts are sufficiently short (according to Section 2, shorter than 5 ms). The average number of bits discarded per talkspurt is given by $r_s E\{B\}$, with r_s the nominal data rate of an active voice source, whereas the average number of bits generated per talkspurt is given by $r_s E\{T_t\}$. Consequently,

$$\mathcal{F}_{MSC} = \frac{E\{B\}}{E\{T_t\}}. \tag{5}$$

$E\{B\}$ can be determined by the use of Little's formula. Let N denote the number of talkspurts in the system. Then,

$$P(N = n) = p_n = \binom{l}{n} \gamma^n (1-\gamma)^{l-n}, n = 0, \ldots, l. \tag{6}$$

The average number of talkspurts in excess of c, whose presence results in mid-speech clipping affecting all talkspurts, is given by $E\{(N-c)^+\}$, where $X^+ = \max(0, X)$. The arrival rate of talkspurts is $l/(E\{T_t\}+E\{T_s\})$. Therefore,

$$E\{B\} = \frac{E\{(N-c)^+\}}{l}(E\{T_t\}+E\{T_s\}) \tag{7}$$

and

$$\mathcal{F}_{MSC} = \frac{E\{(N-c)^+\}}{E\{N\}} = \sum_{n=c+1}^{l} \frac{(n-c)}{l\gamma} p_n. \tag{8}$$

5 Numerical results

Figure 3 depicts the fraction of talkspurts $\mathcal{F}_{FEC}$ that experience FEC in excess of $b = 50$ ms for the reservation method, as a function of the number of voice sources in the system, with an activity factor $\gamma = 0.4$, a mean talkspurt duration $1/\mu = 1.5$ s and a number of voice channels $c = 25$. We see that 46 voice sources can be supported while keeping $\mathcal{F}_{FEC} \leq 0.02$.

The corresponding performance curve for the demand-assignment method is shown in figure 4, where we plot the total fraction of speech clipped $\mathcal{F}_{MSC}$ as a function of the number of voice sources in the system, for the same set of parameters $\gamma = 0.4$ and $c = 25$ (the values of $\mathcal{F}_{MSC}$ do not depend on the mean talkspurt duration). In this case, 55 voice sources can share the transmission channel while assuring $\mathcal{F}_{MSC} \leq 0.02$. These first results show the superiority of the demand-assignment method as compared to the reservation method.

In order to investigate the influence of the various system parameters on the performance of the multiplexing methods considered, we adopt as the basic performance measure the normalized multiplex capacity, defined as the number of voice sources per voice channel that can be supported while satisfying the speech quality requirements of Section 2.

5.1 Speech activity factor

Figure 5 depicts the normalized multiplex capacity of both the reservation method, η_R, and the demand-assignment method, η_D, for $1/\mu = 1.5$ s, $c = 25$ and a speech activity factor, γ, ranging from 0.35 to 0.45. Clearly η_R and η_D decrease with increasing γ. The superiority of demand-assignment over reservation is kept along the values considered for γ, although it is slightly more pronounced for lower values of γ. At $\gamma = 0.35$, $\eta_R = 2.08$ and $\eta_D = 2.48$ which corresponds to a 19% enhancement of the demand-assignment method over the reservation method, whereas at $\gamma = 0.45$, $\eta_R = 1.68$ and $\eta_D = 1.96$ which corresponds to a 17% enhancement.

5.2 Mean talkspurt duration

The influence of the mean talkspurt duration on the multiplex capacity is shown in figure 6, where $\gamma = 0.4$, $c = 25$ and $1/\mu$ ranges from 0.5 to 2.5 s. The capacity of the demand-assignment method does not depend on $1/\mu$ whereas the capacity of the reservation method decreases with increasing $1/\mu$. The longer $1/\mu$ the longer the process $N(t)$, describing the evolution of the number

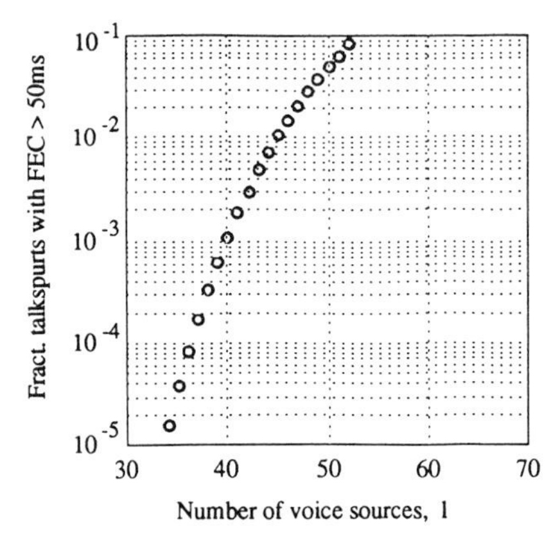

Fig. 3 - Fraction of talkspurts incurring FEC in excess of 50 ms as a function of the number of voice sources in the system, for γ=0.4, 1/μ=1.5 s and c=25.

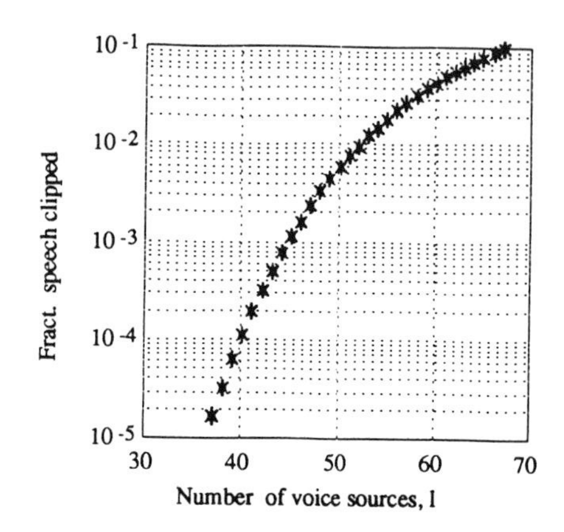

Fig. 4 - Total fraction of speech clipped as a function of the number of voice sources in the system, for γ=0.4, 1/μ=1.5 s and c=25.

of talkspurts in the system over time, spends in each state and the greater the probability that a talkspurt, arriving during an overload period, has to discard more than 50ms of speech before being assigned reserved access to the transmission channel. The asymptotic value of η_R for large $1/\mu$ corresponds to the case in which all talkspurts arriving during overload periods will experience FEC in excess of 50ms.

5.3 Number of voice channels

The effect of the number of voice channels on the multiplex capacity is illustrated in figure 7, where $\gamma = 0.4$, $1/\mu = 1.5$ s and c varies from 15 to 35. We see that the greater the number of voice channels, the greater the multiplex capacities η_R and η_D. Nevertheless, the advantage of demand-assignment over reservation is kept aproximately constant as the number of voice channels varies from 15 to 35.

6 Conclusions

We presented a comparative study between a reservation and a demand-assignment multiplex method for use in packet voice communication systems with slow speech activity detection. These two types of multiplexing strategies affect the perceived speech quality in different ways and set different performance requirements to the underlying transmission mechanism. The results show that about 20 % more voice sources can be accommodated in the demand-assignment method as compared to the reservation method. We found that this advantage is kept for a wide range of values of variation in the speech activity factor, mean talkspurt duration and number of voice

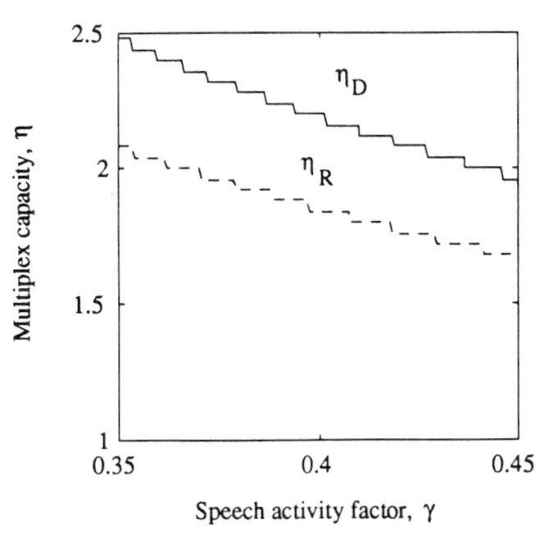

Fig. 5 - Multiplex capacity as a function of the speech activity factor, for $1/\mu$=1.5 s and c=25.

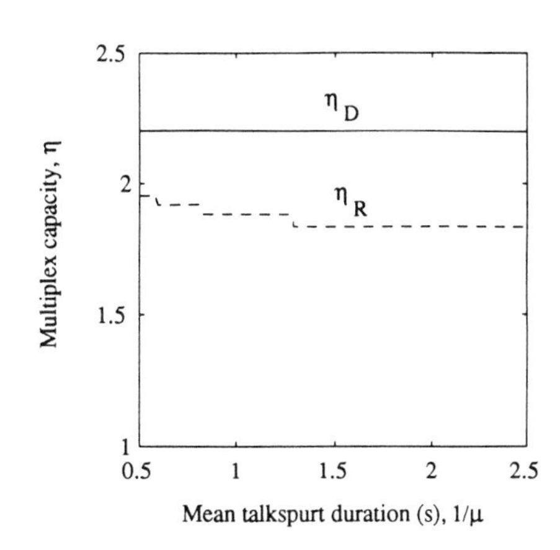

Fig. 6 - Multiplex capacity as a function of the mean talkspurt duration, for γ=0.4 and c=25.

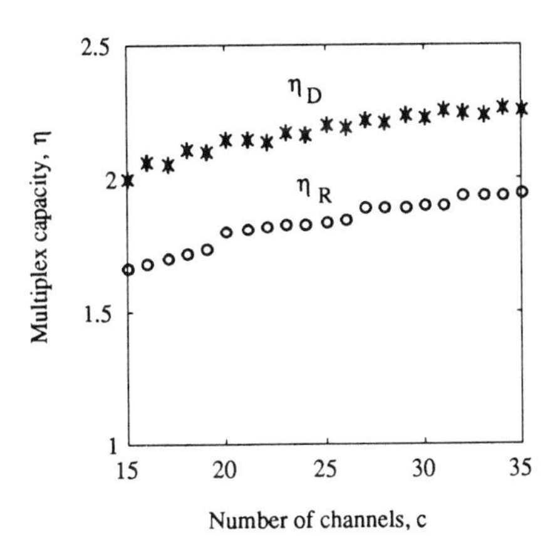

Fig. 7 - Multiplex capacity as a function of the number of voice channels, for γ=0.4 and $1/\mu$=1.5 s.

channels.

These conclusions suggest consideration of demand-assignment methods in the planning of third generation wireless telephony systems, where only reservation strategies have been proposed so far.

Acknowledgements

J. L. Sobrinho acknowledges the financial support of the Portuguese Research Council for Science and Technology (JNICT).

References

[1] Michael Fine and Fouad A. Tobagi. Packet voice on a local area network with round robin service. *IEEE Trans. on Commun.*, 34(9), September 1986.

[2] Timothy A. Gonsalves and Fouad A. Tobagi. Comparative performance of voice/data local area networks. *IEEE Trans. on Commun.*, 7(5), June 1989.

[3] D. J. Goodman, R. A. Valenzuela, K. T. Gayliard, and B. Ramamurthi. Packet reservation multiple access for local wireless communications. *IEEE Trans. on Commun.*, 37(8), August 1989.

[4] N. M. Mitrou, TH. D. Orinos, and E. N. Protonotarios. A reservation multiple access protocol for microcellular mobile-communication systems. *IEEE Trans. Veh. Technol.*, 39(4), November 1990.

[5] K. Bullington and J. M. Frazer. Engineering aspects of TASI. *Bell Syst. Tech. J.*, 38, March 1959.

[6] R. L. Easton, P. T. Hutchison, R. W. Kolor, R. C. Mondello, and R. W. Muise. TASI-E communications system. *IEEE Trans. on Commun.*, 30(4), April 1982.

[7] John G. Gruber. A comparison of measured and calculated speech temporal parameters relevant to speech activity detection. *IEEE Trans. on Commun.*, 30(4), April 1982.

[8] Paul T. Brady. A statistical analysis of on-off patterns in 16 conversations. *Bell Syst. Tech. J.*, 47, January 1968.

[9] David J. Goodman and Sherry X. Wei. Efficiency of packet reservation multiple access. *IEEE Trans. Veh. Technol.*, 40(1), February 1991.

[10] S. J. Campanella. Digital speech interpolation. *COMSAT Tech. Rev.*, 6, Spring 1976.

[11] John G. Gruber and Leo Strawczynski. Subjective effects of variable delay and speech clipping in dynamically managed voice systems. *IEEE Trans. on Commun.*, 33(8), August 1985.

20

THE FIRST ANTENNA AND WIRELESS TELEGRAPH, PERSONAL COMMUNICATIONS SYSTEM (PCS), AND PCS SYMPOSIUM IN VIRGINIA

G. H. Hagn, Senior Staff Advisor
Information and Telecommunications Sciences Center
Information, Telecommunications and Automation Division

E. Lyon, Assistant Director
Systems Technology Center
Systems Technology Division

SRI International
1611 North Kent Street
Arlington, Virginia 22209

ABSTRACT

This is the story of the early history of wireless communications using antennas. The first antenna was conceived by Dr. Mahlon Loomis, a Washington, D.C. dentist, in 1864--at the same time that Prof. James Clerk Maxwell first presented his equations to the Royal Society in London. Loomis used 600-ft wires, suspended by kites, a hand-held key to transmit and a galvanometer detector to receive on-off-keyed (OOK) signals at 0.033 baud in October 1866 between two Virginia mountain tops separated by 18 miles. These two-way communications, which were witnessed, provided the impetus for Loomis' U.S. Patent No. 129,971, "Improvement in Telegraphing (Wireless)," dated 30 July 1872. Loomis' patent was the second *U.S. wireless patent, by 3 months, to that of his friend, William Henry Ward, of Auburn, NY. It also inspired HR 772, "Act of Incorporation of Loomis Aerial Telegraph Company," during the 3rd Session of the 42nd Congress in 1873. This law established the first U.S. wireless company, but the company failed due to losses by Loomis' venture capitalists in the great Chicago fire of 1871 and the "Black Friday" financial crises of the 19th century. Whether Loomis' system utilized electromagnetic (Hertzian) waves, of the type first demonstrated definitively for the first time by Heinrich Rudolph Hertz in Karlsruhe, Germany, in 1886 (the year of Loomis' death), or whether his system functioned totally on the basis of electrostatic conduction is unresolved. Loomis' system did radiate electromagnetic waves when his key closed and generated sparks, but it is not clear whether his DC galvanometer could detect these waves due to some hysteresis effect or other nonlinearity. Loomis used the word* wireless *and he apparently coined the term* aerial. *Loomis also achieved communications between two ships on the Chesapeake Bay which were separated by 2 miles. This represented the first maritime mobile wireless communications, and it occurred before the birth of Marchese Guglielmo Marconi in 1874. But it remained for Marconi to recognize and successfully exploit the commercial aspects of wireless communications using Hertzian waves to provide service to governments and to the general public.*

INTRODUCTION

Dr. Mahlon Loomis, a Washington, D.C. dentist, was born on 21 July 1826, in Oppenheim, near Gloversville, Fulton County, New York. He was a member of the seventh generation of descendants of Joseph Loomis, who came to Massachusetts from England in 1638. The Loomis family was moved to Springvale, Virginia (near Dranesville, about 20 miles west of Washington,

D.C.), in about 1840. Loomis began the study of dentistry in Cleveland in 1848. He subsequently moved to Cambridgeport, Massachusetts, where he became world famous for his invention of mineral-plated false teeth, which were patented in the United States, Great Britain and France. During the academic year 1853-54, Loomis attended the lectures on electricity given by Professor Joseph Lovering at the Lowell Institute in Boston. He subsequently employed a battery and 9 3/4 pounds of No. 18 AWG copper wire (purchased for $4.87), to electrically fertilize his garden and vineyard, apparently with successful results.

Loomis conceived the first communications antenna (see Figure 1) in Washington, D.C. on Thursday, 21 July 1864. This was the same year that the Scottish physicist James Clerk Maxwell (1831-1879) first presented his ideas on electricity and magnetism to the Royal Society in London. A description was given in his notes (now in the Library of Congress) which accompanied the sketch in Figure 1: "a plan may be matured by which to remit shocks." His notes also stated:

> . . . disturb the electrical equilibrium, and thus obtain a current of electricity, or shocks or pulsations, which traverse or disturb the positive electrical body of the atmosphere above and between two given points.

He called his invention an *aerial telegraph* because it electrically coupled into the air, into what he believed to be an electrified conducting layer that also was the source of lightning and the northern lights. The term *aerial* became associated with the elevated wire, which we now call an antenna. He was inspired by the famous 1752 kite experiment of Benjamin Franklin and by the successful transition from Samuel F. B. Morse's two-wire telegraph (which Morse demonstrated experimentally in the winter of 1835-36, and which began commercial operation in 1844) to the one-wire telegraph of Carl August von Steinheil (and later Morse) using the ground as the return wire. In the fall of 1866, Loomis performed the remarkable feat of wireless telegraphic communications, in both directions at the rate of two symbols per minute (0.033 baud), over an 18-mile path in the Blue Ridge Mountains of Virginia. He used kite-borne 600-foot copper wire aerials of his own design (flown from mountain tops at Bear's Den on Mt. Weather and Furnace Mountain, opposite Point of Rock, Maryland on the Potomac about 60 miles west of Washington, D.C.), a hand-operated switch to "key" the transmitter, and a galvanometer for a detector. After receiving the second U.S. patent (by only 3 months) for wireless telegraphy in 1872, he formed the Loomis Aerial Telegraphy Company to exploit this new technology. His company was authorized by an act of Congress signed by Secretary of State Hamilton Fish on 19 January 1873, and later signed into law by President Ulysses S. Grant. The company was funded by venture capitalists, but it was not successful. The negative financial impact of the 1871 Chicago fire on his first set of investors and the financially turbulent times (e.g., the series of "black Fridays" in the second half of the 19th century—especially the panic of 1873) on his other investors proved decisive, and he could not personally fund the company for a long enough period. Nevertheless, Loomis did coin the term *aerial*, and he did design and successfully demonstrate the first wireless telegraph that used antennas. He also demonstrated the first wireless personal communications system (PCS) that achieved a range greater than 10 miles—a direct ancestor of the modern wireless PCS. If the "distinguished witnesses" of his successful October 1866 wireless demonstration at Bear's Den toasted his success with wine (or some other alcoholic beverage), then the event might have legitimately been the original Virginia wireless PCS *symposium*.

A *Boston Post* editorial in 1865, the year before Loomis' first wireless telegraph demonstration, noted that "Well informed people know it is impossible to transmit the voice over wires and that were it possible to do so, the thing would be of no practical value." On 10 March 1876, Alexander Graham Bell (1847-1922) said over the first telephone, "Mr. Watson, come here, I want to see you," thus disproving the *Boston Post's* technology forecast only a little over a decade after it was made. Only two years later (in 1879), the *Hartford Times* mentioned that Loomis had successfully achieved wireless telephony over a distance of about twenty miles. Bell used his *photophone* in 1880 to send the human voice via electromagnetic (albeit optical and not radio) waves to his Washington, D.C. L Street laboratory from the Franklin School. In 1881, William Henry Preece (1834-1913) invented the duplex telegraph; and, he successfully transmitted wireless telephonic speech to Mr. A. W. Heaviside over a span of 1000 yards using square horizontal loops one-quarter of a mile on a side. It remained for Guglielmo Marconi (1874-1934) to employ "Hertzian" electromagnetic waves and, shortly after the turn of the century, make wireless telegraphy and telephony a financial success. Thus, these pioneers (and many others) paved the way for modern-day wireless analog and digital PCS—the topic of the annual modern-era Virginia symposia on wireless PCS begun by Professor Theodore Rappaport of the Virginia Polytechnic Institute and State University (Virginia Tech) in Blacksburg in 1991.

THE EVOLUTION OF LOOMIS' AERIAL (WIRELESS) TELEGRAPH

Benjamin Franklin (1706-1790), the scientist, statesman and philosopher of Philadelphia, made several transatlantic crossings while studying the Gulf Stream. In 1726, he met Dr. Henry Pemberton, Secretary at the Royal Society (who had helped Isaac Newton edit the 3rd edition of *Principia*) and other English scientists. Franklin was aware of an incident that occurred in July 1731 in Wakefield, England, when a tradesman's knives were struck by lightning. Some of the knives melted, but their sheaths (which were insulators from an electrical standpoint) were not damaged. When the damaged knives were dumped onto a counter on top of some nails, the knives "took up the nails." In Boston, Franklin met Dr. Archibald Spencer in 1743 and saw some of Spencer's demonstrations of static electricity. Franklin's interest in lightning derived in part from his desire for better weather forecasts for his *Almanac*. In 1746, Franklin obtained assistance from the English botanist, Peter Collinson, who sent him an "electric tube," and with whom Franklin had much scientifically significant correspondence. Franklin experimented with Leyden jars (which then was the only storage devices for electronic charge) from 1747 on. In 1749, he mentioned the use of conductors to safely discharge low thunderclouds. In July 1750, Franklin wrote of his ideas on lightning and atmospheric electricity to the Royal Society in London. He suggested an experiment using a grounded rod with a sharp tip pointed upward during a thunderstorm to try to draw an electric charge to earth in an attempt to determine whether the clouds contained atmospheric electricity, as suggested by the occurrence of lightning. The letter was not published by the Royal Society, but was translated into French, printed in 1751 as a pamphlet, and circulated in France. At the request of Louis XV, King of France, Thomas Fransçois D'Alibard prepared for the experiment Franklin had suggested by installing in his garden at Marly-la-Ville a 30- or 40-foot grounded conducting rod with a sharp point on the top end. D'Alibard then waited for the next storm. He wasn't home when it occurred in May 1752; however, in his absence, his servant successfully performed the experiment and drew electricity out of the thundercloud—and in a dramatic fashion. Franklin learned of this a few months later (via sea mail), but he did not think the French experiment was definitive. Therefore, he set out to continue the investigations using a different approach. Accompanied by his son, he conducted his own classic experiment in an open Pennsylvania field in 1752. Franklin demonstrated the existence of atmospheric electricity by flying a kite up into a thunderstorm. The kite he used was a cross type, with a wire projecting upward from the upright stick. It was tethered by a silk string, said to have been about 500 ft long and about 1/16 inch in diameter, one end of which he held wrapped in a dry cloth. After the silk string became wet from the rain (and hence conductive), he used a brass key on the ground end to draw sparks to his knuckles. In this manner, Franklin revealed to the world a larger and more powerful source of electric charge than the friction machines of the day (which were impractically small as sources of energy for signalling over significant distances).

Friction machines had been around since 1672, when Otto Von Guericke of Amsterdam (who had demonstrated the force of the atmosphere in the famous Magdeburg experiment performed for Emperor Ferdinand) had rotated a sulphur ball (one of Dr. William Gilbert's "electrics," *materials which could be electrified**) on a shaft to collect a charge by being stroked with a dry palm. These machines were mostly used for parlor tricks, sometimes by charlatans, but serious English scientists like Charles Darwin(1809-1882), James Wimshurst (1832-1903), and William Thompson (Lord Kelvin, 1824-1907), also tried their *hand* with it.

The lightning rods resulting from Franklin's research were first recommended to the public in 1753. They were the first commercially successful products designed for a mass market which were based on electricity. Franklin survived his dangerous experiment to become world famous. This was long before he signed "A Declaration by the Representatives of the United States of America" (popularly known as the Declaration of Independence) in his native city of Philadelphia on 4 July 1776. He was made a Fellow of the Royal Society for his work on lightning, and he became an American hero for his role in the history of electricity, the founding of the United States, for his *Poor Richard's Almanac*, and for inventing bifocals and the Pennsylvania fireplace (also known as the Franklin Stove).

A Spanish physicist named Silva (spelled Salva by Mary Texanna Loomis, a distant cousin of Mahlon), read a paper before the Academy of Sciences in Barcelona on 16 December 1795: "On the Application of Electricity to Telegraphy." According to Collins [1909], Silva advocated that "a given area of earth be positively electrified at Mellorca and that a similar area of earth be charged to the opposite sign at Alicante; the sea connecting these two cities would then act as a conductor when the electric difference of potential would be restored, and by a proper translating device the transfer of energy could be indicated." Presumably, the translating device would consist of an energy coupler of some type suitable for this *conduction* method of wireless communication and a device that could indicate a transient current (e.g., a galvanometer with small enough mass so that it could produce a temporary deflection).

---

*Gilbert (1544-1603), the physician of Queen Elizabeth (1601) and James I (1603), and noted natural philosopher, coined the term *electric* for materials which could be electrified--in about 1600 (before the Mayflower sailed for America). Gilbert died in November of 1603, the same year that Shakespeare penned Hamlet. It wasn't until 1675 that the word electricity was used in print by Robert Boyle, in a manuscript published in Oxford: "On the Mechanical Production of Electricity." Gilbert showed that other materials than amber could be made to attract light objects when rubbed, and he called this group of substances *electrics*. He used the terms electric force, electric attraction, and magnetic pole; and he is considered to be the "father of electricity." Gilbert also discovered the dip of the earth's magnetic field and noticed that the dip also occurred near a spherical magnet.

In the winter of 1819, a Danish physics professor, Hans Christian Oersted (1777-1857), discovered the relationship between electricity and magnetism after a 12-year quest. Oersted was aware of Franklin's observation that electricity seemed to have a mysterious effect on a compass needle. In 1751, Franklin had tried to magnetize (with some success) sewing needles with sparks from Leyden jars. In his classroom, Oersted used a battery to generate a current in a wire, and noted with a compass that the current produced a circular magnetic field around the wire. Oersted, for whom the unit of magnetic field strength (in the CGS system, a force of 1 dyne exerted on a unit magnetic pole) is named, motivated other scientists to study the phenomenon, including Sir Humphrey Davy (1778-1829) at the Royal Institution in London. Davy, a Fellow of the Royal Society, hired, tutored, and mentored Michael Faraday. Faraday made the next major advance in the evolutionary process of wireless communications.

Michael Faraday, (1791-1867) the English chemist and physicist, followed up on the work of Oersted and André Marie Ampère (1775-1836). Faraday discovered magnetic induction on 29 August 1831 in his laboratory in the Royal Institution in London. He showed that a current in one conductor "induced" currents in neighboring inductors. Faraday next connected a coil to a galvanometer and a battery, and he noted a deflection in the galvanometer when the current flowed in the coil, indicating that the current had produced a magnetic field. He reported the discovery to the Royal Society on 24 November 1831, and published the results in the classic text, *Experimental Research on Electricity*, which documented his work from 1831 through 1855. Faraday was subsequently honored by having the Faraday (the charge required in an electrochemical reaction involving a given amount of a substance; 96, 500 Coulombs per gram equivalent weight of an element or ion involved in the electrolysis) and the Farad (the unit of capacitance corresponding to the retention of 1 Coulomb of charge when 1 Volt is applied to a capacitor) named after him to recognize his contributions to both chemistry and physics. Only a few days after Faraday's discovery, Ampère (for whom the unit for measuring current is named: 1 Ampère is the current that flows when 1 Volt is applied to a circuit with a resistance of 1 Ohm) demonstrated that two unrestrained parallel wires carrying current attracted each other when the currents were in the same direction and repelled each other when the current flows were in the opposite direction. Magnetic induction was later used by Joseph Henry and others to achieve short-distance wireless communications. Henry (for whom the unit of inductance was named) was the first to use insulated wires, formed into coils, to make *inductors* and electromagnets which were used to lift iron.

While sailing from Le Havre, France, on the packet ship *Sully* in 1832, Samuel Finley Breese Morse (1791-1872) heard an explanation of Ampère's experiments in electromagnetism presented by Dr. Charles Thomas Jackson (1805-1880) a chemist and physician from Boston. Morse is said to have asked Jackson if electricity could be made to pass over many miles of wire almost instantaneously, and to have received a positive reply. Morse, then 41, is said to have made the following prescient comment: "If the presence of electricity can be made visible in any part of a circuit, I see no reason why intelligence may not be transmitted instantaneously by electricity." In the winter of 1835-1836, Morse successfully sent signals over 1700 feet of wire strung around his room at the University of the City of New York, where he was on the faculty. Morse applied for a patent on his two-wire telegraph in 1837; and in that same year, Sir Charles Wheatstone (1802-1875), who had been elected a Fellow of the Royal Society the year before, applied for a British patent on a similar system. In 1845, Wheatstone, more famous for his bridge circuit, developed the single-needle telegraph (which he had designed with William F. Cooke).

Karl Friedrich Gauss (1777-1855), the German mathematician from Göttingen and Fellow of the Royal Society (1804), suggested to his colleague Carl August von Steinheil the use of railway tracks as conductors for the outgoing and return legs of a telegraph circuit. In 1838, Steinheil found that approach to be impractical because he could not sufficiently insulate the iron rails from the ground. He converted this problem to an economic advantage, however, by using the ground for the return conductor of an above-ground telegraph wire circuit, thus reducing the cost of the wire per circuit mile by a factor of two. Thus, the one-wire telegraph was discovered and first demonstrated in Germany by Steinheil six years after Morse first began his consideration of telegraphy. Steinheil (for whom the transparent mineral Steinheilite, or iolite is named) also invented the printing telegraph (used later for stock ticker tapes) in 1836, the *sounder* (which was the first nonvisual indicator for a telegraph signal), the electric clock, and an electromagnetic motor. He became Head of the Telegraph Department of the Austrian Ministry of Commerce in 1849. Steinheil also dreamed about wireless communications: "Had we means that could stand in the same relation to electricity that the eye stands to light, nothing would prevent our telegraphing through the earth without conducting wires; but it is not probable we shall ever attain this end." Thus, Steinheil may have started the trend of pessimistic predictions about PCS which was continued by the *Boston Post* in 1865.

In 1842, Morse is said to have communicated for short distances under water and, in 1843, Morse was experimenting with communication across water. He concluded that water also could replace one wire in a telegraph circuit. That same year, he devised the code that bears his name, in collaboration with Alfred N. Vail. He also was successful in getting $30,000 from the U.S. Congress to construct a telegraph line along the right of way of the Baltimore and Ohio Railroad. The first commercial telegraph was put into service between Washington and Baltimore by Morse on 24 May 1844.

In 1854, William Henry Preece of England was assigned to investigate a "wireless" system, based on earth conduction, that recently had been developed by James Bowman Lindsay (1799-1862). Scottish philologist Lindsay had communicated over short distances near Dundee, Scotland, using an earth conduction system powered by artificial batteries. Lindsay used a galvanometer as a detector for these experiments during the period 1844-1853. He was awarded British Patent No. 1242 for this system on 5 June 1854. Earlier (in 1835) he had invented a lamp that utilized an electric current to improve its performance. In 1885, Preece transmitted telegraphic speech over 1000 feet by ground conduction. In 1896, he used the induction method to achieve a range of 4.5 miles between the English mainland and the Scottish Island of Mull by laying wires on the two shores. At the time, this range was greater than the ranges achieved the year before by Alexander Popov in Russia and Guglielmo Marconi in Italy, using wireless equipment with "coherers" for detectors.

Morse's company evolved into Western Union, which completed the first transcontinental telegraph line in 1861, at the beginning of the Civil War. The following year, John Haworth was awarded British Patent No. 843 for a method of conveying electric signals "without the intervention of an artificial conductor" between buried copper and zinc plates and buried coils supplied with weak battery power. Others also experimented with various wireless conduction and induction communications systems.

The next evolutionary step was a no-wire (i.e., wireless) telegraph, with the air providing the upper conductor through atmospheric electricity (which was sufficient on cloudy days) and the ground providing the return path. The first wireless telegraph using antennas (which we believe he called aerials), was conceived toward the end of the Civil War on 21 July 1864 by a Washington, D.C. dentist, Dr. Mahlon Loomis. The first wireless that did not rely exclusively on ground conduction (or on magnetic induction) utilized antennas and what we believe to be an electrostatic effect to achieve a range of 18 miles between two northern Virginia mountaintops (see Figure 2), in 1866, the year after the Civil War ended. Some Loomis' supporters (e.g., Appleby, 1967; Marriott, 1925) claim that he used electromagnetic waves, but this is controversial since Loomis' description of his detector is somewhat "sketchy" (see Figures 3 and 4).

Professor James Clerk Maxwell, a young Scottish physicist from Edinburgh who was at Kings College, London, presented his ideas on electricity and magnetism to the Royal Society in 1864. In 1867, Maxwell published his still-famous mathematical equations that described electrostatic and magnetostatic effects and predicted the existence and behavior of electromagnetic waves propagating through a medium he called an "ether."

Dr. Loomis, unaware of Lindsay's or Maxwell's equations, was inspired by the experiment of Franklin, the successful evolution of Morse's two-wire telegraph to a one-wire telegraph, and the phenomena of aurora borealis causing currents to flow in unkeyed telegraph lines. Loomis was an inventive man, and he had patented an improved version of false teeth that used a kaolin mineral plate process in 1854. His friend, William Henry Ward, took some examples to England at Loomis' request. Loomis was to acquire 6 more patents, including the second U.S. Patent for wireless telegraphy in 1872:

1. U.S. Patent No. 10,847, dated May 2, 1854; for *Artificial Teeth*, and issued to him at Cambridgeport, Massachusetts;
2. British Patent No. 1175 dated 1854; for *Artificial Teeth*, and issued to him at Cambridgeport, Massachusetts;
3. French Patent No. 10261 dated May 31, 1854; for *Artificial Teeth*, and issued to him at Cambridgeport, Massachusetts;
4. U.S. Patent No. 129,971 issued July 30, 1872; for *Improvements in Telegraphing (Wireless)*, and issued to him at Washington, D.C.;
5. U.S. Patent No. 241,387 issued May 10, 1881; for *Convertible Valise*, and issued to him at Washington, D.C.;
6. U.S. Patent No. 250,268 issued November 29, 1881; for *Cuff or Collar Fastener*, and issued to him at Lynchburg, Virginia; and,
7. U.S. Patent No. 338,090 issued March 16, 1886; for *Electrical Thermostat*, and issued to him at Terra Alta, West Virginia, the year he died.

He conceived the first antenna (for which we believe he coined the term *aerial*) in 1864 as a long wire held aloft by a metal-gauze-covered balloon or kite and grounded at its base with copper radial wires and a copper plate, as recorded in his remarkable 1864 sketch (Figure 1). Loomis was aware of observations during the period of sunspot maximum around 1860 (when daily sunspot numbers exceeded 200) that on some occasions currents flowed in unkeyed telegraph circuits when aurora borealis (northern lights) were visible in the sky. He dreamed of a no-wire telegraph that used atmospheric electricity to substitute for the overhead wire in a wireless telegraph circuit. He also used the term *wireless*; and he conceived the first wireless telegraph, based on the use of antennas and natural conductors: atmospheric electricity for an overhead conductor and the ground for a lower conductor (see Figures 1 and 2). Loomis successfully demonstrated his aerial (wireless) telegraph in October 1866 before "distinguished" witnesses. The names of these witnesses were not recorded by Loomis or others, with the exception of a note in his diary regarding attendance by two Congressmen (Senator Pomeroy of Kansas and Representative Bingham of Ohio). Appleby (1967) quoted the *Western New England Magazine*, Vol. 3, pp. 27-29, January 1913: ". . . of demonstrations by Loomis in the presence of prominent scientists and electricians," and also, ". . . who demonstrated to Professor Joseph Henry and other scientists from the Smithsonian Institution in the Blue Ridge mountains in 1868." It seems reasonable that Loomis would have invited Henry to witness one of his demonstrations. The date implies a later demonstration than the original October 1866 demonstration. This view is partially substantiated in a report which included comments on the Mahlon Loomis collection and biography, published by the Reference Department of the Library of Congress in 1955, (*Guide to the Special Collections of Prints and Photographs in the Library of Congress*, compiled by Paul Vanderbuilt), which mentioned:

> Dr. Mahlon Loomis (1826-1886) succeeded in a demonstration before Members of Congress and eminent scientists in 1868, in transmitting telegraphic signals from one electrically grounded kite on a mountain in Virginia to another grounded kite 18 miles away. Loomis produced sparks and caused deflection of the galvanometer of one kite when the wires of the other kite were touched to the ground.

Loomis successfully communicated in both directions over an 18-mile path (mislabeled as a 14-mile path in Loomis' sketch, given in Figure 2) in Northern Virginia's Blue Ridge Mountains. He personally operated the terminal at Bear's Den, Virginia, about 5 miles north of the U.S. Weather Bureau's site on Mount Weather (see Figure 3). The attendance at this demonstration by Senator Pomeroy was partially confirmed by his statement on Loomis' aerial telegraph in the 13 January 1873 *Congressional Globe*: "I believe in it. I have seen two or three experiments, and I think there is something to it. I have seen it tested in a small way, and I am inclined to think it will succeed." Loomis' associate (whose name was not recorded) operated the terminal at Furnace Mountain, Virginia—the Catoctin Spur site near the Potomac River approximately opposite Point of Rocks, Maryland (see Figure 4). Furnace Mountain has an elevation of 889 ft; whereas, Mt. Weather's peak is 1760 ft above sea level. Five miles north of the peak, however, the elevation is only 1450 feet above sea level. Opening and closing the antenna circuit at the transmit end for 30 seconds each time caused a DC galvanometer to change its deflection at the receive end, thus conveying two bits of information every minute (i.e., wireless communication at the rate of 0.033 baud). Loomis' distant cousin, the Washington, D.C. radio educator Mary Texanna Loomis, quoted Dr. Elisha Loomis of Berea, Ohio, in her textbook (which sold for $3.50 in 1928 as the 4th edition, revised). Elisha noted Mahlon's success with the first fixed and maritime-mobile wireless communications in a speech on 7 October 1914 at the dedication of the Loomis Institute in Windsor, Connecticut: ". . . he demonstrated [wireless telegraphy] beyond controversy in 1866, in the presence of eminent scientists and electricians, by sending many messages between two stations, in Virginia, eighteen miles apart, and at sea, on Chesapeake Bay, between two ships two miles apart."

THE FIRST TWO U.S. WIRELESS PATENTS AND THE FIRST WIRELESS TELEGRAPH COMPANY

On 15 January 1869, Loomis petitioned the 41st Congress for funds for a telecommunication system, as had Morse before him. He asked the Congress for $50,000 to start a Federal aerial (wireless) telegraph company, but the petition was denied on 19 May 1870. Also, in 1869 (the same year that Senator Leland Stanford pounded the golden spike at Promontory, Utah, to complete the first transcontinental railroad), Loomis had interested Bostonian venture capitalists in funding his proposed company, but one of the "black Fridays" wiped them out. In 1871, Loomis went to Chicago and found new backers, but the famous Chicago fire that year wiped out the resources of those investors. Loomis seemed to be a technical success but financially jinxed.

A supportive congressman, the Honorable John A. Bingham of Ohio (who evidently was present at the October 1866 demonstration), suggested that Loomis file for a patent instead. With the assistance of his friend, Attorney General Edward Washburn Whitaker, Loomis filed a handwritten application on 7 February 1872 for his 1864 invention: "Improvement in Telegraphing (Wireless)." No sketch was required and none was filed. Loomis received U.S. Patent No. 129,971 for his wireless telegraph on 30 July 1872.

Loomis' patent was preceded three months earlier by the patent of one of his acquaintances, William Henry Ward of Auburn, New York. Ward was issued U.S. Patent No. 126,356, "Improvement in Collecting Electricity for Telegraphing," on 30 April 1872. Ward's patent did include a sketch of a tall tower on a mountain top (see Figure 5); and he also sought to use atmospheric electricity to complete a telegraph circuit. Ward's patent described his concept of an electrical tower for connecting aerial currents with earth currents to complete communications circuits between North America (New York City) and South America (Buenos Aires), and he also proposed communications across oceans. Ward envisioned a chain of such towers, interconnected by wire telegraph circuits, to go from New York to Denver and from Denver to Pike's Peak via wire, and via wireless from a tower on Pike's Peak in Colorado to a tower in the Andes. From there, the circuit would continue via wire telegraphy to Quito, Ecuador and on to Buenos Aires (which he spelled Ayres). A system implementing Ward's concept was not built. Nevertheless, Loomis' patent was the second patent (to Ward's) of many in the wireless field, most of which proved to be contentious (and a great source of frustration for radio scientists and engineers, but a great source of income for lawyers).

On 19 December 1871, Representative Bingham presented a new bill to the House of Representatives to incorporate Loomis' company. During the discussion of H.R. 772 on 20 May 1872, Representative Omar D. Conger of Port Huron, Michigan (later Senator Conger) provided the following technical explanation of the operation of Dr. Loomis' aerial telegraph, which undoubtedly had been provided to him by Loomis:

> MR. CONGER: If we look to the Committee for the District of Columbia, who have in charge the very birthplace of this invention, we shall find the legitimate objects of their supervision just at this time "earthy and superficial."
>
> It is proposed by the inventor to do this (Aerial Telegraph) from the highest attainable physical elevations across the aerial spaces from point to point, without wires, and by simple vibrations of electrical forces spreading out like waves across the continent.

This theory assumes that the earth itself, the atmosphere surrounding it, and the infinite depths of space encompassing this aerial world, contains a succession of concentric circles or planes of electricity—which may be affected by an interpenetrating galvanic force from beneath, causing electrical vibrations, or waves, to pass from that point within such electrical plan around the world, as upon the surface of some quiet lake one wave circlet follows another from point of disturbance to the remotest shores. So that from any other mountaintop upon the globe any conductor which shall pierce this plane and receive the impression of such vibration may be connected with an indicator, which shall mark the length and duration of such vibration, and indicate by any agreed system of notation convertible into human language, the message of the operator at the point of first disturbance; and this not only from one, but from many mountaintops, piercing far above circumambient atmosphere, the devotee of science and the solemn student of nature may gather the unwritten messages of interest or affection from the silent solitudes of nature and the cerulean depths of heaven with unerring accuracy, and transmit them to the denizens of all lands by the mundane machinery of telegraphic instrumentalities.

Such, Mr. Speaker, in brief, is the outline of this simple but marvelous theory which the Committee on Commerce have the honor of submitting for your consideration.

[This account was printed in the *Congressional Globe*, the forerunner of the *Congressional Record*.]

The press of the day was enthusiastic (but sarcastic) about the idea of an aerial telegraph, and Loomis (not unlike pioneers in other fields) was the object of some public ridicule. For example, the next day (22 May 1872), the *New York World* stated: "who will not be edified to know that the time of the House of Representatives was taken up for an hour or more Monday night by an elaborate oration by Mr. Conger of Michigan in support of a proposition for establishing an aerial telegraph."

Some months later, the Philadelphia Press quoted several other sources (Appleby, 1967):

Friday, November 29, 1872. The New York Herald urges government aid for ocean telegraphs, which, we think is all proper; but if it is to be done, why not make an appropriation to test the plan of Professor Loomis' aerial line, which, if his theory is correct, can be carried out without the aid of wire or cable. Professor Loomis has the true idea, and it will eventually revolutionize the whole system of telegraphing. It is not only within the probabilities of science, but is accepted by some of the best scientists in the world—Washington Sunday Morning Chronicle.

As we understand it, Dr. Loomis proposes to make his demonstration from our Western mountain-peaks to the Eastern-shore mountain-peaks of Japan—the very line we now need to have direct communication with China and Japan, and complete the entire communication around the earth. The only hiatus is now across the Pacific Ocean.

> If this system, which has been tested on short experimental lines with complete success, be once inaugurated, the principle will apply to all distances in any direction, whether across the seas or over land, wherever salient points can be secured. The practical working of the present Atlantic Cable was at first utterly scouted by scientific men and ridiculed by all; but this system offers such inconceivable advantages over the cable line, that, judging from past experience through the history of all new and astounding propositions—whether Fulton's steamer or Morse's telegraph be sufficient, capital should be supplied to make the experiment, either by Congress or individuals. This will decide whether we can communicate through the air as we now do through the earth. It will reduce the expense of transoceanic messages to a trifle

The Scientific American Supplement of February 1911, as quoted by Appleby (1967), who was quoting Dr. Elisha Loomis' 1914 speech, stated:

> A description of the actual conditions under which wireless telegraphy is today commercially successful would differ little from the theory advanced in the debates on (Loomis Aerial) wireless telegraphy to which the House of Representatives listened 'with dreamy indifference' thirty-eight years ago.

The 1911 *Scientific American* article, according to Appleby (1967), continued:

> But such is the fate of the man with a big idea who is born too soon. It has been so, is so now, and will continue to be so. Morse with his Magnetic Telegraph, in 1838, was laughed to scorn, and in place of receiving help he received sneers and ridicule. In 1873 Loomis's important discoveries were called "Moonshine" and "airy nothing" and he was made the 'butt of ridicule, the target of merciless arrows of wit, and dubbed a dreamer.'

On 21 January 1873, Bingham and Loomis' other supporters in the 42nd Congress passed H.R. 772 (Act of Incorporation of Loomis Aerial Telegraphy Company) during Session III: This first U.S. wireless telegraph act established his company to exploit the then-new wireless communications technology. Secretary of State Hamilton Fish signed the act on 29 January 1873; but apparently it was not signed into Law by President Ulysses S. Grant. No government funding resulted from the act (in contrast to the funding that earlier benefitted Morse, but only after several attempts by Morse). Loomis' company was authorized capital stock of up to $200,000 (with authority to increase that amount by an order of magnitude, should circumstances warrant such an increase). Loomis successfully attracted venture capitalists with his impressive mountaintop demonstrations. His attorney, Whitaker, purchased ten shares in the Loomis Aerial Telegraph Company, Inc., but Loomis raised only about $10,000 through the sale of stock at $100 per share.

On 19 January 1873, Loomis gave an interview to a reporter from the *Washington* (D.C.) *Chronicle*. This interview indicated that a wireless experiment may have been performed in the Blue Ridge Mountains of Virginia during the summer of 1872. The distance between the mountain peaks for this test was about 15 miles. In the interview, Loomis speculated on transcontinental wireless communications and on evidence to support his theory of how it could work.

On 5 February 1873, the *N.Y. Journal of Commerce* noted the passage of the legislation and observed: "We will not record ourselves as disbelieving in the Aerial Telegraph, but wait meekly and see what the Doctor will do with his brilliant idea now that both Houses of Congress have passed a bill incorporating a company for him. Congressmen, at least, do not think him wholly visionary; and it is said that the President (Grant) will sign the bill; all of which is some evidence that air telegraphy has another side than the ridiculous." A few days later, on 24 February 1872, the *Buffalo Express* concluded that "It would be a fatal objection to the popularity of the system if people had to go to the top of Mt. Hood, Chimborazo, Popocatapetl, or to the crests in the Himalayas and Andes to send their dispatches." They evidently were aware of both the Loomis and Ward patents.

LOOMIS' EXPERIMENTS AND APPARATUS

Loomis established two-way communications in 1866 using on-off keying (OOK) from mountaintop to mountaintop. Identical kite-borne 600-feet copper wire antennas were employed at both ends of the path. These antennas, which were probably fabricated from No. 18 American Wire Gauge (AWG) uninsulated annealed copper, were well grounded into ponds with loose coils of similar copper wire (see Figures 3 and 4). The No. 18 AWG wire used earlier by Loomis in his Massachusetts garden was 40.3 mils in diameter, weighed only 4.917 lbs per 1000 ft, and had a DC resistance of only 6.152 ohms per 1000 ft at 77°F. Six hundred feet of this wire would weigh only about 3 pounds, well within the lift capability of a kite. The kite was flown with a string (see Figures 3 and 4), so there was no stress on the wire other than that from gravity and any wind loading. The DC resistance was only about 4 ohms, exclusive of any resistance in the galvanometer and grounding (which was only a few ohms at most). A key to open and close the circuit and a DC galvanometer (the only sensitive electrical indicating device available to him to use as a detector) were placed in series with the elevated portion of the antenna and its ground connection on both ends of the link. The "transmit key" originally employed by Loomis was only an open circuit in the hand-held wire (see Figure 3). The galvanometer deflection on the receive end (or lack of a deflection) provided the indication of what had been sent during a given time interval when the circuit was functioning properly. The receive-end key was always closed, and the transmit-end key was either open (for a zero) or closed (for a one) to send the information bits in a time-synchronized manner. This scheme was first tried near Mt. Weather, Virginia, but it had weather problems. On clear days no deflection could be produced on the receive end when the transmit key was closed (see Figure 2). On most cloudy days, however, the operation was (according to Loomis) "indistinguishable from that of a standard wire telegraph." This weather limitation may have been overcome by using tall telescoping towers topped with steel rods, as used later by Loomis in tests near Lynchburg, Virginia.

We believe that Loomis' first system operated on an electrostatic effect using conduction rather than by employing a propagating electromagnetic (Hertzian) wave. Any sparks caused by opening or closing the "key" in series with a triboelectrically charged antenna wire could have caused low-level spark-gap-type radiation of electromagnetic waves. The DC galvanometer at the receive end was (in our view) an inappropriate device for detecting such radio-frequency waves—despite some speculation on hysteresis modification by radio frequency (RF) currents. The basic laws of magnetic hysteresis were not established until 1891, by Charles Proteus Steinmetz (1865-1922). Therefore, we believe that he was not using radio waves. His supporters, such as Appleby (1967), have credited him with using a kite with copper-mesh panel loading capacity to achieve a top-loaded vertical monopole, a keying spark-gap arrangement for transmitting and a galvanometer indicator for receiving (see Figure 6, redrawn from Appleby's Plate 7 on his page 21). Loomis kept a "Memorandum or Book of Domestic and Public Occurrences in the Life of Mahlon Loomis," commenced at Cambridgeport, Mass., in the year 1852, the centennial of Franklin's kite experiment. Some time after 15 August 1858, Loomis wrote: "It must not be the primary electricity as we find it in the atmosphere that must be used, but modified or a secondary current used (Dream)." This implies using coils, transient waveforms, and (perhaps) a transformer. There are sketches of coils among the materials in the Mahlon Loomis Collection in the Manuscript Division of the Library of Congress (see Figure 7). The interest of one of the authors (Hagn) in atmospheric electricity was inspired by his colleague Dr. Edward T. (Ted) Pierce, a former student of C.R.T. Wilson (who pioneered cloud chambers at Cambridge). Pierce had earlier written on atmospheric electricity and lightning for the Encyclopedia Brittanica in order to earn "pen money." He also measured with a field mill the decrease in the earth's fair weather electrostatic field (from about 100 V/m) at the top to a negative value of -60 V/m (at midnight) or +10V/m (at noon) at the bottom of Yosemite Falls [Pierce and Whitson, 1965], to document the "Lenard effect." A year earlier, Pierce and Whitson (1964) had noted the effects of gamma rays from atomic weapons tests in Nevada on the earth's static electric field. The earth's field increases during cloudy weather (as documented at Cape Canaveral, to judge when it is safe to launch satellites with lower risk of a lightning strike), and this increase evidently was needed to provide enough current in Loomis' receive antenna to cause his galvanometer to deflect (see Figure 2). This fact can be used to estimate crudely the sensitivity of Loomis' receiver. A fair weather field of 100 V/m, when multiplied by the approximately 200 m height of his aerial, yields an open-circuit voltage of about 20 kV. Accounting for the fact that his kites held the wire up at an angle, this voltage could have been only about 10 kV. Presumably, Loomis used one of Henry's high-resistance type of galvanometers. Otherwise, the charge on his wire would have been drained to earth too quickly through the antenna's equivalent circuit. If the resistance was, say, 10 Mohms, then 10 mA would have flowed through the galvanometer to indicate the signal. The sensitivity of mid-19th century galvanometers could be determined by careful laboratory measurements on devices in the Smithsonian Institution.

Although he was plagued by atmospheric noise from lightning discharges (the first occurrence of interference to wireless communications was caused by natural noise, QRN), and from lightning itself (which damaged the equipment), he did benefit from the absence of man-made noise and interference—and he had no spectrum allocation or assignment problems to solve! He was as fortunate as Franklin to have avoided serious injury from a lightning strike while operating his transmitter in the manner shown in Figure 3! The Russian professor, George Wilhelm Richmann of St. Petersburg, was not so lucky in 1753; he was the first scientist killed studying artificially induced lightning.

LOOMIS' LATER YEARS

As noted earlier, the *Hartford Times* mentioned in its 1878 *Washington Letter* that Loomis had succeeded in wireless telephony over a distance of 20 miles. The following year, on 1 March 1879, the *Electrical Review* of London reported "with telephones in this aerial circuit he (Loomis) can converse a distance of twenty miles." These articles apparently were the first (from the US and UK, respectively) to mention wireless telephony. Loomis' use of wireless telephony evidently preceded Bell's photophone by two years and the 1885 demonstration of Preece by about seven years.

The series of "black Fridays" and the 1871 Chicago fire had financially impacted his investors and doomed his venture before it began to provide a commercial service, and he never got to attempt transoceanic wireless communications. His own personal finances also were depleted from trying to keep his company going. As a result, he was almost penniless when he moved from Washington, D.C. to live with his brother (Judge George Loomis) in Terra Alta, West Virginia. The high terrain that gave the town its name also was the type of terrain he sought for his mountaintop-to- mountaintop wireless communications, and he did further successful wireless telegraph experiments and demonstrations in West Virginia. On 20 September 1882, Loomis wrote a caveat description of his invention, quite similar to Lindsay's 1854 patent description, presumably with the intention of filing it with the U.S. Patent Office. The caveat, however, never was filed.

Loomis' accomplishments were recognized in England in 1904 by A.T. Story in his *Story of Wireless Telegraphy*: "With the experiments of Mahlon Loomis we hear of the application of vertical conductors, or antennae, as they are sometimes called for the transmission of signals to a great distance." Mary Texanna Loomis (1928) commented on the definition of the term antenna: "thus the word antenna, from antennae or feelers of insects, is more apt than aerial, which conveys the sole idea of wires elevated in the air." The IEEE [1993] definition of antenna is: "That part of a transmitting or receiving system that is designed to radiate or to receive electromagnetic waves." Since Loomis' transmitting antenna undoubtedly radiated electromagnetic waves produced from the spark that occurred when he closed his "key," he actually did conceive, build and radiate from the first antenna.

Loomis earned his living by practicing dentistry until his death on 13 October, 1886, from a heart attack at age 60, at his brother's home in Terra Alta. He died a frustrated and impoverished innovator and entrepreneur. Some of his admirers say he died of a broken heart, but they were incorrect. According to Appleby (1967), he said to his brother, George, shortly before his death:

> My compensation is poverty, contempt, neglect, forgetfulness. In the distant future, when the possibilities of the discover, as I see them, are more fully developed, public attention will be directed to its originator; and the congressional records will furnish the indisputable evidence that the credit belongs to me. But what good then?
>
> Still, there is a present satisfaction in knowing that some time the proper credit will be given. In the meantime, others will reap the benefit in worldly wealth and worldly honors. Monuments will be reared to their memory, costly monuments, in token of the world's appreciation of their genius. I ask but a rose-bush to mark my grave, affording a brief resting place for passing song-birds, and I have a feeling that I shall even then be conscious of their carolings.

His wife had a memorial headstone, reading "In Memory of Mahlon Loomis, M.D., Died at Terra Alta, West Va., 1826-1886, Achsah Ashley Loomis, his wife," erected for Dr. Loomis, near his family home, in the Meeting House Hill Cemetery in West Springfield, Massachusetts. He actually was actually buried in Terra Alta near the grave of his brother George. A Virginia

Conservation Commission highway marker was erected in 1948 on Leesburg Pike (Route 7) at the junction of Loudoun County highway 601, just to the west of the Appalachian Trail at Snicker's Gap (near Bluemont, Virginia), about 5 miles north of Bear's Den (see Figure 8). This monument reads: "Forerunner of Wireless Telegraphy. From nearby Bear's Den Mountain to the Catoctin Ridge, a distance of fourteen miles, Dr. Mahlon Loomis, Dentist, sent the first aerial wireless signals, 1866-73, using kites flown by copper wire. Loomis received a patent in 1872 and his company was chartered by Congress in 1873. But lack of capital frustrated his experiments. He died in 1886." As noted earlier, the actual distance was closer to 18 miles.

Snicker's Gap, on the boundary between Loudin and Clarke counties, played a prominent role in Civil War history just prior to Loomis' 1866 experiment. This gap is a key low (1070 feet above sea level) passage through the Blue Ridge Mountains to the Shenandoah River Valley. For example, General Jubal Anderson Early, while returning to the Shenandoah Valley on Saturday, 16 July 1864, after a remarkably successful raid on the perimeter defenses of Washington, D.C., was attacked by the troops of Union General George Crook (who destroyed a few wagons while General Early captured a Union cannon). Early's raid on Washington on 11 and 12 July was less than 2 weeks before Dr. Mahlon Loomis drew the historic sketch of the first antenna, reproduced as Figure 1, in his dentist office at 907 Pennsylvania Avenue, Northwest. Two months later, on 15 September 1864, Confederate Cavalry General John Singleton Mosby (of Mosby's Rangers) engaged in his historic fight at nearby Mt. Airy. Mosby had met with Early in the "Gap" on 16 July and he had many skirmishes in the area. Roadside markers for all three events are nearby.

CONCLUSION

In conclusion, Loomis (see Figure 10, and note the resemblance to President Abraham Lincoln) can be credited with coining the term aerial (which is still in use today as a synonym for antenna, especially in the English-speaking world outside the U.S.); conceiving and using the first antenna (which today we would call a Marconi-type top-loaded, grounded vertical antenna, as noted by Appleby); using the term wireless; demonstrating the first wireless communications using antennas over the operationally significant distance of 18 miles (which was a much greater range than the earlier conduction, induction and other electrostatic methods achieved); demonstrating the first maritime mobile wireless telegraphy, holding the first "wireless symposium" in Virginia (and perhaps in the world) in Bear's Den, Virginia in October 1866; obtaining the second U.S. wireless patent; starting the first wireless company; and (arguably) using the first wireless personal communication system (PCS) employing antennas. Regrettably, not all of Loomis' apparatus was preserved; however, some of his magnets and coils are in the collection of the Smithsonian Institution's Division of Electricity and Modern Physics. The lack of preservation of his complete system is especially unfortunate, because his otherwise complete writings and sketches did not contain a very comprehensive description of the galvanometer he used for his detector. Without such a description it is not possible to determine with certainty exactly how Loomis' system worked (and especially whether or not he did employ electromagnetic waves--as claimed by some of his supporters) without a careful replication of his experiment. In spite of these lacks, the originals of his 1864 sketch of the first aerial and wireless telegraph terminal (Figure 1 in this paper) and his sketches of the 1866 Blue Ridge Mountain experiment (Figures 2 through 4 of this paper), along with sketches of his other antenna and communications ideas, and his notes on his experiments and diary fortunately have been preserved and are held in the Loomis Collection in the Manuscript Division of the Library of Congress.

George Loomis, the brother of Mahlon Loomis who buried him in Terra Alta, West Virginia, in 1886, wrote to Loomis' widow (Achsah) on 12 April 1901 (the year of Marconi's famous transatlantic transmission of the letter "S") about some of Loomis' last thoughts. The advent of the technology of low-earth-obit satellites (LEOS), and the almost instantaneous low-cost global communications that will result, may prove true Loomis' 1886 prediction that his discovery of wireless communications using antennas ". . . will be regarded as of more consequence to mankind than was Columbus's discovery of a New World. I have not only discovered a new world, but the means of invading it, not with little frail boats of human build, but with the 'invisible chariots of the Almighty'." His 1865 dream of low-cost transoceanic wireless communications (see Figure 11) will indeed come true.

POSTLOGUE

Appleby (1967) notes: "In a letter dated December 26, 1936, Dr. George Beerbower, Bowermaster Building, Kingwood, West Virginia, writes:

> In 1906 I purchased the Judge Loomis home in Terra Alta. On a trash pile on the back of the lot were dozens of little nickel-plated boxes about four inches long and an inch square, with a spiral spring in each one. On inquiring of a neighbor what they were, he told me that 'Judge's brother used them in trying to send a wireless message.'

These spiral springs probably were coils.

The Weather Bureau apparently performed wireless experiments with kites in the vicinity of Bear's Den in 1909 and Reginald Aubrey Fessenden (another wireless pioneer) had consulted with them in 1900-1902. The Weather Bureau acknowledged the lead role played by the U.S. Navy Department regarding government work pertaining to wireless communications. An excellent account of the history of the Navy's role has been provided by Howeth (1963). Other useful accounts are provided in the bibliography. *The Washington Star* (1 February 1909) carried the following article on the Weather Bureau's 1909 attempt at a replication of Loomis' 1866 demonstration:

> After lapse of about forty years the government intends to try out a system of wireless telegraph invented by a Washington man long before the advent of Marconi. Dr. Mahlon Loomis carried on his experiments in the Blue Ridge Mts. not far from where the Weather Bureau will duplicate them.
>
> The facts in the case have been transmitted by Prof. Moore to Dr. Blair, who is carrying on a series of kite flying experiments at Mt. Weather, and he will try to reproduce the results.

The results of the Weather Bureau's 1909 attempt at replication currently are unknown to the authors. We believe it is now timely, with the current momentum of PCS, to try replication of Loomis' 1866 experiment again; and we hope to do so using calibrated equipment—both mid-1860s equipment and modern equipment—while recording the geophysical conditions using field mills, ground-current meters, magnetometers, and meteorological instruments.

ACKNOWLEDGMENTS

The authors acknowledge the helpful review comments of Mr. Elliot Sivowitch of the Smithsonian Institution, Mr. Thomas Doeppner of Mount Vernon, Virginia, and Professor Duncan McNair Porter of the Virginia Polytechnic Institute and State University. They also appreciated the assistance of Mr. Wallace Burns and Ms. Susan Smith of SRI International in helping locate some of the items in the bibliography. Finally, they appreciate the editorial assistance of Ms. Jean Stockett of SRI and the draft and final manuscript preparation by Ms. Rose Hagn, and by Ms. Elaine Matthews and Ms. Cynthia Newton of SRI, respectively.

BIBLIOGRAPHY

Appleby, T. Cdr., USNR (Ret.), Commander, *Mahlon Loomis, Inventor of Radio*, Washington, D.C., 1967. (Library of Congress Call No. TK5739. L7 A6).

Boyle, R., *On the Mechanical Production of Electricity*, Oxford, 1675.

Clarkson, R.P. *The Hysterical Background of Radio*, J.H. Sears & Company, Inc., New York, 1927. (Library of Congress Call No. TK6550).

Collins, A.F., "Wireless Telegraphy," in *Modern Engineering Practice*, F.W. Gunsaulus, Ed., American School of Correspondence, pp. 331-368, Chicago, 1906. (Library of Congress Call No. TJ163.A53).

Debus, A.G., ed., *World Who's Who in Science*, Marquis Who's Who, Inc., 1st Edition, Chicago, 1968.

DeSota, C.B., *Two Hundred Meters and Down--The Story of Amateur Radio*, the American Radio Relay League, West Hartford, CT, 1936.

Dunlap, O.E., Jr., *Radio's 100 Men of Science, Biographical Narratives of Pathfinders in Electronics and Television*, Harper & Brothers, New York and London, 1944. (Library of Congress Call No. TK6545.A1 D8; Repr. 1970 by Books for Libraries Press).

Franklin, B., *Philosophical Transactions*, Philadelphia, 1786.

Henry, J., "Scientific Writings of Joseph Henry," Vol. I, Washington, D.C. 1886.

Howeth, Capt. L.S., USN (Ret.), *History of Communications-Electronics in the United States Navy*, U.S. Government Printing Office, Washington, pp. 15-16, 515, 1963. (Library of Congress Catalog No. 64-62870; Call No. VG77.H65).

IEEE, IEEE Dictionary, IEEE Std 100-1992, *Institute of Electrical and Electronics Engineers*, New York, 1993.

Jenkins, C.F., "The First Radio Channel, Vision by Radio—Radio Photographs," Jenkins Laboratories, Inc., Washington, D.C., pp. 67-69, 1925 (Library of Congress Call No. TK6600.J4).

Lewis, Tom, *Empire of the Air, The Men Who Made Radio*, Edward Burlingame Books, Harper Collins Publishers, New York, 1991.

Loftin, E.H., "Marconi—Father of Radio?" Radio Craft, Radcraft Publications, Inc., Springfield, MA, p. 426, January 1939.

Lenard P., "Über der Elektrizität der Wasserfälle," *Ann. Phys.*, Vol. 46, pp. 584-636, 1892.

Loomis, M.T., *Radio Theory and Operating, for the Radio Student and Practical Operator*, 5th Ed. (Rev.), Loomis Publishing Co., Washington, D.C., 1930. (Library of Congress Call No. TK5741.L6). See also earlier editions.

Mahlon Loomis Collection, Manuscript Division, Library of Congress, Washington, D.C.

Marriott, R.H., "How Radio Grew Up," *Radio Broadcast*, Vol. VIII, No. 2, December 1925.

Maxwell, J.C., *A Treatise on Electricity and Magnetism*, 3rd Ed., Dover Publications, Inc., 1954. This is an unabridged, unaltered repubication of the third edition of 1891. The first edition was published in 1867.

Picard, J.G.W., "How I Invented the Crystal Detector," *Electrical Experimenter*, August 1919.

Pierce, E.T. and A.L. Whitson, "The Variation of Potential Gradient with Altitude Above Ground of High Radio Activity," *Journal of Geophysical Research*, Vol. 69, No. 14, pp. 2895-2898, 15 July 1964.

Pierce, E.T. and A.L. Whitson, "Atmospheric Electricity and the Waterfalls of Yosemite Valley," *Journal of the Atm. Sci.*, Vol. 22, No. 3, pp. 314-319, May 1965.

Pierce, E.T., "Water Falls, Bathrooms and—Perhaps—Supertanker Explosions," *Proceedings of the Lightning and Static Electricity Conference*, Wright Patterson AFB, Ohio, 9-11 December 1970.

Rhees, W.J., "Mahlon Loomis and the Wireless or Aerial Telegraph," 1899. (Library of Congress Call No. TK5739.L7 R5).

Scientific American, Supplement, February 1911. Note: The authors were unable to find the quote by Dr. Elisha Loomis in Nos. 1831-1834.

Sievers, M.L., *Crystal Clear*, The Vestal Press, Ltd., P.O. Box 97, Vestal, NY, 1991.

Sivowitch, E.N., "A Technological Survey of Broadcasting's Prehistory, 1876-1920," *Journal of Broadcasting*, Vol. XV, No. 1, pp. 1-20, Winter, 1970-1971.

Smyth, A.H., ed., *The Writings of Benjamin Franklin*, Vol. 1-10, Macmillan, New York, 1905-1907.

Solari, M.L., "The Development and Latest Achievements of the Wireless Telegraph Employed by the Italian Government," *Transactions of the International Electrical Congress*, St. Louis, 1904, Vol. III, J.G. Lyon Co., Albany, New York, pp. 531-554, 1905.

Story, A.T., *Story of Wireless Telegraphy*, D. Appleton and Co., New York, New York, pp. 46-47, 1904. (Library of Congress Call No. TK5742.58).

Wright, E., *Franklin of Philadelphia*, Harvard University Press, Cambridge, Massachusetts, 1986.

Yates, R.F., and L.G. Pacent, *The Complete Radio Book*, Century Company, New York, 1922.

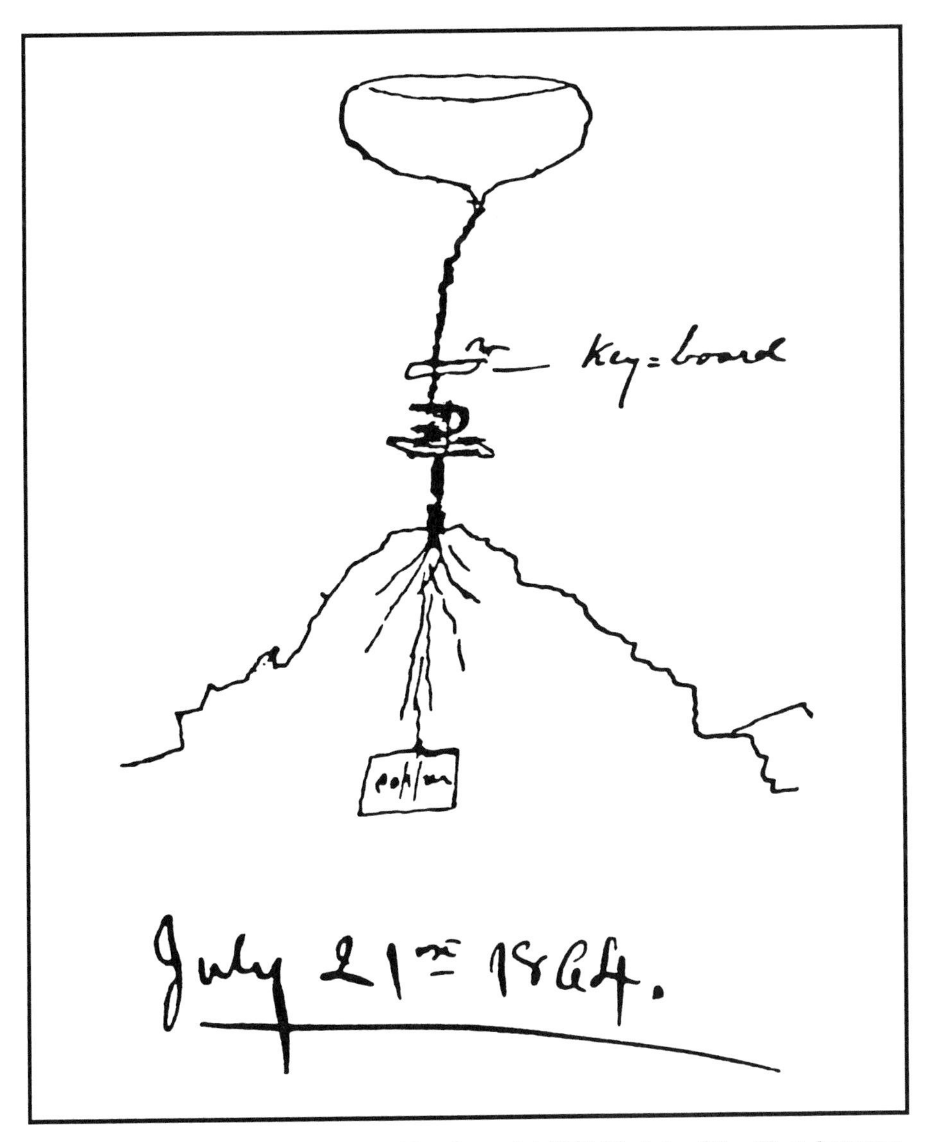

Figure 1 Reproduction of Dr. Mahlon Loomis' 1864 Sketch of the First Antenna as Component of the First Wireless Telegraph Terminal

Original in Library of Congress

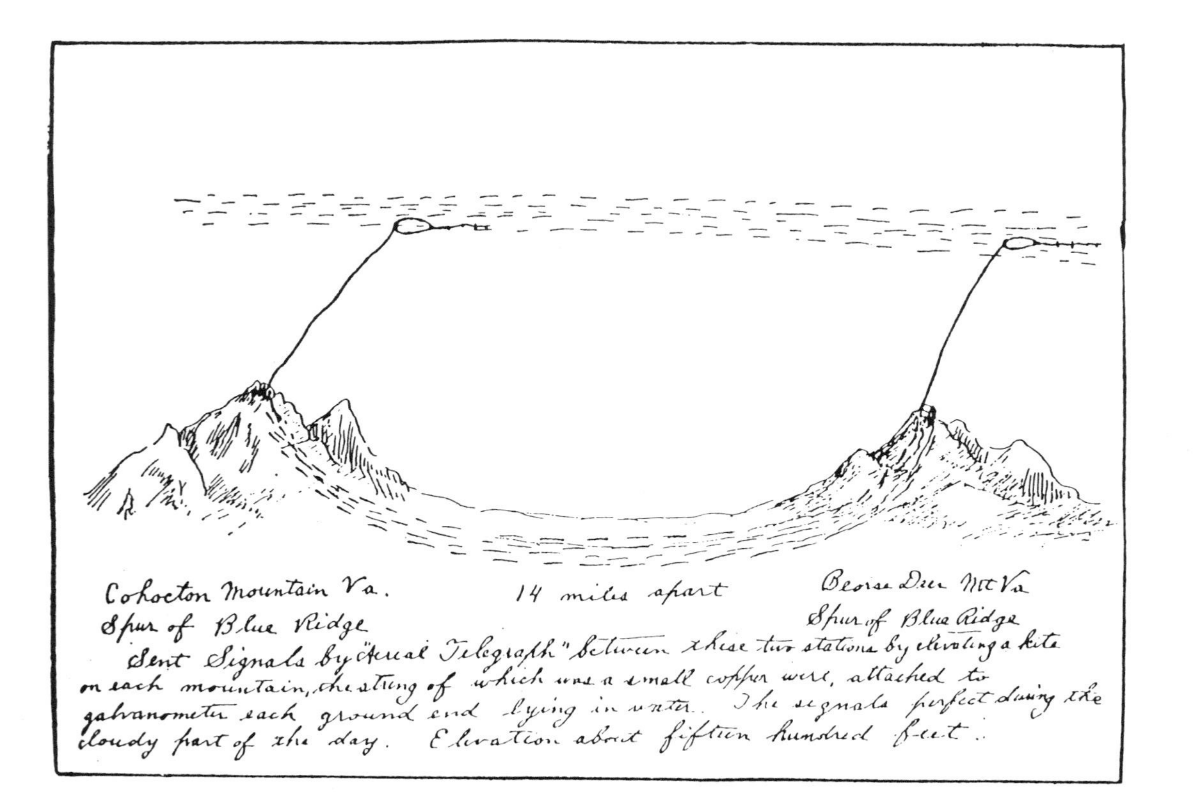

Figure 2 Reproduction of Dr. Mahlon Loomis' Sketch of His Successful 1866 Demonstration of a Wireless "Aerial Telegraph" Between Catoctin Mountain and Bear's Den, Virginia
(Note: The stations actually were 18 miles apart)
Original in Library of Congress

Figure 3 Reproduction of Dr. Mahlon Loomis' Sketch of Himself at the Bear's Den, Virginia Transmit Site During His 1866 Wireless Telegraph Demonstration
(Note: Grounding wire is in pond and antenna wire is handheld for on-off keying)
Original in Library of Congress

Figure 4 Reproduction of Dr. Mahlon Loomis' Sketch of His Assistant at Furnace Mountain, Virginia Receive Site During 1866 Wireless Telegraph Demonstration in Virginia's Blue Ridge Mountains
(Note: Ground wire is in pond and key is closed)
Original in Library of Congress

WILLIAM H. WARD.
Improvement in Collecting Electricity for Telegraphing. &c.
No. 126,356. Patented April 30, 1872.

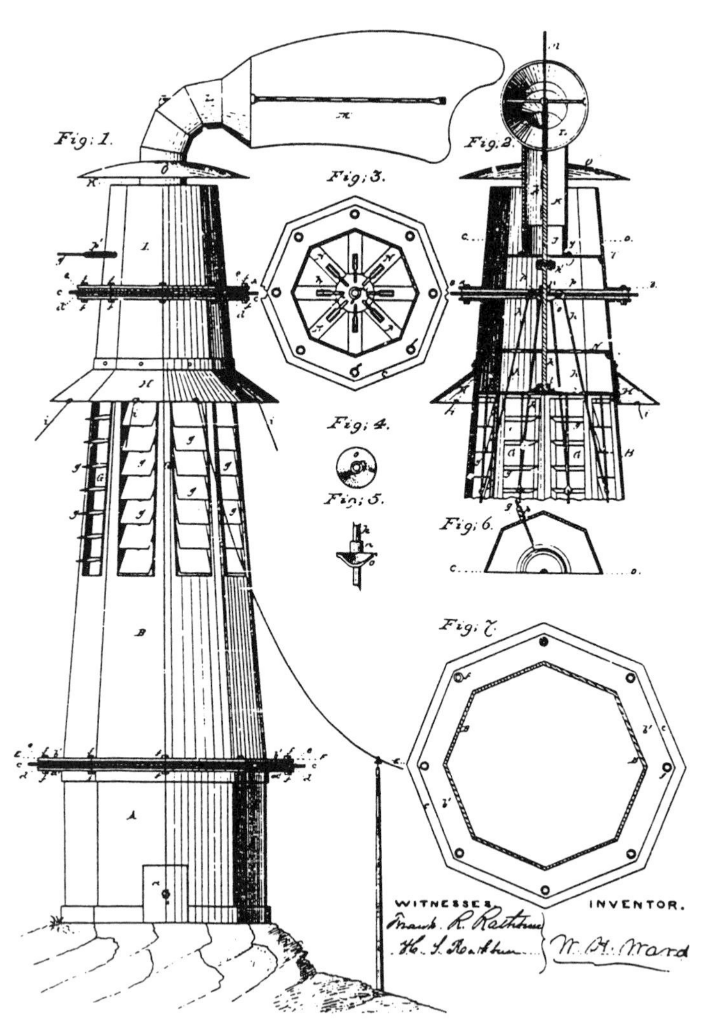

Figure 5 William Henry Ward's Proposed Electrical Tower for Accumulating Natural Electricity for Telegraphic Purposes

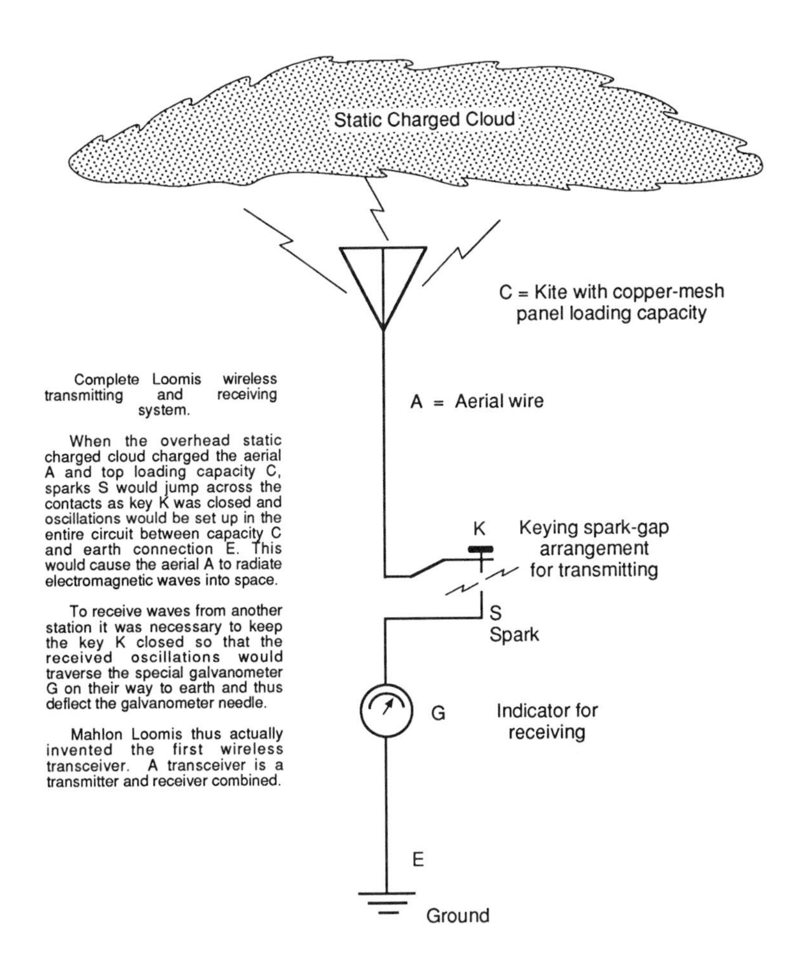

Figure 6 Loomis Wireless System of 1864 (in modern symbols)
(A reproduction of Plate 7, Appleby, 1967)

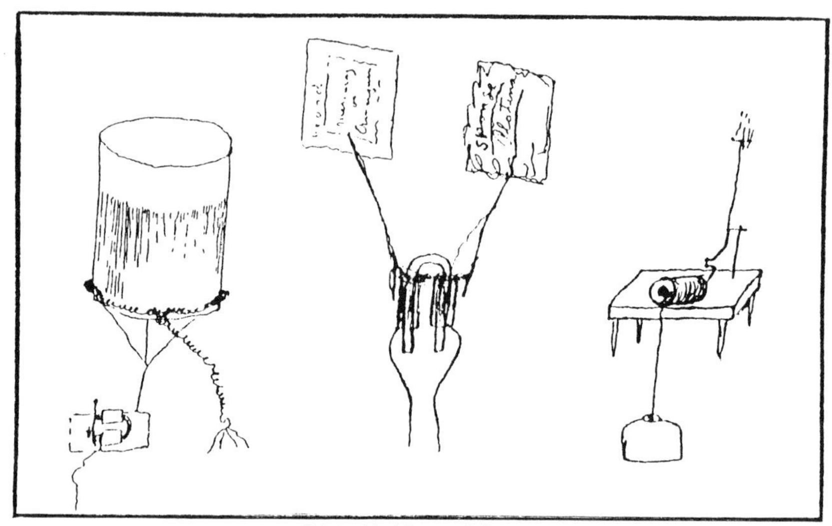

Original in Library of Congress

Figure 7 Reproductions of Sketches by Mahlon Loomis, Apparently Showing Coils

Photo Courtesy of David Hagn

Figure 8 Photo of Snicker's Gap, Virginia Roadside Marker for Loomis

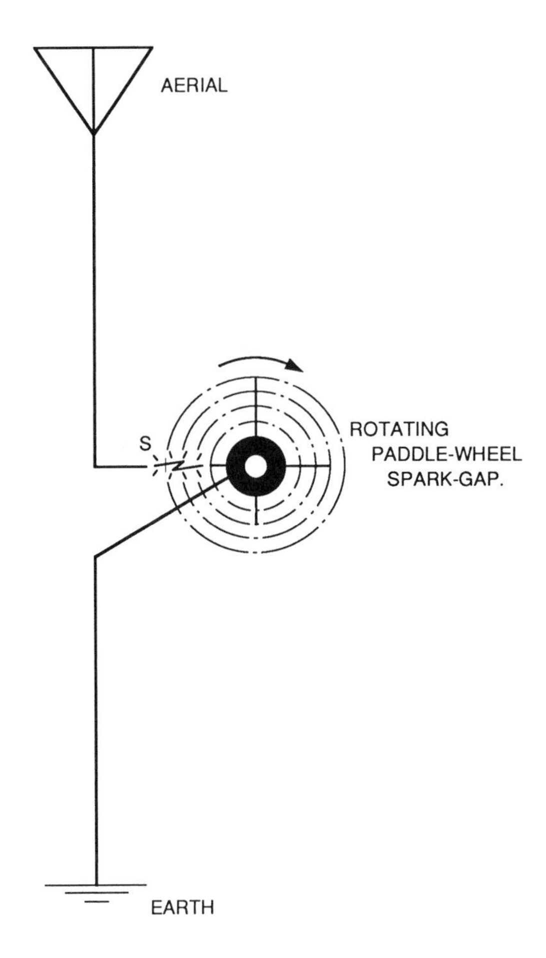

Figure 9 Loomis Wireless Transmitting System as Used by Robert H. Marriott, First President of the Institute of Radio Engineers (IRE) (from Appleby, 1967)

(Ref: *Radio Broadcast*, Vol. VIII, No. 2, Dec. 1925.)

Figure 10 Photo of Dr. Mahlon Loomis

From Mary Texanna Loomis', *Radio Theory and Operating for the Radio Student and Practical Operator, 1928*

Figure 11 Reproduction of Dr. Mahlon Loomis' 1865 Sketch Depicting His Dream of a Trans-Pacific Wireless Telegraph Circuit Between San Francisco, California and Yeddo (Tokyo), Japan
(Note: Antenna wires suspended with kites above mountain peaks and ground wires in the Pacific Ocean)

Original in Library of Congress

21

Mobile Communications: An IC Designers Perspective

Michael Schwartz
National Semiconductor
Santa Clara, CA

Abstract

The projected growth rates of mobile communication products has created some very enticing large volume markets for integrated circuit manufacturers. At the same time advancements in semiconductor techniques have allowed communication system designers to design products more suited to these larger volume markets. Both integrated circuit and communication system manufacturers recognize this and are working together to design more highly integrated, smaller, lower cost, and low power products. This paper will concentrate on these markets, advancements, and integrated circuit solutions and what can be expected in the future.

1. Markets

"Telecommunications: A $3 trillion market in 2010, $600 billion will be wireless." (Motorola Inc.)

There currently is a trend of shifting paradigms: from wired to untethered, from mobile to portable, and from business use to consumer / business use. These shifts and the projected market size make it easy to understand why the wireless market is so appealing to semiconductor and other manufacturing companies (including battery developers, antenna technologists, board developers, etc.). New wireless communications products are becoming available for all sectors of the market. However the biggest market is the one geared for the everyday consumer: small PBX systems, residential cordless telephones, cellular telephones, and ultimately PCS, personal communication services.

To be able to capture a portion of this market you must get consumer acceptance. To get consumer acceptance you must meet many criteria. The products must be low power which means longer battery life, longer talk time, and longer standby time (less charging). They must be small and cheap which is generally accomplished by reducing parts count. The consumer handsets must cost on the order of $200 or less (see Table 1) and be small enough to carry comfortably in the pocket. The products must be full featured, easy to use and secure. They must be designed to operate at lower supply voltages to reduce the required batteries and associated weight. The products must be upward compatible allowing new generations and innovations. They must also be state of the art in regards to quality and reliability. The voice and data quality must be superior. There must be no maintenance or servicing required. The packaging must be of the highest standards.

For semiconductor companies these acceptance criteria take on a few added dimensions. The company must be able to respond quickly with new products and enhanced performance and features. The semiconductor designers must become radio system designers to better understand the tradeoffs necessary to produce the most cost effective solution. There must be reliable low cost production. There must also be a willingness to do custom designs for unique circumstances or unique standards, all geared to provide extra value to the customer. Semiconductor companies must also try to predict which standards will be available and when, so they can be prepared when the appropriate market windows open up. Today there are more than 10 significant cellular standards and at least 8 cordless standards. The standards are also divided between analog and digital. In the future these standards will be reduced to 2 or 3 in each category and ultimately only 2 or 3 total. Thus the manufacturers must track the standards accurately and hopefully pick the winners for tomorrow's products.

Presently analog phones dominate the market. Relative to digital phones they are generally cheaper, use existing standards, and use well known technologies. The digital phones are more secure, have higher capacity, better quality, but few standards. See Figure 1 for a comparison of expected analog versus digital cellular growth. It is apparent that digital phones will grow at a faster rate and overtake analog phones, but analog phones will still be available as a market to address.

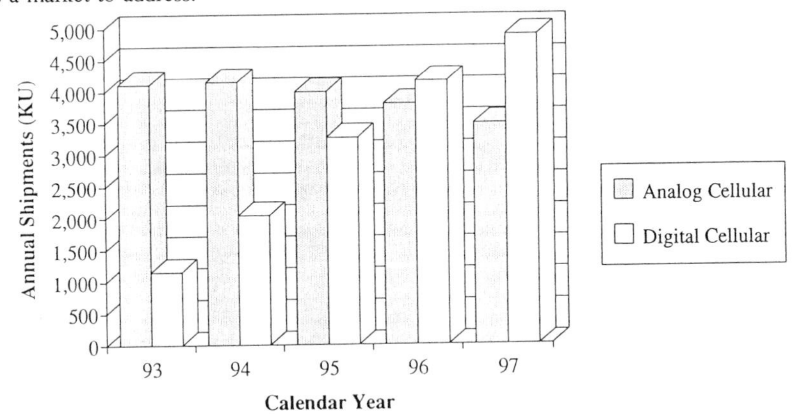

Figure 1: North American and European Cellular Shipments

As the standards become more firmly established and suppliers can produce phones for these standards the digital phone market will grow very rapidly. Presently the growth is mostly in Europe where standards such as GSM (Groupe Speciale Mobile) have been finalized. The standards effort in the United States is still well behind, keeping the unit shipment number down for the next few years. IS54 (US Interim Standard - 54) is still not a released standard, but many companies have extensive efforts in that area. See Figure 2 for projected GSM and IS54 unit shipments.

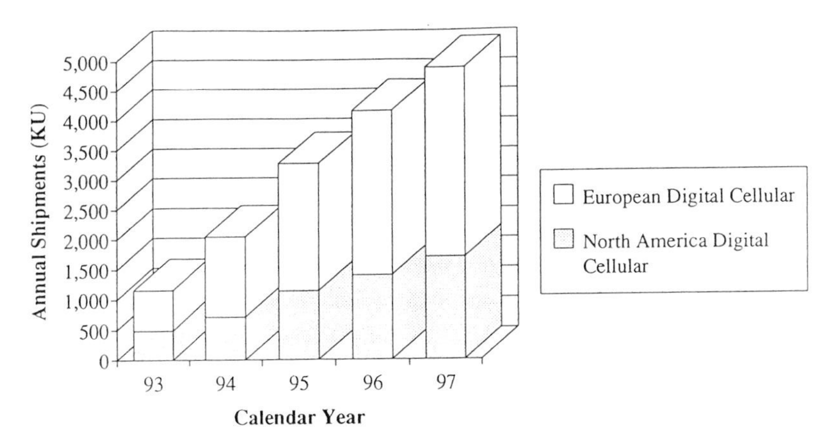

Figure 2: North American and European Digital Cellular Shipments

The digital cordless market will grow even faster than the cellular market. There are several standards presently in Europe that suppliers are addressing including DECT (Digital European Cordless Telecommunications), CT2 (Cordless Telephone - 2nd generation), and DCS1800 (Digital Communication System - 1800MHz). In the US the digital cordless shipments again will lag Europe awaiting some standardization, however ISM (Instrumentation, Scientific, and Medical) and ET (Emerging Technology) band phones will be developed opening a large new market. See Figure 3 for projected US and European digital cordless shipments.

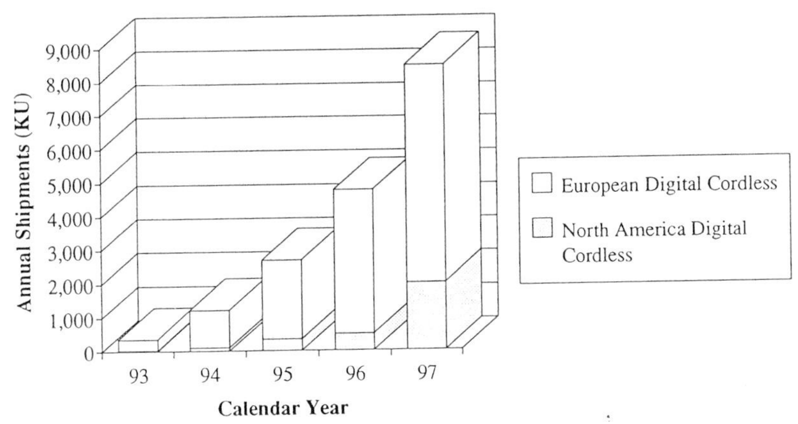

Figure 3: North American and European Digital Cordless Shipments

All of this leads to this very appealing market for semiconductor manufacturers. Figure 4 shows the North American and European total available market for silicon for analog and digital, cordless and cellular products. These numbers do not include Japan, Southeast Asia, or eastern Europe. All of these numbers still don't include the ultimate market which is expected to be the PCS market. This is the ultimate one phone, one number, one person, personal communicator. All of the current and near term standards will eventually be replaced by PCS. This is the market that is most enticing of all to the manufacturers and is the one everyone is trying to prepare for.

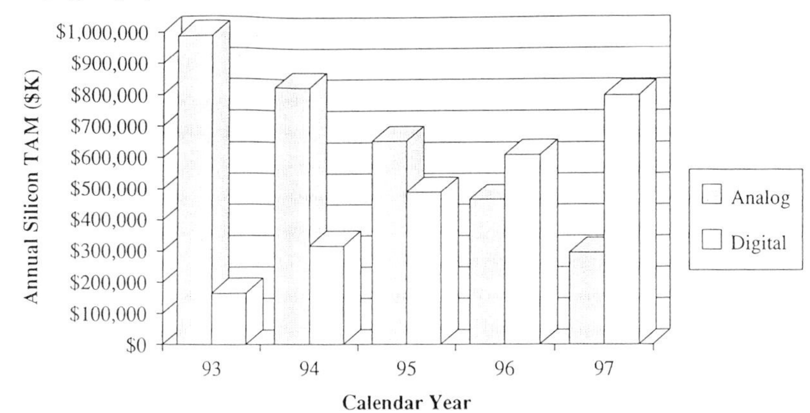

Figure 4: North American and European Silicon TAM

Table 1 shows the retail price versus semiconductor content for cellular phones in the US and Europe. The low analog cellular price in the US in 1996 is due to service providers subsidies. This shows the drive for lower consumer prices and the pressure that will (and is) being applied to manufacturers to reduce their component costs. Current projections show silicon costs of the simpler digital cordless phones to be on the order of $30 in the last half of the 90's.

Retail Price vs Semiconductor Content for Cellular Telephones: US & Europe, 1991-1996				
	1991		1996	
	Handset price	Semi. Content	Handset price	Semi. Content
Europe				
Analog cellular	$1474	$98	$788	$54
GSM	$1500	$155	$862	$69
US				
Analog cellular	$217	$85	$42	$61
Digital	$1380 (1992)	$131	$181	$71

Table 1: Retail Handset Prices

2. Standards

A brief review of the standards shows a very confusing picture when trying to identify the proper markets to address. Table 2 is a summary of a few of the standards and some of the key requirements of each.

	Analog		Digital						
Standard	AMPS	ETACS	IS54	GSM	CT2	DECT	DCS1800	JDC	PHP
Freq (MHz)	824-894	872-950	824-894	890-960	864-868	1880-1900	1700-1880	800/1500	1900
Access method	FDMA	FDMA	TDMA/FDMA	TDMA/FDMA	FDMA/TDD	TDMA/TDD	TDMA/FDMA	TDMA/FDMA	TDMA/TDD
Modulation	NBFM	NBFM	pi/4DQPSK	GMSK	MSK	GMSK	GMSK	pi/4DQPSK	pi/4DQPSK
Bit Rate (Kb/s)	NA	NA	48.6	270.8	72	1152	270	42	384
Voice Coding	NA	NA	VSELP	RPE-LPT	ADPCM	ADPCM	RPE-LPT	VSELP	ADPCM
Max Power Out	3W	3W	3W	20W	10mW	250mW	1W	2W	10mW
Carriers	832	1320	832	125	40	10	375	-	-
Carrier space	30K	25K	30K	200K	100K	1728K	200K	25K	300K
Chan/carrier	1	1	6	8	1	12	8	3	4

Table 2: Standards Summary

A few comments on each standard helps to alleviate some of the confusion when selecting standards to address. AMPS (Advanced Mobile Phone System) is a narrowband FM analog cellular standard. It's quality is fair, but capacity is becoming an issue. NAMPS (Narrowband AMPS) improves the capacity of AMPS by 3 times by reducing the channel spacing. ETACS (Extended Total Access Communications System) is very similar to AMPS and is also of limited capacity. IS54 is not completed yet and is somewhat restricted by backward compatibility issues. It will have improved quality and increased capacity. GSM is a pan-European digital standard. It was one of the first approved digital standards. It is relatively high priced and high performance. CT2 is a digital standard gaining acceptance in Europe as a residential and small PBX standard. It's telepoint application was not very well received in the UK. DECT is another recent digital standard. It is relatively simple and should be low cost. It will be used for residential and small PBX phones, as well as some data connections. DCS1800 is the second generation GSM. It has been designed for both cordless and cellular use. It occupies a less used area of the spectrum. JDC (Japanese Digital Cellular) is very similar to IS54. Japan was following the lead of the US, but has now moved ahead with PHP (Personal Handy Phone). PHP is the first implementation of a PCS-like system.

As can be seen the newer standards are all moving toward digital designs. The analog systems are running low on capacity, their quality is poor, and they are very unsecure. Digital systems will address all of these issues. Capacity is increased by architecting the systems differently, quality is improved through the use of digital signal processing, and security is introduced by coding and encrypting the transmitted data. Digital systems also lend themselves to further integration. More of the phone design, feature sets, and signal processing can occur in lower cost, lower power, integrated circuits.

3. Technology

The key technologies to service these standards include: semiconductor technology, PCB technology, plastics / casing, battery technology, display technology, and antenna / RF technology. All of these technologies are being improved in response to the consumer requirements reviewed earlier: low cost, light weight, small, and easy to use. PCB's must be light weight, small, and RF compatible. The casings must be small, unbreakable, light weight, and non-interfering. The displays must be easy to read and cheap. Antennas must be small and / or hidden. RF and IF filters must be specially designed to be very low loss and very small. Battery technology determines talk and standby time. Semiconductor technology is a cost and size driver.

The battery technology improvements and semiconductor improvements are working closely to reduce weight and improve talk and standby time. The battery is the major source of weight in portable phones. Battery technology has moved toward longer life batteries, on the shelf and in a product, and lighter weight. Some batteries have increased voltage levels to allow less cells for a given power supply voltage requirement. The longer life is derived from the improved capacity during discharge. Lighter weights and higher voltages are from newer material batteries. The most commonly used batteries today are NiCd. Nickel-metal-hydride and lithium batteries will be the batteries most likely used in the future. Both types are being designed into new products now.

Battery Type	Voltage / cell, volts	Capacity, Amp-Hours*
Alkaline	1.5	2.0
NiCd	1.2	0.7
NiMH	1.2	1.0
Lithium	2.7	1.3

* - typical value; capacity is a function of load, voltage, etc.

Table 3: Typical Battery Parameters

With the batteries, the semiconductor components form the most important technology areas in cordless and cellular phones. Semiconductor content is close to 50% of the total parts cost of present phones. They have been the largest contributor to cost savings over the past decade. The semiconductor improvements come from several sources. One is the power supply voltage required to run these devices. Most new designs are aimed at 3v supplies. This eliminates the need for an additional battery cell. Present phones work with a supply range of 5 to 7.5 volts. By redesigning the semiconductors to work at 3 volts, unnecessary cells are removed. Also with semiconductor process improvements devices can achieve higher bandwidths and higher levels of integration. Referring to Table 2 it can be seen that the frequencies of most interest are in the 800MHz to 2.5GHz band. This frequency band has been addressable using GaAs solutions. GaAs however tends to be expensive, difficult to handle, and not highly integratable. Over the past several years bipolar silicon processes have improved to the level where they can accomplish performance levels good enough to meet the requirements of the band defined in Table 2. Bipolar processes tend to be cheaper, but still their integration level is not very high. The bipolar performance is still lower than GaAs, however it is good enough to meet most requirements of the phones to be designed in the frequency range of interest. (There will always be room for GaAs where performance is all important, but in the consumer market where price is critical, GaAs will be lost unless a significant price reduction occurs.) CMOS processes allow high levels of integration and extremely low power. They can meet all requirements for the baseband and digital signal processing that occurs in most phones. However they cannot meet the RF performance requirements of the frequency band. Thus the biggest improvements have come from processes where CMOS and bipolar have been combined. These BiCMOS processes have the performance level necessary to address the band of interest and the integration level to allow future miniaturization.

The BiCMOS processes presently have performance characteristics similar to those listed in Table 4. Table 4 identifies key parameters of the ABIC4 process used by National Semiconductor to design its present line of phase locked loop ICs and DECT chipsets. Processes with this type of performance are well suited to be used in the design of semiconductors for use in the digital cordless and cellular areas. These processes were first introduced in the mid to late 1980's, but have only been reliably manufacturable in the last 2 to 3 years. Newer processes are pushing the transistor performance level to greater than 2 times this level. This will open even more markets as higher bandwidths and improved performance characteristics are achievable.

Parameter	Value
Minimum dimension	0.8u
Minimum emitter area	0.8u x 1.6u
ft	15GHz
Beta	90
Ic for ft,max	150uA
Cbe	5.3fF
Cbc	2.0fF
Ccs	6.8fF
Re	100ohms
CMOS gate delay	~100ps
speed power product	55fj
Leff	0.6u
Vtn	0.75v
Vtp	-1.0v

Table 4: ABIC4 Key Parameters

The present processes have allowed an early level of integration that reduces device count in the RF portions of present phones by nearly a factor of four. Up until recent designs (in the past 2 years) most RF boards were designed using all discrete components. With the improved processes the discrete components have been replaced with integrated circuits. The newest boards have about one hundred components versus four hundred components in discrete designs. This obviously reduces size and improves manufacturability. Cost has decreased as the component count diminishes, board size shrinks, and complexity is reduced. As integration continues the component count and thus size and cost will proceed to be reduced. Presently the second generation of digital phone components are emerging. Complete receivers on a chip are now available. The third generation will allow complete transceivers to be built on a single chip. All circuits from the antenna to digital data (including digital controllers) will be integratable using BiCMOS processes. The ultimate digital phone may have only three IC's and possibly 25 or fewer discrete components.

As an example of the present state of the art in IC's for cordless phones refer to Figure 5. In Figure 5 National Semiconductor's first generation DECT architecture is shown. In this single conversion receiver architecture there are about 10 integrated circuits and less than 100 other components, including filters, VCO, switches, and discretes. Four of the ICs are done in National's ABIC4 BiCMOS process. These parts are the 2GHz LNA and mixer, the 150MHz limiter and discriminator, the 2GHz PLL, and the 1MHz baseband processor. The remaining ICs are regulators and amplifiers.

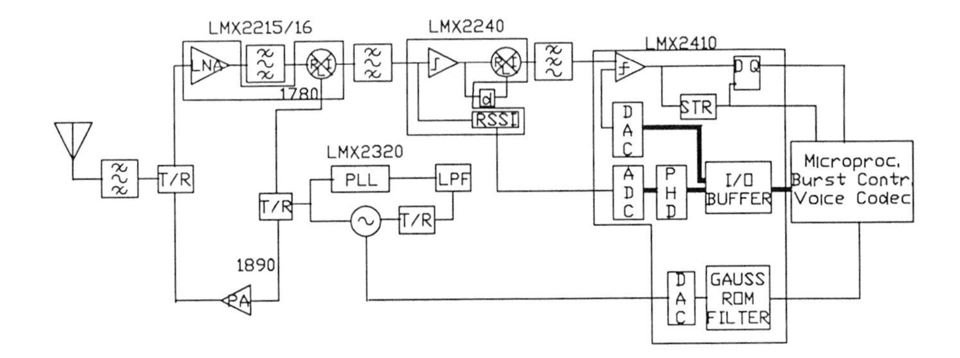

Figure 5: DECT single conversion receiver architecture

The LMX2216, LNA / mixer chip, provides the front end to the single conversion receiver. With the initial band limiting filter and an image reject filter after the LNA, this chip receives and down converts the desired signal to an IF frequency of 110.592MHz (standardized for DECT). The IF signal is then passed to the LMX2240, a limiting amplifier and discriminator. The signal is limited by the amplifier (gain of 70dBm) and frequency discriminated using a 90 degree phase shifted version of itself to provide the instantaneous frequency deviation. The instantaneous frequency deviation is the original modulating signal. This signal is low passed filter to limit the baseband frequency seen by the LMX2410, baseband processor. The baseband processor has a high performance comparator to convert the analog modulation signal to a digital pulse stream. Also in the baseband processor is a symbol timing recovery system that locks onto the incoming data and converts the raw digital data from the comparator into a sampled data stream with corresponding clock. The DC slicing level of the comparator can be controlled by an internal DAC for microprocessor control of DC compensation circuitry. A peak hold detect circuit can be used to help monitor the received signal strength for best use of the available channels. On the transmit side a ROM and DAC provide the Gaussian shaped modulation signal to be applied to the direct modulated VCO. Thus the LMX2410 contains all of the circuitry to effectively convert the received analog radio signal to a digital bit stream and to convert a digital bit stream to an analog modulation signal for up conversion and transmission.

The frequency conversions are controlled by the LMX2320, phase locked loop. With a loop filter and VCO, the LMX2320 provides all of the components necessary to build a 2GHz frequency synthesizer. This synthesizer provides the LO (local oscillator) for the LMX2216 mixer. It is also directly modulated by the LMX2410 during transmission to provide the modulated carrier to the output power amplifier. See Table 5 for a brief summary of the single conversion performance characteristics.

Parameter	Value*
Battery Voltage	3v
Sensitivity	-87dBm
Input Intercept (IIP3)	-23dBm
Overall Gain	78dB
Receiver Noise Figure	12.6dB
Band Filter Loss	1dB
LMX2216 LNA Gain	10dB
LMX2216 LNA Noise Figure	4.7dB
Image Filter Loss	2dB
LMX2216 Mixer Gain	6dB
LMX2216 Mixer Noise Figure	17dB
IF SAW Filter Loss	4dB
LMX2240 Gain	70dB
LMX2240 Noise Figure	8dB
Required Eb/No	13.6dB
BER	1e-3

* - all values are at DECT frequencies (1.88GHz to 1.9GHz), filter values are typical

Table 5: DECT single conversion performance

4. Summary

The growth of the cordless telephone market, enabled by improving technologies, provides a huge opportunity for manufacturers of products for these phones. As improvements continue the phones will become smaller and cheaper. Using BiCMOS processes to allow integration to reach high levels the IC content of these phones will be reduced to the order of three chips. There are many problems which must be overcome to achieve this level, but work is progressing and the future is promising.

22

Network Connection & Traffic Interchange Agreements--
A Wireless Personal Communications Opportunity

James D. Proffitt
PacTel Corporation
Walnut Creek, California

ABSTRACT

Wireless personal communications service providers must build a network infrastructure which will enable them to satisfy the needs of their customers for tetherless access. Interaction with other networks will often be required to complete the end-to-end connections necessary to deliver the desired telecommunications service. This need will grow as more and more networks proliferate today's telecommunications environment.

The connection of these networks and the subsequent interchange of traffic which originates and terminates therein must be accomplished in a manner which is agreeable to the parties involved. The specific details of the business relationship between any two network providers must be represented in a connection and traffic interchange agreement. This "contract" must capture the cost value incurred by each network provider to perform required functions such as transport, switching, and other network services such as providing intercept announcements; the negotiated "settlement" between the parties must describe how actual payments for the performance of these actions will occur.

The wireless personal communications service providers must establish fair and equitable relationships with other network providers in order to survive. This paper describes procedures which have been useful in agreements between cellular service network providers and wired local exchange network providers; suggestions for improvements in the process and other considerations are described.

1. INTRODUCTION

Interconnection, in its most basic form, is a reasonably simplistic concept. If a customer of one network service provider wishes to communicate with a customer of another network service provider, the networks must somehow be connected in order for this to happen. Assuming that both network service providers are equally motivated for this to happen, the connection will be made and the customers of both service providers will communicate. A problem can arise when one network service provider is less motivated than the other to seek interconnection; usually this carrier will be the dominant carrier in terms of size and will try to dictate the terms and conditions of any interconnection arrangement.

Wireless communications service providers must build a network infrastructure which will enable them to satisfy the need of their customers for tetherless access. These customers will wish to interact with other networks, thus forcing the wireless network service provider to interconnect accordingly.

Representing the largest segment of wireless network service providers, the cellular carriers have led the charge to seek interconnection with the wired network community.

2. GOOD FAITH NEGOTIATION

The FCC has recognized that because the terms and conditions of interconnection depend upon numerous local factors, it is best that these arrangements be directly negotiated in good faith between the parties. (In fact, the FCC has asserted that it has plenary jurisdiction to require that the terms and conditions of cellular interconnection must be negotiated in good faith.) This is the approach which most cellular carriers have tried to take in their service areas, where they essentially operate as a co-carrier with the local telephone company in the local exchange network.

A good faith negotiation process requires that both carriers openly discuss all terms in the connection and traffic interchange agreement. The FCC has noted that a carrier is not negotiating in good faith unless that carrier is negotiating in good faith with regard to the entire interconnection agreement.

Telephone companies typically refuse to negotiate those terms which they claim are bound by state regulation. This practice often leaves very little to be "negotiated"--some landline companies may even offer the cellular carriers the same "standard contract."

A good faith negotiation process should not be dominated by one carrier. A genuine good faith negotiation will include, as a minimum, open discussion regarding the following issues:

- Point of Demarcation
- Mutual Compensation
- Co-Carrier Interconnection
- Telephone Numbers
- Billing and Collection
- Volume and Term Discounts

When wireless carriers are forced to interconnect with the landline carriers under conditions which fall short of good faith, an adversarial relationship is established which will often worsen with time.

3. POINT OF DEMARCATION

Historically, co-carriers in the local exchange network have established connection and traffic interchange arrangements which include the identification of a physical point of demarcation where one service provider's network ends and the other's begins. This inter-carrier interface, or boundary, could be a pedestal where the two carriers' cable pairs are physically cross-connected in the right-of-way, or the point of demarcation could be an imaginary point in space where a mid-air microwave meet conceptually occurs (see Figure 1).

The significance of this approach is that each party is responsible for providing the facilities required to reach the point of demarcation, and the parties work together to coordinate this activity in order that it be accomplished in a cost effective manner. For example, in the case of a mid-air microwave meet, the two local exchange carriers work closely together to determine what frequencies, terminal equipment, and antenna systems are required; moreover, a detailed joint testing and cutover plan is established and carried out. This kind of activity has been going on for years, and it is business-as-usual for true co-carriers.

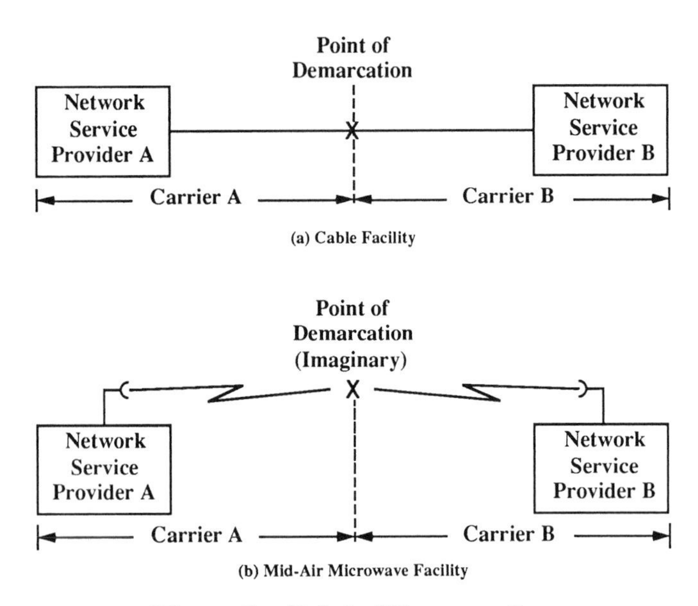

Figure 1. Point of Demarcation

Unfortunately, cellular network service providers have generally been unable to establish mid-air microwave meets with the telephone companies because of policy issues, even though it remains to be an extremely expeditious and cost effective means of interconnection for both parties.

In fact, most connection and traffic interchange agreements between the cellular carriers and the local exchange telephone companies require that the wireless carrier pay for the wired carrier's facilities needed to connect to the point of demarcation. This often forces the cellular carrier to uneconomically build facilities to reach a point of demarcation close to the telephone company's switching office in order to minimize this charge which is ordinarily mileage sensitive.

There are times when the most economical alternative would include placement of the wireless carrier's facility terminal in the wired carrier's central office, but such physical co-location of equipment is also generally not permitted. Wireless carriers need such opportunities for cost effective interconnection.

4. MUTUAL COMPENSATION

An equitable connection and traffic interchange agreement between any two carriers recognizes the mutual actions and associated costs incurred by these carriers as they originate outgoing calls and terminate incoming calls in their respective networks. The basis for these costs is well founded. The landline telephone companies, for example, record costs pursuant to the Uniform System of Accounts or a similar accounting system and further separate these costs into intrastate and interstate categories.

Even though these telephone companies are extremely diligent in pursuing all aspects of their costs in support of their tariff filings and contracts, they are often quite reluctant to recognize that cellular carriers likewise incur costs when they perform the same actions within their wireless networks. In fact, one would be hard pressed to find any reference to mutual compensation in connection and traffic interchange agreements between cellular carriers and telephone companies.

The telephone companies are quick to point out that because they are entwined in complex separations agreements with other telephone companies and further encumbered by "provider of last resort" obligations, they are not obligated to fully compensate cellular carriers for the performance of the functions required to originate and terminate calls. The FCC has refuted this claim and specifically stated that in the case where a cellular mobile switching center (MSC) performs the identical function that a telephone company end office performs when originating and terminating a call, the carriers are equally entitled to just and reasonable compensation for their provision of access.

With more than 10 million people using cellular services nationwide, it is time that the mutual compensation issue be addressed.

Failure to receive mutual compensation costs the wireless carriers in other ways as well. With no concern for mutual compensation claims from the wireless carriers, the telephone companies are free to liberally mandate usage charges. For example, some landline carriers will include call set-up time in their measurement of the minutes of network usage which they bill to the wireless cellular carriers for mobile-originated calls. This quantity of time cannot be measured or in any way controlled by the cellular carrier since it is entirely dependent upon the efficiency of the operation of the landline carrier's network; moreover, there is no motivation for the landline carrier to minimize call set-up time on these mobile-originated calls since they are fully compensated regardless.

Without mutual compensation, the wireless carrier is nothing more than an "end-user" of the landline network. If the telephone companies are successful in establishing end-user status upon the wireless carriers, they will force the wireless carriers to "order" interconnection services from tariffs. This situation must be avoided; wireless service providers must achieve and maintain co-carrier status.

5. CO-CARRIER INTERCONNECTION

Figure 2 represents a co-carrier interconnection arrangement amongst an interLATA carrier, a wireless network service provider, and the local exchange telephone company. The wireless carrier's MSC and the wired carrier's end office have equal status in this network configuration.

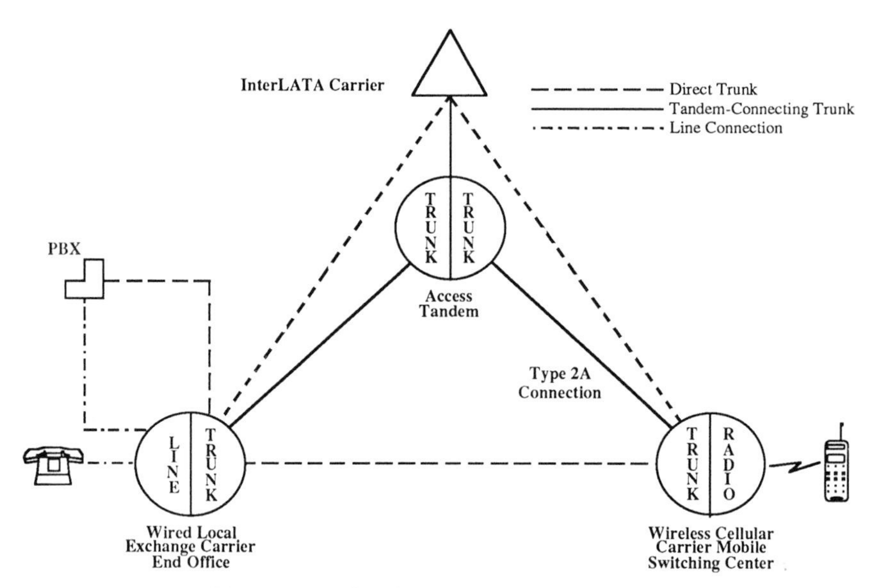

Figure 2. Co-Carrier Interconnection

The direct connection between the cellular carrier's MSC and the telephone company's access tandem switching office is known as a Type 2A Connection; this connection typically provides access to all NXX codes residing in end offices within the Local Access and Transport Area (LATA). However, some telephone companies require that the wireless carrier obtain a direct connection to every tandem within the LATA to reach every NXX even though they do not require this of their own end offices.

It took years and intervention on the part of the FCC before the wired telephone companies would provide a physical connection of co-carrier status to the cellular network service providers. Even today, many telephone companies inappropriately utilize tariffs instead of negotiated connection and traffic interchange agreements to characterize these arrangements. Such tariffs improperly base costs upon multiple market averages which ignore the unique network characteristics of each wireless carrier's system. As noted earlier, wireless carriers must constantly be wary of the telephone companies' attempts to restore the customer/vendor relationship which existed when cellular carriers were restricted to end user status in the network with PBX-like connections to end offices.

6. TELEPHONE NUMBERS

Wireless carriers need access to telephone numbers to provide the features and services that their customers require. Because they are tetherless, these customers' needs can only be met with multiple options for telephone number availability. At a minimum, wireless carriers must have access to the following:

- Geographic Numbers
- Non-Geographic Numbers
- N00 Service Access Codes

Geographic Numbers

Bellcore, in its role as North American Numbering Plan Administrator (NANPA), assigns "area codes" or NPAs. The local exchange telephone companies assign the NXX codes within a specific NPA. This approach to number administration served the wired community well for a number of years; however, problems occurred in 1983 when the cellular carriers requested NXX codes for their MSCs.

In their role as administrators of the NXX codes, many of the telephone companies confused the authority delegated to them by NANPA with "ownership" and imposed recurring charges on the wireless carriers for the use of the telephone numbers. The FCC intervened and indicated that only a reasonable non-recurring charge was permitted to compensate the telephone company for its administrative costs. Furthermore, the FCC acknowledged that the wireless carriers were afforded the same entitlement to these numbers as the telephone companies.

But NXX codes remained difficult to get because the telephone companies believed that "code conservation" was their top priority; wireless cellular carriers were forced to meet restrictive "fill requirements" before additional numbers could be obtained. These restrictions often prevented the cellular carriers from providing new features and services.

Once Type 2A Connections were available, the cellular carriers were able to obtain a dedicated NXX code (10,000 telephone numbers) which they could fully administer; however, the telephone companies continued to impose a non-recurring charge for their administrative costs. Some cellular carriers have paid as much as $35,000 for a dedicated NXX code, but the telephone companies generally refuse to mutually compensate the cellular carriers for their code administration costs. These costs can be quite significant when the cellular carriers must recall all mobile units for chip changes or reprogramming when the telephone company does a code split.

Non-Geographic Numbers

Wireless service providers' customers travel beyond the geographic boundaries of telephone company local exchanges, LATAs, and NPA boundaries. Many of these customers do not require a number with geographic significance. For them a telephone number which covers a broad region of the country, or perhaps the entire country, is important.

A non-geographic NPA which "overlays" two or more geographic NPAs will serve this need. In reality, these non-geographic numbers are still geographic in nature, but they are no longer restricted to boundaries which are meaningless to the wireless community.

Wireless carriers have achieved little success in obtaining non-geographic NPAs from NANPA thus far; however, wireless carrier associations, such as the Cellular Telecommunications Industry Association (CTIA) and Telocator, have begun to increase the level of awareness regarding this issue. Strong demands have been made for a substantial allocation of non-geographic NPAs from the pool of 640 new NPA codes which become available in 1995.

N00 Service Access Codes

NANPA also administers the N00 Service Access Codes, e.g., 500-NXX-XXXX. Many of these codes have not been assigned, and one was requested on behalf of the wireless service providers in August, 1992 for personal communications services.

7. BILLING AND COLLECTION

Conventional telephony has required that the originator of a call incur the total cost for the call; this concept has become generally accepted by the public since, intuitively, the "cost causer" should pay. Through separations and pooling arrangements, the network providers are able to recover their "costs plus" for the services they perform in completing the call.

With few exceptions the cellular service providers have been unable to persuade the wired local exchange carriers to include the wireless airtime charges on the originator's telephone bill when calls are made to wireless customers. This service option, often referred to as Calling Party Pays (CPP), has been denied to most cellular customers for almost 10 years despite constant pleas by the wireless carriers. The justifications for this exclusion are often based upon the telephone company's unwillingness to "modify their billing system" and/or to "deal with the high volumes of customer complaints" which they anticipate.

It is interesting to note that, historically, the telephone companies have not encountered problems with their ability to make billing system modifications and to notify their customers of billing changes; moreover, it is expected that these problems will be quickly solved when the telephone companies prepare to implement cellular-like personal communications services of their own.

8. VOLUME & TERM DISCOUNTS

Volume and term discounts are a proven means of recognizing the efficiencies associated with traffic interchange regarding larger volumes and the certainty of longer term deals, respectively. Telephone companies generally refuse to include either of these considerations in connection and traffic interchange agreements with wireless carriers. Opportunity exists for both carriers to reach mutually beneficial arrangements which acknowledge volume and term discount considerations.

9. BEYOND 1993

The revolution in technology, regulation, financing, public policy, competition, and customer service will result in a telecommunications network evolution which will force the landline carriers to fully acknowledge the wireless co-carriers. Technological advances are creating a network structure which is obliterating the local exchange as we know it today.

The cellular network service providers reached the 10 million-users plateau in less than a decade; whereas, the landline service providers required 20 years to reach one million users. It is quite likely that the number of wireless users will double well within the next five years.

As the local exchanges fragment and disaggregate, more communications nodes will be deployed in the network. As more network intelligence is distributed in these network nodes, new features and services will be provided more quickly; moreover, customer-designed billing distribution will become commonplace. Connections between the nodes will shorten, and competition for making those connections will grow rapidly.

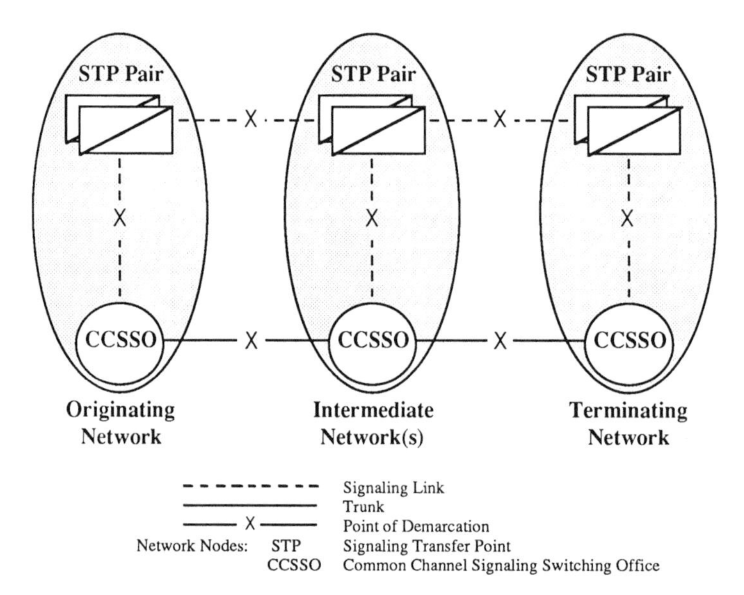

Figure 3. Future Network Connectivity

The architecture of the telecommunications network connectivity will eventually look something like that shown in Figure 3. Unlike in the past, when signaling resided in the voice band, separate communications paths will be utilized to carry the signaling messages and the voice/data traffic.

A switch with Common Channel Signaling (CCS) capability will interconnect with Signaling Transfer Points (STPs) to complete call control and transaction (signaling) messages and with other switches to complete the voice/data calls. Additional points of demarcation will be required to define network boundaries. As shown in Figure 3, with a single intermediate network, at least five (5) additional points of demarcation are included in the network configuration which would have contained only two (2) points of demarcation without CCS. Connection and traffic interchange agreements will establish the relationships between network service providers. Competition will provide multiple alternatives for network access and transport.

Competition will erase the current local exchange carrier intercompany compensation arrangements and create an environment where mutual compensation considerations will be mandatory and available to all network providers. Those telephone companies which fail to change will be passed by as the wireless carriers utilize other alternatives for connection and traffic interchange. It is time for the wired and wireless carriers to work together to achieve mutually beneficial goals in the fast-changing telecommunications industry.

REFERENCES

1. "North American Numbering Plan Administrator's Proposal on the Future of Numbering in WZ1--Second Edition", **Bellcore IL-93/01-008**, January 8, 1993.
2. Peter W. Huber, Michael K. Kellogg and John Thorne; **The Geodesic Network II: 1993 Report on Competition in the Telephone Industry**; The Geodesic Company, 1992.
3. "Local Competition May Strain Regulators to Breaking Point", **State Telephone Regulation Report**, December 3, 1992.
4. Kurt C. Maass, "Interconnection: A Challenge to the Wireless Industry"; **Telocator**; August/September, 1992; Pp. 42-44.
5. James D. Proffitt, "Calling Party Pays" Presentation, **CTIA Winter Meeting & Exposition**, January 17, 1991.
6. FCC Memorandum Opinion and Order on Reconsideration, Cellular Interconnection Proceeding, **FCC Docket 89-60**, Released March 15, 1989.

7. FCC Declaration Ruling, Cellular Interconnection Proceeding, **FCC Docket 87-163**, Released May 18, 1987.
8. Jan D. Jubon, "Cellular Interconnection: A Technical Primer for 1984"; **Proceedings of the Industry Conference, Cellular Communications '84**; November, 1984; Pp. 111-127.

23

Whither Personal Communications

by
Dr. Larry U. Dworkin and Louis L. Taylor
The MITRE Corporation
145 Wyckoff Rd.
Eatontown, NJ 07724

ABSTRACT

Concern is expressed that ideas and designs abound for personal communications systems (PCS) and equipment, but there is no comprehensive long range plan for a world-wide personal communications architecture. Before arguing about different modulation techniques, security codes, satellite designs, etc. we must realize that the Cheshire Cat was correct--if we don't know where we are going, it doesn't make much difference which way we go. Now is the time for U.S. academia, government, and industry to unite and develop an integrated PCS architecture that can support government, military (especially tactical communications), and civilian requirements; and focus our energies in the direction of this architecture. If we do not unite our efforts we are going to see personal communications of the future follow the television receiver, camcorder, tape recorder, children's electronic toys, etc., overseas.

This paper discusses the requirements and limitations of a long range plan for a personal communications system ten to twenty years hence and provides a strategy for the U.S. to maintain the lead in striving toward this goal. This strategy includes cooperation between system and equipment developers, interested technical societies and standards committees, and the U.S. government (including the FCC and military planners) to develop standards for wireless personal communications in the context of the public switched telephone system.

INTRODUCTION

Occasionally, rents appear in the veil over the future that permit glimpses of the way some aspects of our world could, or will, change. When we are especially fortunate, we can foresee major technological changes ten or fifteen years hence. A good example is the foresight of BG "Billy" Mitchell in his advocacy of an independent air force, an advocacy which led to his court martial and suspension from active service. Too bad he didn't live long enough to see himself vindicated.

When opportunities for change are not exploited the future unfolds without their benefit. Such a case is solid state versus vacuum tube technology. If the reasons for the odd behavior of galena crystals had been fathomed earlier, Lee De Forest may never have put the control grid in a vacuum tube.

This paper is not a discussion of futurology, we refer those interested in a rigorous discussion of this discipline to De Jouvenel and to Martino. This paper is a plea for coordinated, long range planning to develop less expensive, more effective personal communications. Some foundations for this long range planning are provided.

While the commercial networks evolve to support high speed integrated voice, data, and video services, a migration has been taking place in military communications from the development of specialized military systems towards the use of commercial backbone networks. Reliance on the public switched telephone infrastructure for more military applications will offer both advantages and disadvantages. On the one hand, it represents a cost effective solution since developmental costs for specialized military networks can be reduced significantly. Satellite channel capacity, for instance, can be leased from commercial systems that are operational now instead of procuring new privatized versions which would provide the same transmission means tomorrow. A major disadvantage, on the other hand, is the necessity to correlate the guidelines of future military communications with the developments occurring in the commercial sector. Secure communications requirements must remain compatible with the designs in consideration for the commercial networks.

Due to competition and easing regulatory statutes by federal communications agencies, a myriad of technologies and services are expected to exist for next generation wirelike and wireless systems. Little consideration has been given to compatibilities and interpretability, only to what has been fair for market competitiveness. These facts represent a critical need to define a secure communications roadmap that can adapt to commercial industry developments.

TRENDS

The following represent an overview of some current military trends that have been considered as baseline to the development of a fully interoperable military communication environment. The Defense Information System Agency's (DISA's) vision of the goal architecture is one that offers DoD opportunities to lower communication costs, reduce the inventory of disparate networks and numerous dedicated facilities, and provide advance services. The emerging concept is one of a multigigabit global grid permitting warfighters located anywhere in the world to quickly and reliably exchange information with fellow soldiers and their commanders. Many of the trends are obviously security related. However, some of the trends, listed below, are similar to characteristics expected in the commercial market (Weissman):

Use of commercial networks as supplement to existing military communications infrastructure, i.e., digital cellular and commercial satellite

Higher data rates

Higher quality voice coding (under 9.6 kbps)

Integrated secure voice, data, and video transmission

Secure teleconferencing and net broadcast modes

Fixed networks are expected to converge to B-ISDN and asynchronous transfer mode (ATM)

Strategic and tactical communications interoperability

Development cycles are continuing to lengthen yet life cycles of advanced technological products and systems continue to get shorter and shorter. A cessation to this divergence of trends has been much discussed, but is not yet readily apparent.

Communication components are continuing to shrink and coalesce as their capabilities increase. A good example is the modem which used to be a separate unit with its own power supply, is now available as a chip, and soon will be incorporated on a chip along with other communication functions. Communication security (COMSEC) devices have gone from separate boxes to plug-in cards, and are now plug-in chips. Another example is the size of telephone switching units. They used to occupy entire floors of buildings, the military had to squeeze to make them transportable. Now cellular telephone switches hang on telephone poles.

Military communications have continually lagged behind civilian communications in the efficient utilization of bandwidth. Civilian users have used 25 KHz channel spacing for FM ever since the end of World War II, while the military has just recently gone from 50 to 25 KHz channel spacing. Now the civilian field is starting to use 5 KHz channel spacing with the use of amplitude compandored single sideband, and 6.25 kHz channel spacing is being contemplated for the near future by the FCC to pack more communications channels in a given spectrum allocation. There is no indication that the military could ever go to such narrowband voice channels, mainly because of susceptibility to jamming; however, the same motivation to pack more users per spectrum allocation should be applied to develop more efficient spread spectrum techniques. For example, digital voice civilian communications appear to be quite promising at 1,250 bits per second (BPS), while military system planning is based upon the continued use of 16,200 BPS. Agile filters and other transmitter noise and spurious response reduction techniques coupled with improved voice digitization could permit more efficient spectrum usage by spread spectrum radios, and at the same time decrease the probability of intercept and probability of detection. These same benefits would be helpful to the radios used in the civil sector.

It often appears that we design military hardware and software as if we had a priori knowledge of where our next conflict is going to occur. The fact is we don't know where our military forces will need communications in the future and we must design communications facilities that will work anywhere in the world. This means that a military operation that requires echelons of radio nets in one theater might need a switched cellular system in another.

Personal communications which provide individuals with voice and data terminals which are not location specific are becoming more universal, and more ubiquitous. Today it is possible to have a telephone with world-wide access, a computer connected to a wide-area network, and a facsimile in your automobile. Soon these three types of user terminals will be embodied in one unit which can be carried in a briefcase from home to automobile to office to airliner. Later, the same functions will be embodied in a pocket-sized terminal and will provide more enhanced universal services, if standards are developed to achieve uniformity at the user end (Weissman).

CONJECTURE

If we take a hard look at these facts it appears that it may be no longer necessary, or practical, for the military to embark on the development of large discrete, stand-alone, communication systems. Large system procurements for communication systems have usually resulted in the systems being fielded long after they are outmoded. In the future, data communications can be obtained as components of the system needing the communications support. With a major proviso, voice communications can be provided by modifying and extending civilian personal communications. Furthermore, with a little long range planning and cooperation, the data communications and voice communications would coalesce into a single system serving all civilian, government, and military users and would be interconnected with a world-wide, public switched voice and data communications system.

To some, the foregoing might bring up visions of every soldier with a personal communication device (PCD) that would provide access to every other telephone in the world. In effect this is correct, but in reality, hierarchical order, security, and discipline would have to be maintained. This is the proviso. A removable insert would be used to implement transmission security (TRANSEC) and COMSEC variables to convert the unit from one that could access only the public telephone system to one that also could access secure military communication systems. Maybe a soldiers "dog tags" could contain this insert? Every soldier could have a PCD furnished as a pocket device, or incorporated into the helmet, armament, or other tools. Every PCD would be a cellular telephone under the control of the cell in which it happened to be situated. The access and networking of a particular PCD would depend upon the identity and security clearance of the bearer and the strategic/tactical situation prevailing at the time access is requested.

For example, the PCD of a soldier in a squad on patrol would be operating in a cell and the squad leader would carry the cell control (or if you prefer, call it "network control") station and switching unit. The squad leader's control station and switching unit would be a PCD in the next higher hierarchy, perhaps a vehicular mounted system or a satellite system. Such a tree structure would extend all the way to the military headquarters in the United States and the public switched telephone system, and would extend downward to all United States armed forces and the armed forces of our allies. In a similar manner, computers used by military personnel would have embedded cellular telephones. The access and privileges would be dependent upon the identity of the computer, the computer operator, and the tactical/strategic situation. During off-duty hours, in garrison, the soldier could use his PCD as a personal telephone. When on patrol in enemy territory it could only be used to contact his squad leader. If the squad leader were disabled it could assume the control station and switching duties of the squad leader's radio. In another extreme, a commanding officer could use the same PCD to contact military headquarters in the field or in Washington. Such contacts could be by voice or by digital device if the digital device (facsimile, computer, sensor, readout, etc.) were available. In a field headquarters the PCDs would be incorporated into the digital equipment employed for command and control.

It should be apparent that the communication and networking concepts presented here are extensions of some of the concepts of intelligent networks presented by Chorafas and Steinmann in their recent book INTELLIGENT NETWORKS coupled with extensions of the concepts of the future of personal communications services (PCS) presented by Sam Ginn in the February 1991 issue of the IEEE Communications Magazine. The lack of comments concerning the extension of intelligent networks and PCS to military applications in either of these references illustrates the dichotomy existing between civil sector and military communications. Such a dichotomy is an expensive burden upon the nation and is not necessary. Communication equipment that can withstand the rigors of construction sites, factories, and traveling salesmen's briefcases don't need much modification to withstand military environments. Security precautions that protect multimillion dollar fund transfers are not really so remote from the security needs of the military. Equipment that satisfies the radio/telephone intelligent networking needs of the civil sector should be extended to the military instead of attempting to design new systems for the military.

In an earlier paper on networking I pointed out that, "The most overlooked need today is the need to provide interfaces between ... networks and their human users that are based upon sound human factors engineering. Much human factors engineering is yet to be completed, but a lot that has been completed is being overlooked by the network developers." This same lament applies to communication systems in general. Competition is going to force better design of personal communication end user equipment and military procurement should take advantage of such designs, and try to nourish them to include features useful to military users, instead of trying to re-invent the wheel with attendant waste of resources and potential for errors. Those that heed such tenets will have a high probability of success in the communication marketplace in the future. Those that chase after large contracts for military communication systems will probably end up rather hungry.

SUGGESTIONS

Civilian, government and military planners should participate in the decisions that are now being made on a global scale to implement personal communications systems. Of particular interest should be the contract to LM Ericsson for the cellular telephone system in New York City, General Magic's Personal Communicator, Motorola's Iridium PCS, and Anterior Technology's wireless electronic mail service. The effect of the implementation of these three systems upon the future of PCDs will be similar to the effect twenty years ago of the Chicago trials upon cellular radio. Many of the requirements for military users are coincident with civil sector requirements. Examples are:

Utilization of spread spectrum techniques to make it difficult to intercept and compromise information being transmitted

Direct communication link between subscriber and orbiting satellite

Efficient handling of the roamer location problem

Numbering plans

Elimination of the government-military-civilian labeling of the frequency spectrum (and the further lower-level divisioning) to enable more efficient sharing of the spectrum by all concerned, i.e., denote portions of the spectrum by type of use, not by the allegiance of the user

Adaptable (self-adjusting) transmitter power to conserve batteries and to minimize interference

Solar recharge of batteries from ambient light

Voice control of most functions

Automatic operator identification to minimize compromise

Digitized vocoder for speech processing, as opposed to simply digitizing the audio waveform, to enable conservation of spectrum

Modular construction, as envisioned in the SPEAKEASY system, so a desired unit can be constructed from basic modules, and to enable ease of incorporating improvements and supporting specialized applications. Typical modules would be:

Speech processing (vocoder and digitizer)

Modulator, rf (radio frequency) control and low level rf amplifier

High level rf amplifier

Receiving unit (rf, detectors, processors, and audio amplifier)

Automatic and manual control

Switching module to enable PCD to be assembled to create cell control unit and switch

Bandwidth adjustment to suit desired information rate

Built-in-test capabilities

Ability to communicate point-to-point (half-duplex or full duplex) as well as operate in a cellular mode

Last, but not least, is a switching and networking capability that will enable the PCDs to be used either as subscriber telephone units or as radio units that will provide integrated voice and data communications in a tactical network. Such a feature would have many uses in the civilian sector (law enforcement, construction industry, delivery services, and most of the other places that mobile and portable radio units are used today), and in military tactical networks.

CONCLUSION

The reader will note that such details as advocation of code-division multiple access (CDMA), frequency-division multiple access (FDMA), time-division multiple access (TDMA), or the Digital European Cordless Telecommunications (DECT) standard; use of military waveforms such as SATURN, HAVE QUICK, etc.; frequency bands of operation; protocols; speech-bandwidth compression techniques; etc., are omitted. Before problems at this level of detail are addressed, the problems at the next higher level must be resolved: Are we going to develop integrated communications or not? Are we going to pack more communications into our fixed spectrum or are we going to continue to squander spectrum resources? Are we going to continue to sacrifice long-term gains for all to obtain short-term profits for a few? Are we going to persist in developing fractionated military communication systems that get implemented long after they become outmoded, or are we going to work together and develop personal communications that will serve both civil and military sectors? Equipment and system developers must cooperate and develop standards for public switched systems (including local exchanges and wide-area networks) that include the needs of military users. Military planners must make their future requirements known to equipment and system developers years before surprising them with the requirements in *Commerce Business Daily*.

We recommend the following four-part strategy to enhance information flow, coordination, and common standards between commercial and military developers:

(1) A review of current government and industrial PCS standards to ensure appropriate and effective linkage between them, e.g., Federal Wireless-Services Users Forum (FWUF) and Telecommunications Industry Association (TIA).

(2) Leveraging of the Commercial Communication Technology Testbed (C2T2) established by the Advanced Research Projects Agency (ARPA) to determine which architectures and protocols should be recommended to the standards groups.

(3) Coordination of the DISA Goal Architecture and Transition Strategy with the above and other initiatives.

(4) Expansion of present efforts to reassign portions of the frequency spectrum to a more general overhaul of the spectrum to provide contiguous allocations for integrated systems that will serve government, military, and civilian users.

(5) The creation of a PCS government/industry working group, if it appears necessary, to identify gaps in existing PCS technology, standards, and appropriate tests.

PLAN

We plan to explore the four-part strategy outlined above. The MITRE Corporation, in conjunction with the Signal School at Fort Gordon and DISA, has begun an examination of the various standards groups associated with government and commercial wireless communications (Rudisill). The purpose of this program is to identify the various standards groups, their roles, and their method of developing standards for current and future systems. This will facilitate contact between appropriate government and commercial standards groups. The objective will be to develop common standards that permit PCS and land mobile systems integration, and reduce the need for extensive military investment in new wireless/PCS equipment.

In order for the government to make wise standard recommendations and fully define their requirements for wireless/PCS battlefield systems, the use of a proposed ARPA testbed, the C2T2, will be a primary vehicle. An evolutionary paradigm will be followed. Commercially available mobile satellite and ground-based cellular systems will be used to determine an accurate set of requirements and evaluate the ability of existing and prototype communications systems to support secure military C3I applications. In addition, NDI user-friendly processor hardware and software will be used. A set of desired and required services will be established and the associated interfaces will be defined. (This paradigm was successfully followed in the Army/DARPA Distributed Communication and Processing Experiment to define C3I upper echelon requirements and services in the 1983-1988 time period.)

Finally, if it is necessary, the creation of a joint government/industry working group will formalize the process of identifying gaps in standards and recommend appropriate tests and field trials.

BIBLIOGRAPHY

Chorafas, Dr. Dimitris N., and Steinmann, Heinrich, *Intelligent Networks*, CRC Press, Inc., Boca Raton, FL, 1990.

De Jouvenel, Bertrand, *The Art of Conjecture*, Basic Books, NY, 1967.

Ginn, Sam (February 1991), "Personal Communication Services: Expanding the Freedom to Communicate," *IEEE Communications Magazine*, Vol. 29, No. 2, pp. 30-32 and p. 39.

Martino, Joseph P., *Technological Forecasting for Decision Making*, North Holland, NY, 2nd Ed. 1983.

Rudisill, Richard E. III, Powel, John W., and Toy, David H., *An Overview of Digital Wireless Services and Associated Standards Organizations*, (Draft), The MITRE Corporation, February 1993.

Taylor, Louis L. (October 1992), "Whither Tactical Communications," *Conference Record, IEEE Military Communications Conference*, Vol. 2, pp. 0667-0671.

Taylor, Louis L. (November 1989), "Whence Networking," *Proceedings of International Conference on Systems, Man, and Cybernetics*, pp. 977-980.

Weissman, Dave, (April 1993), GTE Government Systems Corporation, private correspondence.

INDEX

O

P

Q

R

S